DNA Barcodes for Evolution and Biodiversity

DNA Barcodes for Evolution and Biodiversity

Editor

Stephan Koblmüller

MDPI

Basel • Beijing • Wuhan • Barcelona • Belgrade • Novi Sad • Cluj • Manchester

Editor
Stephan Koblmüller
University of Graz
Graz
Austria

Editorial Office
MDPI
St. Alban-Anlage 66
4052 Basel, Switzerland

This is a reprint of articles from the Special Issue published online in the open access journal *Diversity* (ISSN 1424-2818) (available at: https://www.mdpi.com/journal/diversity/special_issues/bar_bio).

For citation purposes, cite each article independently as indicated on the article page online and as indicated below:

Lastname, A.A.; Lastname, B.B. Article Title. *Journal Name* **Year**, *Volume Number*, Page Range.

ISBN 978-3-7258-0487-0 (Hbk)
ISBN 978-3-7258-0488-7 (PDF)
doi.org/10.3390/books978-3-7258-0488-7

Contents

diversity

MDPI

Editorial

DNA Barcodes for Evolution and Biodiversity

Stephan Koblmüller

Institute of Biology, University of Graz, Universitätsplatz 2, 8010 Graz, Austria; stephan.koblmueller@uni-graz.at

Questions centered around how biological diversity is being generated and maintained, as well as how this biodiversity can be conserved/protected, are being frequently asked in basic and applied evolutionary biological and biodiversity research. However, identifying the entities of biodiversity, i.e., the species, by means of traditional morphological methods is often anything but trivial and is time-consuming. Our ability to identify and assess biodiversity has been enhanced by the establishment of DNA barcoding, which had, has, and will continue to have a great impact on many fields of basic and applied research.

DNA barcoding is a method used for identifying specimens (ideally to species level) and involves employing an expert-based reference system (an open-access database) that drastically increases the number of people who are able to identify organisms down to the species level and reduce the number of misidentifications among morphologically similar taxa. Specifically, DNA barcoding is a standardized approach used for identifying organisms based on specific sections of their DNA [1]. Depending on the taxonomic group, different genes have been established as the standard DNA barcoding markers, even though also other genes may be used for certain applications or taxa. Consequently, DNA barcodes should (in most cases) allow for an unambiguous specimen identification, as well as of morphologically unidentifiable life stages/sexes or parts of organisms, once a reliable DNA barcode reference database is available. Thus, DNA barcoding has become an important tool in basic and applied biodiversity and evolutionary biology research. Indeed, since the onset of large-scale DNA barcoding initiatives, researchers have aimed to increase the time and cost-efficiency of this method [2–5], obtain reference data from samples with suboptimal DNA quality (e.g., from older museum specimens) [6,7], provide comprehensive reference DNA barcode libraries for certain taxa and regions [8–11], or characterize entire communities via (eDNA-)metabarcoding [12,13].

This Special Issue includes a collection of 14 papers that use DNA barcodes to answer questions in basic and applied biodiversity and evolutionary biology research. Many of the key aspects of DNA barcoding are addressed by these studies which provide some important new insights in their respective fields of research.

Typically, in any biological study, species identification and delimitation is the first and often most important step. For a long time, this has solely been based on morphological characteristics, but with the establishment of molecular genetic methods, and especially with the advent of DNA barcoding, DNA data have been increasingly used for this purpose within an integrative taxonomical framework. DNA-based methods are a particularly useful supplement for delimiting species in taxa that comprise several phenotypically similar or indistinguishable species (=cryptic species complexes) (e.g., [14–16]). Five papers in this Special Issue focus on species delimitation among, in part, morphologically very similar taxa [17–21]. The first paper explores species diversity in dogfish sharks (genus *Squalus*) from the Pacific and western Atlantic Oceans, a taxon notorious for its conserved morphology, by means of DNA barcodes and a variety of molecular species delimitation methods [17]. The study shows that all samples analyzed represent species that are already known. The presence of obviously misidentified samples in databases, however, makes drawing inferences on the real distribution and diversity of species that belong to this genus difficult. The second paper [18] characterizes the diversity of invertebrates in Croatian

Citation: Koblmüller, S. DNA Barcodes for Evolution and Biodiversity. *Diversity* **2023**, *15*, 1003. https://doi.org/10.3390/d15091003

Received: 1 September 2023
Accepted: 7 September 2023
Published: 8 September 2023

olive orchards and vineyards by means of DNA barcoding and by comparing the obtained DNA barcodes with available reference data. With their protocol, which uses standard barcoding primers for animals (LCO1490/HCO2198 [22]), the authors managed to obtain data for only slightly more than half of their samples, a finding that is in line with other studies that show that these standard primers do not work well for all taxa. A finding that is particularly interesting and relevant for many taxa, especially within Collembola and Oligochaeta, which are considered major players in (soil) ecosystems, is that many samples cannot be assigned to particular species, which indicates that there is a lack of reference data in the two largest databases (BOLD and GenBank). The third paper [19] establishes a DNA barcode library for mealybug species from Espírito Santo, a major coffee-producing region in Brazil, by combining newly generated DNA barcodes with barcodes available on BOLD. The study shows that, in principle, molecular species delimitation works well in the relevant taxa, if obviously misidentified samples on BOLD are taken into account/excluded. The fourth paper [20] focuses on the diversity of Lepidoptera from Greece. DNA barcodes of ~600 morphospecies were generated and assigned to molecular taxonomic units (MOTUs) on BOLD. A large number of these could be assigned to MOTUSs/species that are also present in other parts of Europe (including new records for Greece). However, about one-sixth of the identified MOTUs had no references in BOLD, despite the generally good coverage of European Lepidoptera on BOLD (e.g., [10]). Hence, these MOTUs may be restricted to Greece (or southeastern Europe) and may potentially include a large number of undescribed species. The fifth paper [21] focuses on the agaricoid mushroom genus *Cortinarus* s.l. in Romania. By means of an integrative taxonomical approach, morphological analyses and DNA barcoding data were combined. Of the 109 *Cortinarius* s.l. species identified in this study, only 43 were previously reported for Romania, while 66 species were new to the country. Collectively, these five papers show the potential of DNA barcoding for species delimitation, species discovery, and general biodiversity assessment, but also highlight obvious problems/difficulties associated with erroneous species identification for some samples on BOLD and a lack of reference data for some important taxa, which makes it difficult to for taxonomic laymen (one of the alleged huge advantages of DNA barcoding used for species identification) to characterize their (local) diversity.

Two further studies [23,24] used DNA barcodes to identify samples to species level. The first study [23] used DNA barcoding to identify a cichlid fish population from a freshwater reservoir in the Ouémé Basi, Benin, as an invasive and primarily estuarine/brackish species that is only rarely found further upstream in pure freshwater habitats. Whether this represents a case of natural range expansion or human introduction remains unclear. The second study [24] used DNA barcoding to identify species of single-drug herbal powders collected from markets in Tamil Nadu, India. As herbal powder is more prone to adulteration than intact plant parts, its authentication is essential to ensure the safety and efficacy of herbal drugs. The study shows that of the 107 herbal powders analyzed, a surprisingly large portion of samples (46%) were adulterant. In 59% of these adulterant samples, the authentic species were entirely replaced with taxonomically unrelated, but sometimes phenotypically similar, species. This low rate of authentic plants in the investigated herbal powders is alarming and calls for thorough training centered around the correct identification of relevant plants and routine validation, e.g., by means of DNA barcoding, to minimize potential health risks for consumers.

High taxonomic coverage is crucial for the applicability of reference DNA barcode databases like BOLD. Although there is comprehensive coverage for certain taxa and regions [8–11], this is not the case for other taxa and regions. These are typically understudied taxa and regions l. In this Special Issue, one paper [25] targets the monogenean fish parasites of gobies in Greece. By conducting morphological analysis combined with the sequencing of three genes, including the DNA barcoding region, the authors provide the first record of *Xenoligophoroides cobitis* (Monogenea: Dactylogyridae) for Greece and the first DNA barcode of this monotypic genus. In addition, the authors proposed some hypotheses regarding the evolution of this monotypic genus.

A combination of DNA barcoding, the sequencing of additional genes, and phenotypic data analysis was also used in another paper included in this Special Issue [26] to characterize population of the invasive colonial ascidian *Botrylloides niger* in the northeastern Mediterranean Sea. Several distinct morphotypes were found, but DNA-based species delimitation methods suggest that these all belong to *B. niger*. In addition, this study provides important information on population dynamics, demographic history, and intraspecific genetic diversity for this invasive species.

The intraspecific diversity of the crambid moth *Maruca vitrata* in India was the focus of another study [27]. This species is of the one of the most destructive pests of grain legumes across the subtropical and tropical regions of the world, and hence knowledge on intraspecific diversity is important for its management. Based on DNA barcoding data, very little intraspecific variation was inferred, indicating the presences of a panmictic population in India. Furthermore, the data show clear signatures of recent population growth. Thus, the study provides very important baseline data for the future management of this pest species.

Recent advances in (eDNA-)metabarcoding methods have resulted in a range of technologies that now can be applied to monitor the occurrence and abundance of diversity in different environments [13,28–30]. A review paper included in the Special Issue [31] focuses on how eDNA-based methods are used in the biodiversity monitoring of protected areas. Specifically, the advantages (and disadvantages), as well as the challenges and limitations, of potential applications are discussed. The paper provides useful information on the use of eDNA approaches in protected areas and also explicitly states what is needed to increase applicability and comparability. Thus, this review may serve as guideline for where to focus in the future development/improvement of eDNA approaches to be applied for monitoring-associated research and answering explicit management questions (not only) in protected areas.

One study included in this Special Issue [32] used metabarcoding of bulk samples to assess the species composition of ichthyoplankton in the Oujiang River estuary in China. The authors compared the performance of 12S and cytb as metabarcoding markers and found that 12S consistently performed better, both in terms of species coverage and detection rates. In total, 145 taxa were identified. This study makes an important contribution to our knowledge about fish diversity in Chinese river estuaries.

The monitoring of pathogens and parasites to identify high-risk-infection areas, enabling disease control, is also facilitated by eDNA methods [33,34]. A study included in this Special Issue [35] provides reference DNA barcodes for the Austrian avian schistosomes of the genus *Trichobilharzia*. Based on these data, an eDNA-based PCR assay was developed to identify *Trichobilharzia* in water samples. Though these parasites typically use birds as final hosts, they may also infect humans as accidental hosts, causing dermatitis symptoms. Thus, these trematodes are of human medical relevance, suggesting that the assay developed in this study will be of great use for the routine monitoring of waterbodies.

Finally, one opinion paper included in this Special Issue [36] presents the reflected opinions of early-career biodiversity researchers regarding questions related to the future of DNA barcoding and whether the currently employed standard barcoding, i.e., the sequencing of short standardized fragments of DNA, will also remain the method of choice for rapid and reliable species identification in the future. From their reflections, it seems to be clear that DNA (meta-)barcoding will also continue to impact biological sciences and environmental management in the future, as long as a focus on data quality is prioritized and the methodological and technological advancements remain aligned.

Conflicts of Interest: The author declares no conflict of interest.

References

1. Hebert, P.D.N.; Cywinska, A.; Ball, S.L.; DeWaard, J.R. Biological identifications through DNA barcodes. *Proc. R. Soc. B Biol. Sci.* **2003**, *270*, 313–321. [CrossRef]
2. Rulik, B.; Eberle, J.; von der Mark, L.; Thormann, J.; Jung, M.; Köhler, F.; Apfel, W.; Weigel, A.; Kopetz, A.; Köhler, J.; et al. Using taxonomic consistency with semi-automated data pre-processing for high quality DNA barcodes. *Methods Ecol. Evol.* **2017**, *8*, 1878–1887. [CrossRef]
3. Hebert, P.D.N.; Braukmann, T.W.A.; Prosser, S.W.J.; Ratnasingham, S.; deWaard, J.R.; Ivanova, N.V.; Janzen, D.H.; Hallwachs, W.; Naik, S.; Sones, J.E.; et al. A Sequel to Sanger: Amplicon sequencing that scales. *BMC Genom.* **2018**, *19*, 219. [CrossRef] [PubMed]
4. Yang, C.; Zheng, Y.; Tan, S.; Meng, G.; Rao, W.; Yang, C.; Bourne, D.G.; O'Brien, P.A.; Xu, J.; Liao, S.; et al. Efficient COI barcoding using high throughput single-end 400 bp sequencing. *BMC Genom.* **2020**, *21*, 862. [CrossRef]
5. Srivathsan, A.; Lee, L.; Katoh, K.; Hartop, E.; Kutty, S.N.; Wong, J.; Yeo, D.; Meier, R. ONTbarcoder and MinION barcodes aid biodiversity discovery and identification by everyone, for everyone. *BMC Biol.* **2021**, *19*, 217. [CrossRef]
6. Hernández-Triana, L.M.; Prosser, S.W.; Rodríguez-Perez, M.A.; Chaverri, L.G.; Hebert, P.D.N. Recovery of DNA barcodes from blackfly museum specimens (Diptera: Simuliidae) using primer sets that target a variety of sequence lengths. *Mol. Ecol. Resour.* **2014**, *14*, 508–518. [CrossRef]
7. Schäffer, S.; Zachos, F.E.; Koblmüller, S. Opening the treasure chest: A DNA-barcoding primer set for most higher taxa of Central European birds and mammals from museum collections. *PLoS ONE* **2017**, *12*, e017449. [CrossRef]
8. Morinière, J.; Balke, M.; Doczkal, D.; Geiger, M.F.; Hardulak, L.A.; Haszprunar, G.; Hausmann, A.; Hendrich, L.; Ragalado, L.; Rulik, B.; et al. A DNA barcode library for 5200 German flies and midges (Insecta: Diptera) and its implications for metabarcoding-based biomonitoring. *Mol. Ecol. Resour.* **2019**, *19*, 900–928. [CrossRef]
9. Zangl, L.; Daill, D.; Schweiger, S.; Gassner, G.; Koblmüller, S. A reference DNA barcode library for Austrian amphibians and reptiles. *PLoS ONE* **2020**, *15*, e0229353. [CrossRef]
10. Dinča, V.; Dapporto, L.; Somervuo, P.; Vodă, R.; Cuvelier, S.; Cascoigne-Pees, M.; Huemer, P.; Mutanen, M.; Hebert, P.D.N.; Vila, R. High resolution DNA barcode library for European butterflies reveals continental patterns of mitochondrial genetic diversity. *Commun. Biol.* **2021**, *4*, 315. [CrossRef]
11. Geiger, M.; Koblmüller, S.; Assandri, G.; Chovanec, A.; Ekrem, T.; Fischer, I.; Galimberti, A.; Grabowski, M.; Haring, E.; Hausmann, A.; et al. Coverage and quality of DNA barcode references for Central and Northern European Odonata. *PeerJ* **2021**, *9*, e11192. [CrossRef] [PubMed]
12. Bohrmann, K.; Evans, A.; Gilbert, M.T.P.; Carvalho, G.R.; Creer, S.; Knapp, M.; Yu, D.W.; de Bruyn, M. Environmental DNA for wildlife biology and biodiversity monitoring. *Trends Ecol. Evol.* **2014**, *29*, 358–367. [CrossRef] [PubMed]
13. Bohrmann, K.; Lynggaard, C. Transforming terrestrial biodiversity surveys using airborne eDNA. *Trends Ecol. Evol.* **2022**, *38*, 119–121. [CrossRef] [PubMed]
14. Dziki, A.; Binford, G.J.; Coddington, J.A.; Agnarsson, I. *Spintharus flavus* in the Caribbean—A 30 million year biogeographical history and radiation of a "widespread species". *PeerJ* **2015**, *3*, e1422. [CrossRef]
15. Schäffer, S.; Kerschbaumer, M.; Koblmüller, S. Multiple new species: Cryptic diversity in the widespread mite species Cymbaeremaeus cymba (Oribatida, Cymbaeremaeidae). *Mol. Phylogenet. Evol.* **2019**, *135*, 185–192. [CrossRef]
16. Pfingstl, T.; Lienhard, A.; Baumann, J.; Koblmüller, S. A taxonomist's nightmare—Cryptic diversity in Caribbean intertidal arthropods (Arachnida, Acari, Oribatida). *Mol. Phylogenet. Evol.* **2021**, *163*, 107240. [CrossRef]
17. Ariza, A.A.; Adachi, A.M.C.L.; Roque, P.; Hazin, F.H.V.; Vianna, M.; Rotundo, M.M.; Delpiani, S.M.; Díaz de Astarloa, J.; Delpiani, G.; Oliveira, C.; et al. DNA barcoding and species delimitation for dogfish sharks belonging to the Squalius genus (Squaliformes: Squalidae). *Diversity* **2022**, *14*, 544. [CrossRef]
18. Anđelić Dmitrović, B.; Jelić, M.; Rota, E.; Šerić Jelaska, L. DNA barcoding of invertebrates inhabiting olive orchards and vineyards accelerates understudied Mediterranean bidodiversity assessment. *Diversity* **2022**, *14*, 182. [CrossRef]
19. Oliveira, P.V.; dos Santos, A.R.; Olive, E.L.; Britto, K.B.; de Almeida, F.A.N.; da Silva, V.C.P.; Machado, C.B.; Fornazier, M.J.; Ventura, J.A.; Culik, M.P.; et al. Molecular species delimitation using COI barcodes of mealybugs (Hemiptera: Pseudococcidae) from coffee plants in Espírito Santo, Brazil. *Diversity* **2023**, *15*, 305. [CrossRef]
20. Huemer, P.; Mutanen, M. An incomplete European barcode library has a strong impact on the identification success of Lepidoptera from Greece. *Diversity* **2022**, *14*, 118. [CrossRef]
21. Szabó, E.; Dima, B.; Dénes, A.L.; Papp, V.; Keresztes, L. DNA barcoding data reveal important overlooked diversity of Cortinarius sensu lato (Agaricales, Basidiomycota) in the Romanian Carpathians. *Diversity* **2023**, *15*, 553. [CrossRef]
22. Folmer, O.; Black, M.; Hoeh, W.; Lutz, R.; Vrijenhoek, R. DNA primers for amplification of mitochondrial cytochrome c oxidase subunit I from diverse metazoan invertebrates. *Mol. Mar. Biol. Biotechnol.* **1994**, *3*, 294–299.
23. Pèlèbè, R.O.E.; Toko, I.I.; Verheyen, E.; Van Steenberge, M. Molecular identification of an invasive Sarotherodon species from the Atchapka freshwater reservoir (Ouémé River basin, Benin) and comparison with S. melanotheron using COI markers. *Diversity* **2021**, *13*, 297. [CrossRef]
24. Balaji, R.; Parani, M. DNA barcoding of the market samples of single-drug herbal powders reveals adulteration with taxonomically unrelated plant species. *Diversity* **2022**, *14*, 495. [CrossRef]

25. Vanhove, M.P.P.; Giakoumi, S.; Zogaris, D.; Kovačić, M.; Huyse, T. First eastern Mediterranean record of Xenoligophoroides cobitis, the only dactylogyrid monogenean infecting Mediterranean gobies: Just arrived or missed the boat? *Diversity* **2022**, *14*, 580. [CrossRef]
26. Temiz, B.; Öztürk, E.; Blanchoud, S.; Karahan, A. Phylogeographic and morphological analysis of Botrylloides niger Herdman, 1886 from the northeastern Mediterranean Sea. *Diversity* **2023**, *15*, 367. [CrossRef]
27. Mahalle, R.M.M.; Chakravarty, S.; Srivastava, C.P. Population genetic differentiation and structure of Maruca vitrata (Lepidoptera: Crambidae) in India. *Diversity* **2022**, *14*, 546. [CrossRef]
28. Ruppert, K.M.; Kline, R.J.; Saydur Rahman, M. Past, present, and future perspectives of environmental DNA (eDNA) metabarcoding: A systematic review in methods, monitoring, and applications of global eDNA. *Glob. Ecol. Conserv.* **2019**, *17*, e00547. [CrossRef]
29. Kirse, A.; Bourlat, S.J.; Langen, K.; Fonseca, V.G. Unearthing the potential of soil eDNA metabarcoding—Towards best practice advice for invertebrate biodiversity assessment. *Front. Ecol. Evol.* **2021**, *9*, 630560. [CrossRef]
30. Ariza, M.; Foulks, B.; Mauvisseau, Q.; Halvorsen, R.; Greve Alsos, I.; de Boer, H.J. Plant biodiversity assessment through soil eDNA reflects temporal and local diversity. *Methods Ecol. Evol.* **2023**, *14*, 415–540. [CrossRef]
31. Pascher, K.; Švara, V.; Jungmeier, M. Environmental DNA-based methods in biodiversity monitoring of protected areas: Application range, limitations, and needs. *Diversity* **2022**, *14*, 463. [CrossRef]
32. Jiang, R.; Lusana, J.L.; Chen, Y. High-throughput DNA metabarcoding as an approach for ichthyoplankton survey in Oujian River estuary, China. *Diversity* **2022**, *14*, 1111. [CrossRef]
33. Bass, D.; Stentiford, G.D.; Littlewood, D.T.J.; Hartikainen, H. Diverse applications of environmental DNA methods in parasitology. *Trends Parasitol.* **2015**, *31*, 499–513. [CrossRef]
34. Sengupta, M.E.; Lynggaard, C.; Mukaratirwa, S.; Vennervald, B.J.; Stensgaard, A.S. Environmental DNA in human and veterinary parasitology—Current applications and future prospects for monitoring and control. *Food Waterborne Parasitol.* **2022**, *29*, e00183. [CrossRef]
35. Helmer, N.; Hörweg, C.; Sattmann, H.; Reier, S.; Szucsich, N.U.; Bulantová, J.; Haring, E. DNA barcoding of Trichobilharzia (Trematoda: Schistosomatidae) species and their detection in eDNA water samples. *Diversity* **2023**, *15*, 104. [CrossRef]
36. Grant, D.M.; Brodnicke, O.B.; Evankow, A.M.; Ferreira, A.O.; Fontes, J.T.; Hansen, A.K.; Jensen, M.R.; Kalaycı, T.E.; Leeper, A.; Patil, S.K.; et al. The future of DNA barcoding: Reflections from early career researchers. *Diversity* **2021**, *13*, 313. [CrossRef]

diversity

MDPI

Article

DNA Barcoding and Species Delimitation for Dogfish Sharks Belonging to the *Squalus* Genus (Squaliformes: Squalidae)

Ailton A. Ariza [1,*], Aisni M. C. L. Adachi [1], Pollyana Roque [2], Fabio H. V. Hazin [2,†], Marcelo Vianna [3], Matheus M. Rotundo [4], Sergio M. Delpiani [5,6], Juan M. Díaz de Astarloa [5,6], Gabriela Delpiani [5,6], Claudio Oliveira [1], Fausto Foresti [1] and Vanessa P. Cruz [1,*]

1 Laboratório de Biologia e Genética de Peixes, Instituto de Biociências de Botucatu, Universidade Estadual Paulista, Botucatu 18618-689, Brazil; aisnimayumi@gmail.com (A.M.C.L.A.); claudio.oliveira@unesp.br (C.O.); f.foresti@unesp.br (F.F.)
2 Laboratório de Oceanografia Pesqueira, Universidade Federal Rural de Pernambuco, Recife 52171-900, Brazil; pollyana_cgr@hotmail.com (P.R.); fhvhazin@terra.com.br (F.H.V.H.)
3 Laboratório de Biologia e Tecnologia Pesqueira, Universidade Federal do Rio de Janeiro, Rio de Janeiro 21941-901, Brazil; mvianna@biologia.ufrj.br
4 Acervo Zoológico da Universidade Santa Cecília, Universidade Santa Cecília, Santos 11045-907, Brazil; mmrotundo@unisanta.br
5 Grupo de Biotaxonomía Morfológica y Molecular de Peces, Departamento de Biología, Facultad de Ciencias Exactas y Naturales, Instituto de Investigaciones Marinas y Costeras, Funes 3250, Mar del Plata 7600, Argentina; matidelpiani16@yahoo.com.ar (S.M.D.); juanastarloa@gmail.com (J.M.D.d.A.); gabriela.delpiani@gmail.com (G.D.)
6 Consejo Nacional de Investigaciones Científicas y Técnicas, CABA AAJ, Buenos Aires 2290, Argentina
* Correspondence: ailton.ariza@unesp.br (A.A.A.); cruzvp@outlook.com (V.P.C.)
† In memorian.

Citation: Ariza, A.A.; Adachi, A.M.C.L.; Roque, P.; Hazin, F.H.V.; Vianna, M.; Rotundo, M.M.; Delpiani, S.M.; de Astarloa, J.M.D.; Delpiani, G.; Oliveira, C.; et al. DNA Barcoding and Species Delimitation for Dogfish Sharks Belonging to the *Squalus* Genus (Squaliformes: Squalidae). *Diversity* **2022**, *14*, 544. https://doi.org/10.3390/d14070544

Academic Editor: Stephan Koblmüller

Received: 30 May 2022
Accepted: 30 June 2022
Published: 6 July 2022

Abstract: The *Squalus* genus comprises a group of small demersal sharks occurring circumglobally, popularly known as dogfish sharks. This genus exhibits a conserved morphology, thus making correct morphological identification difficult. Considering these taxonomic problems and the scarcity of molecular data, the present study aimed to identify *Squalus* genus MOTUs, using DNA barcoding for species delimitation via ABGD (automatic barcode gap discovery), PTP (Poisson tree process), and GMYC (general mixed Yule coalescent) employing the mitochondrial COI gene. A total of 69 sequences were generated from samples obtained from the American coast in both the Atlantic and Pacific Oceans. The ABGD analysis was the most conservative among the three applied delimitations, indicating three taxonomic units, while the PTP analysis revealed nine MOTUs, with two conflicting units noted between *S. clarkae* + *S. mitsukurii* and *S. albicaudus* + *S. cubensis*. The GMYC analysis indicated an excessive division, with *S. acanthias* and *S. mitsukurii* subdivided into six MOTUs each and *S. blainville*, into four. These findings demonstrated that *Squalus* presents a complex of previously defined species, with misidentified samples deposited in databases leading to difficulties in analyzing the real distribution and diversity of species belonging to this genus. Thus, further efforts to highlight possible new species are recommended.

Keywords: dogsharks; Elasmobranchii; mitochondrial DNA; COI; species identification

1. Introduction

Dogfish are cartilaginous fish belonging to the genus *Squalus* Linnaeus, 1758 (Squaliformes, Squalidae), comprising 35 described species [1–4]. Dogfish are small, migratory, demersal sharks living between 100 and 500 m in depth [5], reaching up to 1.5 m in length and presenting a wide global geographical distribution, occurring on continental shelves in the Atlantic, Pacific, and Indian oceans [5–7].

As with many other elasmobranchs, the life history of the *Squalus* genus is characterized by slow growth, late sexual maturation, long life expectancy, and low fecundity [8], with individuals tending to aggregate by sex and size [6,9]. The dogfish is a yolk-sac viviparous species [10–12] with a long gestation period, estimated as lasting up to two years [13–15]. These attributes lead to low population growth rates and limited capacity to withstand fishing pressures, resulting in rapid population declines [16]. Because of their biology and anthropic actions, most *Squalus* species are currently classified as Threatened by the International Union for Conservation of Nature (IUCN) [17], whereas the Chico Mendes Institute for Biodiversity Conservation (ICMBIO), a Brazilian environmental agency linked to the Ministry of the Environment, categorizes only *S. acanthias* as Critically Endangered (CR) [18].

Overall, sharks belonging to the *Squalus* genus exhibit a conserved body morphology, making identification based solely on morphological characters problematic, leading to misidentifications [19]. This complexity is amplified even further by the high overlap of morphological characters among species, as identification is often based on limited and insufficiently consistent characters, such as number of vertebrae and morphometric data [2,4,5,20–22].

Reliable information on species richness is essential for any biodiversity study and conservation policies, although it is often difficult to discriminate a species based on highly similar morphological characters [20]. In this regard, reliable species identification is the first and most important step for the application of conservation policies and sustainable exploitation of natural resources [23], even more so considering the presently accelerated biodiversity crisis induced by human activities [24].

In recent years, different genetic studies have attempted to identify *Squalus* species using mitochondrial COI and NADH2 genes [3,21,25–28]. In general, three well-defined groups within the genus have been reported, namely group I, comprising *S. suckleyi* and *S. acanthias*; group II, comprising *S. blainville*/*S. megalops*/*S. raoulensis*/*S. brevirostris*; and group III, the *S. mitsukurii* complex, comprising *S. edmundsi*, *S. japonicus*, *S. grahami*, *S. clarkae*, and *S. mitsukurii* [21,22,26,29].

Generally, barcoding researchers have used a 2% divergence threshold as a heuristic cutoff value in fish species delimitation [30,31]. Nevertheless, it is already known that elasmobranchs have low evolutionary rates when compared with other fish species [32], which means that they are a more genetically conserved group [33–35]. Thus, in rays and sharks, we found some genera that presented about 1% of genetic distance between species, including among members of the genus of rays *Mobula* [36], the genus *Carcharhinus* [27,37,38], and the genus *Squalus* [3,21,25,26].

Considering the important taxonomic problems that characterize this group and the scarcity of available molecular data, the main goal of this study was to identify molecular operational taxonomic units (MOTUs) in the *Squalus* genus based on the analysis of sampled nominal species collected in different Western Atlantic and Pacific Ocean regions through the DNA barcoding technique employing the cytochrome c oxidase I (COI) genetic marker. The data were compared with available databases using species delimitation approaches including automatic barcode gap discovery (ABGD) [39], the Poisson tree process (PTP) [40], and the general mixed Yule coalescent (GMYC) [41,42], with the aim to add molecular data of the genus *Squalus* through the tools of species delimitation to assist in future works of integrative taxonomy.

2. Materials and Methods

2.1. Sample Collection

Samples were obtained from 69 dogfish shark specimens from the Western Atlantic and Pacific Oceans belonging to the *Squalus* genus (Squaliformes: Squalidae) representing three nominal species, *S. mitsukurii*, *S. albicaudus*, and *S. acanthias*, (Figure S1). The tissue samples were deposited at the LBGP ichthyological collection (Laboratório de Biologia e Genética de Peixes—Fish Biology and Genetics Laboratory) belonging to UNESP in Botucatu, Sao Paulo, Brazil. All samplings were performed in accordance with Brazilian government standards (SISBIO protocol 13843-1) and an Animal Ethical Committee. Small muscle fragments (<1 cm^2) from each sample were obtained and preserved in 96% ethanol.

Total genomic DNA was isolated from muscle tissues of each specimen with a DNeasy Tissue Kit (Qiagen, Hilden, Germany) according to the manufacturer's instructions. Amplification reactions of mitochondrial gene cytochrome c oxidase subunit I (COI) were performed in a total volume of 12.5 μL, with 1.25 μL of 10× buffer (10 mM Tris-HCl + 15 mM $MgCl_2$); 0.5 μL dNTPs (200 nM of each); 0.5 μL each of the 5 mM primers L6252-Asn and H7271-COXI, as described in Melo et al. [43]; 0.2 μL of PHT Taq DNA polymerase (Phoneutria Biotecnologia e Serviços Ltd., Belo Horizonte, Brasil); 1 μL template DNA (12 ng); and 8.7 μL ddH_2O. The PCR reactions consisted of initial denaturation at 95 °C for 3 min; 25 cycles at 94 °C for 30 s, 52 °C for 45 s, and 68 °C for 1 min; and final extension at 68 °C for 7 min. All PCR products were first visually identified on a 1% agarose gel. The purified PCR products were sequenced using a Big Dye Terminator v3.1 Cycle. Sequencing was performed with the BigDye Terminator v3.1 Cycle Sequencing Kit (Applied Biosystems, Waltham, MA, USA). Individual reactions were performed with approximately 30 ng template PCR product, 3.2 pmol primer, 1 μL terminator mix, and 5 μL Better Buffer (The Gel Co., Eden Prairie, MN, USA) in a total volume of 15 μL. PCR sequencing profiles consisted of an initial denaturation step of 4 min at 96 °C followed by 30 cycles of 30 s at 96 °C, 15 s at 50 °C, and 4 min at 60 °C. Sequencing was carried out on an automated ABI 3130xl Applied Biosystems sequencer.

2.2. Barcoding

The COI sequences were edited in the Geneious 6.0 software(Biomatters, Ltd., Auckland, New Zeland) [44], with each sequence manually reviewed for uncalled and miscalled bases and all variable positions confirmed by comparing sequence reads produced by the forward and reverse sequences of each individual. A consensus sequence was produced for each individual, and all sequences were deposited in GenBank under accession numbers ON827418 to ON827486.

Aligned consensus sequences were compared with those deposited in the National Center for Biotechnology Information (NCBI) database (http://www.ncbi.nlm.nih.gov/, accessed on 27 March 2021) using the Basic Local Alignment Search Tool—Nucleotide (BLASTn). Each sequence used in this study is provided in a supplementary table (Table S1) and was later aligned using the Muscle algorithm [45] implemented within the Geneious 6.0 software [44]. All parameters followed the default version of the algorithm.

The sequences obtained herein were compared with 204 COI GenBank sequences [21,46–53] referring *to S. blainville, S. suckleyi, S. acanthias, S. brevirostris, S. clarkae, S. cubensis, S. japonicus, S. mitsukurii, S. grahami*, and *S. edmundsi*. The final dataset comprised 273 sequences from 11 *Squalus* species and 1 sequence for *Cirrhigaleus asper* (MN982926), representing an external group, totaling 274 sequences.

The best-fit model of nucleotide evolution for the data was estimated for the analyzed matrix with the MEGA X program [54], applying the neighbor-joining (NJ) method using the Kimura two-parameter (K2P) model [55]. Bootstrap replicates were assessed by applying 1000 replicates [56]. Trees were visualized and edited using the FigTree v1.4 program (www.tree.bio.ed.ac.uk/software/figtree, accessed on 27 March 2021), (Edinburgh, UK). The mean genetic inter- and intraspecific distances for nominal species were calculated under the K2P model and displayed in a pairwise distance matrix.

2.3. Automatic Species Delimitation Analyses

To infer *Squalus* species delimitation criteria based on a partial COI gene, molecular operational taxonomic unit (MOTU) estimations were performed by employing three molecular tools to delimit species. The first delimitation was conducted using automatic barcode gap discovery (ABGD) [39] run on the ABGD web server (https://bioinfo.mnhn.fr/abi/public/abgd/abgdweb.html, accessed on: 30 March 2021). All parameters followed the default version of the program (model = Jukes–Cantor (JC69) Pmin = 0.001, Pmax = 0.1, steps = 10, X (relative gap width) = 1.5, number of bins = 20).

The second delimitation was performed applying the Poisson tree process (PTP) [40] based on a nonultrametric tree run on the PTP web server (https://species.h-its.org/ptp, accessed on 30 March 2021). The maximum likelihood (ML) tree was used as the input. The best model used in this dataset was selected based on AIC (i.e., had the lowest AIC) as estimated using the MEGA X software (https://www.megasoftware.net/citations, accessed on 27 March 2021) based on the best nucleotide substitution model HKY + G + I (5968.201). The PTP analysis was then performed for 100,000 generations MCMC, with a thinning value of 100 and burn-in of 0.1.

The third delimitation analysis was performed through the general mixed Yule coalescent (GMYC) method [41,42] run on the GMYC web server (https://species.h-its.org/gmyc/, accessed on 27 March 2021). The Elimdupes software (https://www.hiv.lanl.gov/content/sequence/ELIMDUPES/elimdupes.html, accessed on 27 March 2021) was used to group identical sequences and thus reduce the computational analysis time. The tree parameters were selected in the BEAUTI program belonging to the BEAST program package to calibrate the ultrametric tree, uncorrelated lognormal relaxed clock, and coalescence speciation models, where exponential growth was applied employing the HKY + G + I nucleotide substitution model. The MCMC method was performed for 10 million iterations. The Tracer v1.7 software was used to verify convergence (ESS > 200).

The Tree Annotator v1.8 software was used at a 10% burn-in, and the output file was submitted to the Figtree software to detect possible analysis errors such as polytomies or others. The output file was then submitted to the online GMYC version, applying the site's default parameters.

Genetic groups were selected based on a MOTU consensus, and mean genetic interspecific and intraspecific distances were calculated under the K2P model and displayed in a pairwise distance matrix.

To better understand the relationships among the three major *Squalus* groups known in the literature (group I, comprising *S. suckleyi* and *S. acanthias*; group II, comprising of *S. blainville*; *S. brevirostris*, *S. cubensis* and *S. albicaudus*; and group III, comprising *S. edmundsi*, *S. japonicus*, *S. grahami*, *S. clarkae* and *S. mitsukurii*), the number of variable sites, number of haplotypes, and haplotype diversity of each group were evaluated and estimated by the DnaSP v5 software [57] with the median-joining network produced by the PopArt program [58] for mutational analyses.

3. Results

A total of 69 sequences were generated from dogfish shark samples belonging to the *Squalus* genus, representing the nominal species *S. mitsukurii*, *S. albicaudus*, and *S. acanthias*. The amplification of the COI gene resulted in standardized 711 bp fragments, and the nucleotide composition analysis revealed a mean nucleotide composition of 24.8% adenine (A), 33.9% thymine (T), 16.5% guanine (G), and 24.7% cytosine (C). The dataset was submitted to the Basic Local Alignment Search Tool (BLAST) for correct identification by comparing the obtained results with sequences deposited in the NCBI database. Thirty *S. acanthias*, twenty-one *S. albicaudus*, and eighteen *S. mitsukurii* were identified, presenting 98.73 to 100% similarity. The *S. albicaudus* nomenclature was adopted for the sampled individuals, as they occurred on the Brazilian coast, considering that Viana et al. [2] described the occurrence of this species in the southeast Atlantic Ocean.

The matrix was complemented with 204 *Squalus* sequences obtained from GenBank [21,46–53] (Table S1), totaling 273 sequences for the final matrix analysis, representing 11 nominal species from different Mediterranean, Atlantic, and Pacific Ocean regions. The K2P distances of the COI sequence between species ranged from 0.72 to 8.3%, with the smallest and largest interspecific genetic distances identified between *S. albicaudus* and *S. cubensis* (0.0072) and *S. acanthias* and *S. brevirostris* (0.0832) (Table 1). Intraspecific genetic distances ranged from 0.0000 for *S. grahami* to 0.0043 for *S. mitsukurii*.

Table 1. Genetic distances (K2P) based on COI sequences among *Squalus* species (below the diagonal) and standard errors (above the diagonal). The numbers in bold represent the intraspecific K2P genetic distances.

Species	1	2	3	4	5	6	7	8	9	10	11
1—*S. suckleyi*	**0.0008**	0.0031	0.0118	0.0119	0.0108	0.0109	0.0099	0.0105	0.0101	0.0095	0.0097
2—*S. acanthias*	0.0077	**0.0022**	0.0117	0.0123	0.0110	0.0107	0.0099	0.0112	0.0107	0.0101	0.0101
3—*S. blainville*	0.0785	0.0788	**0.0032**	0.0039	0.0050	0.0052	0.0105	0.0109	0.0104	0.0105	0.0106
4—*S. brevirostris*	0.0792	0.0832	0.0116	**0.0018**	0.0053	0.0054	0.0106	0.0109	0.0106	0.0106	0.0107
5—*S. cubensis*	0.0701	0.0748	0.0168	0.0188	**0.0013**	0.0026	0.0099	0.0103	0.0100	0.0096	0.0100
6—*S. albicaudus*	0.0707	0.0729	0.0178	0.0197	0.0072	**0.0016**	0.0100	0.0102	0.0101	0.0102	0.0097
7—*S. edmundsi*	0.0630	0.0636	0.0638	0.0667	0.0614	0.0618	**0.0021**	0.0054	0.0057	0.0047	0.0049
8—*S. japonicus*	0.0646	0.0720	0.0655	0.0663	0.0601	0.0600	0.0190	**0.0009**	0.0051	0.0051	0.0055
9—*S. grahami*	0.0631	0.0703	0.0634	0.0649	0.0603	0.0610	0.0200	0.0169	**0.0000**	0.0045	0.0048
10—*S. clarkae*	0.0589	0.0675	0.0636	0.0659	0.0577	0.0661	0.0169	0.0185	0.0133	**0.0029**	0.0026
11—*S. mitsukurii*	0.0605	0.0659	0.0644	0.0665	0.0638	0.0613	0.0190	0.0213	0.0154	0.0084	**0.0043**

A neighbor-joining tree based on COI gene sequencing identified three main clades (Figure 1), which represented a *suckleyi/acanthias* group (group I), comprising *S. suckleyi* and *S. acanthias;* the *S. blainville/S. megalops/S. raoulensis/S. brevirostris* group (group II), comprising *S. blainville, S. brevirostris, S. cubensis,* and *S. albicaudus*, and the *S. mitsukurii* complex group (group III), comprising *S. edmundsi, S. japonicus, S. grahami, S. clarkae,* and *S. mitsukurii.* These groups represented 11 nominal species, with bootstrap values ranging from 56 to 100%.

The species delimitation results indicated 3 MOTUs by the ABGD method, 9 by the PTP method, 24 by the GMYC method. Three taxonomic groups were identified based on the ABGD analysis, the most conservative among the three applied delimitation analyses, comprising *S. suckleyi* and *S. acanthias* (group I), *S. blainville, S. brevirostris, S. cubensis* and *S. albicaudus* (group II), and *S. edmundsi, S. japonicus, S. grahami, S. clarkae and S. mitsukurii* (group III).

The discrimination based on the PTP analysis identified nine MOTUs. *S. suckleyi, S. acanthias, S. blainville, S. brevirostris, S. edmundsi, S. japonicus,* and *S. grahami* formed one group, with the other two MOTUs formed were by the pairs of nominal species *S. cubensis* and *S. albicaudus* and *S. clarkae* and *S. mitsukurii.*

The GMYC analysis detected a larger number of groups, totaling 24 MOTUs. A single MOTU grouped *S. suckleyi, S. brevirostris, S. cubensis, S. albicaudus, S. edmundsi, S. japonicus, S. grahami,* and *S. clarkae.* The highest number of MOTUs was noted mainly for *S. acanthias* individuals, at six, while *S. blainville* individuals presented four MOTUs, and *S. mitsukurii* individuals, six (see Table S1).

An investigation of pairwise genetic distances based on the PTP analysis in nine MOTUs was carried out, resulting in interspecific genetic variation values of 0.0032 between *S. blainville* and *S. brevirostris* and of 0.0832 between *S. acanthias* and *S. brevirostris* (Table 2). Intraspecific genetic distance values were 0.0000 for *S. grahami* and 0.0055 for *S. mitsukurii.*

The median-joining network was also used for each of the three major *suckleyi/acanthias* lineages forming the *S. blainville/S. megalops/S. raoulensis/S. brevirostris* and *S. mitsukurii* complex groups. This analysis has the advantage of showing the history (step by step) of the mutations connecting nodes between samples or species. A total of 23 haplotypes were identified in the *suckleyi/acanthias* group, with a haplotype diversity of 0.6962 and 31 variable sites. Among the 82 analyzed sequences, the haplotypes were subdivided into two main groups, one composed by S. *suckleyi* haplotypes and the other by *S. acanthias* haplotypes (Figure 2A).

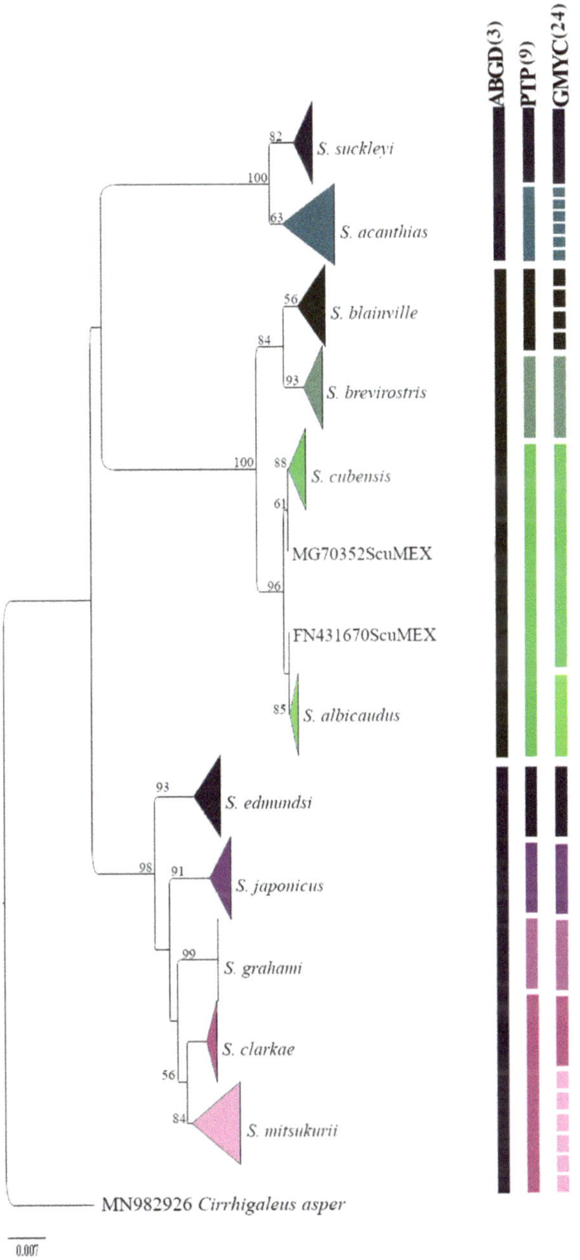

Figure 1. Neighbor-joining tree based on the COI gene from 11 nominal *Squalus* species with bootstrap values on branches. On the right, the vertical bars represent the division into MOTUs (molecular operational taxonomic units) obtained by ABGD (automatic barcode gap discovery for primary species delimitation), PTP (Poisson tree process), and GMYC (generalized mixed Yule coalescent) analyses.

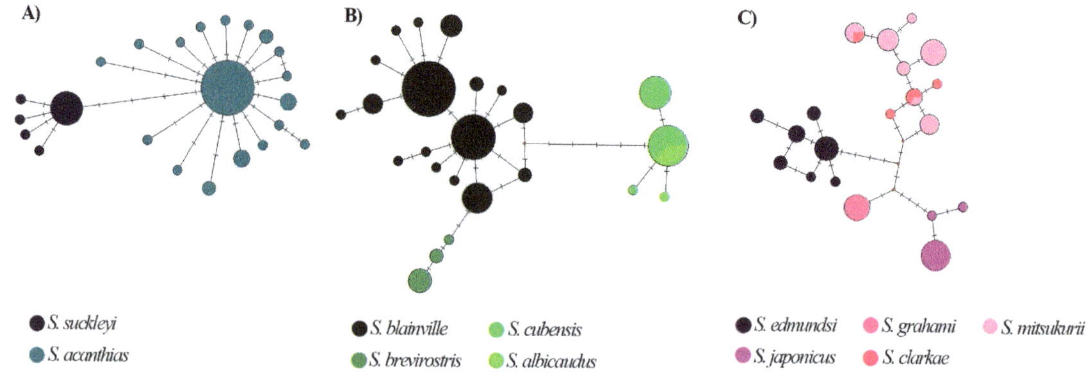

Figure 2. Median-joining network of the three identified groups based on COI gene sequencing from 11 nominal *Squalus* species: (**A**) group I, comprising *S. suckleyi* and *S. acanthias;* (**B**) group II, consisting of *S. blainville; S. brevirostris, S. cubensis,* and *S. albicaudus;* (**C**) group III, grouping *S. edmundsi, S. japonicus, S. grahami, S. clarkae,* and *S. mitsukurii.* The dashes represent mutational steps. The size of the circle representing each haplotype is proportional to the number of individuals within that haplotype.

Table 2. Genetic distances (K2P) based on the division of MOTUs obtained through a PTP analysis among *Squalus* species (below the diagonal) and standard errors (above the diagonal). The numbers in bold represent the intraspecific K2P genetic distances.

Species	1	2	3	4	5	6	7	8	9
1—*S. suckleyi*	**0.0008**	0.0031	0.0121	0.0119	0.0108	0.0106	0.0108	0.0104	0.0100
2—*S. acanthias*	0.0078	**0.0022**	0.0121	0.0122	0.0107	0.0104	0.0114	0.0110	0.0103
3—*S. blainville*	0.0785	0.0788	**0.0038**	0.0041	0.0052	0.0107	0.0112	0.0107	0.0106
4—*S. brevirostris*	0.0792	0.0832	0.0032	**0.0018**	0.0056	0.0110	0.0112	0.0108	0.0107
5—*S. cubensis* +*S. albicaudus*	0.0705	0.0734	0.0176	0.0194	**0.0038**	0.0100	0.0101	0.0100	0.0096
6—*S. edmundsi*	0.0630	0.0636	0.0638	0.0667	0.0617	**0.0021**	0.0053	0.0055	0.0050
7—*S. japonicus*	0.0646	0.0720	0.0655	0.0663	0.0600	0.0190	**0.0009**	0.0051	0.0055
8—*S. grahami*	0.0631	0.0703	0.0634	0.0649	0.0608	0.0200	0.0169	**0.0000**	0.0045
9—*S. clarkae* + *S. mitsukurii*	0.0602	0.0662	0.0642	0.0664	0.0623	0.0186	0.0208	0.0151	**0.0055**

Among *blainville/brevirostris* group individuals, haplotypes were categorized into three main groups among the 128 analyzed sequences. The three groups were composed by *S. blainville, S. brevirostris,* and *S. cubensis* and *S. albicaudus,* the latter two of which shared a total of 24 haplotypes (Figure 2B), with a haplotype diversity of 0.8674 and 25 variable sites. The median-joining network for the *S. mitsukurii* complex group, with 63 analyzed sequences, indicated specific haplotypes for *S. edmundsi, S. japonicus* and *S. grahami,* whereas *S. clarkae* and *S. mitsukurii* shared haplotypes among species (Figure 2C), totaling 20 haplotypes, with a haplotype diversity of 0.9360 and 25 variable sites.

4. Discussion

This study provides a wide genetic analysis aiming at identifying molecular operational taxonomic units (MOTUs) for *Squalus* specimens from the Western Atlantic Ocean. The data allowed for the identification of at least nine *Squalus* lineages among the 11 nominal analyzed species, comprising *S. suckleyi, S. acanthias, S. blainville, S. brevirostris, S. edmundsi, S. japonicus, S. grahami, S. cubensis* + *S. albicaudus,* and *S. clarkae* + *S. mitsukurii.*

Genetic distance employing DNA barcoding is a strong indicator of lineages or species Ward et al. [59], Hubert et al. [60], and Pereira et al. [61] suggested COI distances from

1% [60] to 2% [59,61] as thresholds for fish species. However, as highlighted by Ramirez et al. [62], such values were derived from comparative analyses among phylogenetically diverse groups, whereas DNA barcoding analyses of closely related groups of species may result in lower values [21,25,61,63,64].

Our results coupled to dogfish DNA barcode genetic distances indicated that, among the 54 analyzed comparative values, 41, representing 74.6% of the total data, were higher than 2%, reaching up to 8%. Meanwhile, 12 estimates, representing 21.8% of the total data, exhibited values around 1%. Values were lower than 1% in only 2 estimates, representing 3.6% of the analyzed species. Ziadi-Künzli et al. [25] found similar proportions to those detected herein in an analysis of 27 *Squalus* groups/lineages employing the COI gene, identifying that 66.4% of the estimates displayed genetic distances greater than 2%, while 26% of the estimates were around 1%, and 7.6% of the estimates were below 1%.

An alternative for carrying out species delimitation, especially in cases in which genetic distance values are below 1%, is the application of multiple "automatic species delimitation" methods, which provides an efficient approach in identifying putative species, or MOTUs [65]. The ABGD species delimitation analysis is known to result in conservative delimitation values and be unlikely to partition variations into species [66]. This method identified only three major *Squalus* strains, namely *S. suckleyi* and *S. acanthias* (group I), *S. blainville*, *S. brevirostris*, *S. cubensis*, and *S. albicaudus* (group II), and *S. japonicus*, *S. edmundsi*, *S. grahami*, *S. clarkae*, and *S. mitsukurii* (group III). Verissimo et al. [21] investigated *Squalus* species by employing the COI and NADH2 mitochondrial genes in 19 nominal species and also detected three major lineages. Thus, the present results were in accordance with previous studies [22,25,26].

Formerly, group I was represented only by *S. acanthias*. However, Ebert et al. [9] resurrected S. *suckleyi*, an endemic species from the North Pacific, and allocated it within this group. Our results supported the separation of *S. acanthias* and *S. suckleyi*, despite a genetic distance of below 1%. We identified one MOTU for *S. acanthias* and one MOTU for *S. suckleyi* according to the respective nominal species by means of the PTP analysis. The GMYC analysis also detected one MOTU for *S. suckleyi* but identified six MOTUs for *S. acanthias*. Ebert et al. [9] highlighted that numerous synonyms for *S. acanthias* are in place, with regional subspecies within this subgroup for the North Atlantic Ocean, the Black Sea, and the west coast of Southern Africa. A relationship among the analyzed locations was not, however, detected in the present work. Interestingly, the haplotype network revealed that these species did not share haplotypes.

The four nominal species that were part of group II, *S. blainville*, *S. brevirostris*, *S. cubensis* and *S. albicaudus*, exhibited high COI distance values (>1.9%, Table 1), but group II presented the lowest interspecific genetic distance detected herein, that between *S. cubensis* and *S. albicaudus* (0.72%). PTP analyses indicated one MOTU for *S. blainville*, one for *S. brevirostris*, and only one for *S. cubensis* + *S. albicaudus*. The GMYC analysis identified seven MOTUs, four for *S. blainville*, one for *S. blainville*, one for *S. cubensis*, and one for *S. albicaudus*. Other authors have indicated the existence of more than one species, requiring a taxonomic revision [26], and highlighted the diversity among individuals identified as *S. blainville*. In this context, it is also interesting to note the close relationship with *S. brevirostris* observed in the haplotype network, with only two mutations of difference.

Low levels of genetic variation among species of elasmobranchs with the COI gene have been reported in the literature [21,27,37,38,67]. These were also detected in this study among some members of the genus *Squalus*. The low rates found may have been due to the evolutionary aspects of the group, as it is already known that sharks and rays have lower evolutionary rates than other fish species [32], or even a recent speciation process.

In a taxonomic review of the *Squalus* genus occurring in the Southeastern Atlantic Ocean performed by Viana et al. [2], the authors diagnosed individuals previously identified as *S. cubensis* in the region, with a new species described as *S. albicaudus*. In the present study, the only delimitation analysis able to separate these two nominal species was the GMYC, although the haplotype network analysis revealed haplotype sharing between them. One hypothesis is that *S. albicaudus* may comprise an *S. cubensis* population in the Southeastern Atlantic Ocean currently undergoing a speciation process, which may still be very recent and incomplete, as these species still share haplotypes. As mentioned by other authors, elasmobranch speciation is very common [67,68] and boundaries among populations or species are often difficult to detect.

The complexity detected in group III herein was noted in *S. mitsukurii* and *S. clarkae*, which exhibited low COI distance values (1%); distances were greater than 1.3% for the other species (Table 1). Pfleger et al. [4] reported a 2.8% divergence between *S. clarkae* in the Gulf of Mexico and *S. mitsukurii* in Japan when employing the COI technique [27].

Herein, the GMYC method identified similar results to those of the PTP analysis but divided *S. clarkae* into a single MOTU and subdivided *S. mitsukurii* into six MOTUs. The haplotype network revealed that *S. japonicus*, *S. edmundsi*, and *S. grahami* did not share haplotypes, unlike *S. clarkae* and *S. mitsukurii*, which did. *Squalus mitsukurii* was originally described in Japan by Jordan and Snyder (1903), and despite identification issues due to morphological character overlapping, this species presents a circumglobal distribution [69], and its occurrence has likely been overestimated [4,70,71].

5. Conclusions

Our approach, using molecular tools for species delimitation, presented data to assist in future studies of species delimitation in the genus *Squalus*, since in many cases morphological data by themselves are not decisive. However, molecular data alone do not replace traditional taxonomy in the delimitation of species [72]. This integrative approach has been used over the years and has proven to be quite effective in elasmobranchs [72–76] and in other groups of organisms [76,77].

It is important to emphasize that the use of MOTUs represents an initial approach to support specific integrative analyses aiming for the identification of taxonomic groups [65]. However, because of the difficulty of morphologically defining *Squalus* species, many sequences available in genetic databases, i.e., BOLD and GenBank, indicate misidentifications or identifications only at the genus or family levels, making them not very useful for molecular identification purposes. Incorrect identifications or identifications at a higher taxonomic level often reflect high numbers of BINs, which are generally associated with ghost species but may also indicate undescribed species [5,75,78–80]. We also highlight that the barcode DNA in fish often does not reveal the genetic peculiarities existing in the groups, mainly in species with taxonomic complexity such as that already known to exist in *Squalus* [3,21,25,26], resulting in the need to use other genetic markers [36,81–83], or associations with morphological studies, for an integrated taxonomic approach [73,74,83].

Supplementary Materials: The following supporting information can be downloaded at: https://www.mdpi.com/article/10.3390/d14070544/s1, Figure S1: Samples collected from individuals of the genus *Squalus* in the Western Atlantic Ocean and Pacific Ocean; Table S1. Species, locality, Sequence ID, analyses in this study, and GenBank accession numbers of specimens of *Squalus*.

Author Contributions: Conceptualization, F.H.V.H., M.M.R., J.M.D.d.A., G.D., C.O., F.F. and V.P.C.; methodology, A.A.A., A.M.C.L.A. and M.M.R.; validation, A.A.A.; formal analysis, A.A.A., M.V. and V.P.C.; investigation, A.A.A., S.M.D., J.M.D.d.A., G.D. and F.F.; resources, C.O.; data curation, P.R., M.V., S.M.D. and G.D.; writing—original draft preparation, A.A.A., S.M.D., J.M.D.d.A., A.M.C.L.A., C.O., F.F. and V.P.C.; writing—review and editing, A.A,.A., P.R., M.V., M.M.R., C.O., F.F. and V.P.C.; supervision, F.H.V.H.; project administration V.P.C.; funding acquisition, F.F. All authors have read and agreed to the published version of the manuscript.

Funding: This research was funded by the Conselho Nacional de Desenvolvimento Científico e Tecnológico (CNPq, grants to V.P.C., 107761/2019-0 to A.A.A.). C.O. received financial support from Fundação de Amparo à Pesquisa do Estado de Sao Paulo—FAPESP grants 2018/20610-1, 2016/09204-6, and 2014/26508-3, and from Conselho Nacional de Desenvolvimento Científico e Tecnológico—CNPq proc. 306054/2006-0 (C.O.).

Institutional Review Board Statement: All samples were collected in strict accordance with the regulations of the Brazilian Federal Animal Ethics Committee (SISBIO 13843–1), and the analyses followed the International Guidelines for Animal Experiments, as authorized by CEEAA IBB/UNESP, protocol number 556.

Informed Consent Statement: Not applicable.

Data Availability Statement: The data presented in the present study will be deposited and made available openly through the GenBank genetic sequence database.

Acknowledgments: We would like to thank the UNESP Biosciences Institute for the infrastructure provided,, Fundação de Amparo à Pesquisa do Estado de Sao Paulo—FAPESP and from Conselho Nacional de Desenvolvimento Científico e Tecnológico—CNPq We are grateful for the samples provided by Gavin Naylor, Florida Museum of Natural History, University of Florida, Gainesville, FL, USA.

Conflicts of Interest: The authors declare no conflict of interest.

References

1. Last, P.; White, W.; Pogonoski, J. *Descriptions of New Dogfishes of the Genus Squalus (Squaloidea: Squalidae)*; CSIRO Marine and Atmospheric Research: Hobart, Australia, 2007.
2. Viana, S.T.D.F.; Carvalho, M.R.D.; Gomes, U.L. Taxonomy and Morphology of Species of the Genus Squalus Linnaeus, 1758 from the Southwestern Atlantic Ocean (Chondrichthyes: Squaliformes: Squalidae). *Zootaxa* **2016**, *4133*, 1–89. [CrossRef] [PubMed]
3. Daly-Engel, T.S.; Koch, A.; Anderson, J.M.; Cotton, C.F.; Rubbs, R.D. Description of a New Deep-Water Dogfish Shark from Hawaii, with Comments on the Squalus mitsukurii Species Complex in the West Pacific. *Zookeys* **2018**, *798*, 135–157. [CrossRef] [PubMed]
4. Pfleger, M.O.; Grubbs, R.D.; Cotton, C.F.; Daly-Engel, T.S. Squalus clarkae Sp. Nov., a New Dogfish Shark from the Northwest Atlantic and Gulf of Mexico, with Comments on the Squalus mitsukurii Species Complex. *Zootaxa* **2018**, *4444*, 101–119. [CrossRef] [PubMed]
5. Ebert, D.; Fowler, S.; Compagno, L. *Sharks of the World: A Fully Illustrated Guide*; Wild Nature Press: Plymouth, MA, USA, 2013; ISBN 978-0-9573946-0-5.
6. Compagno, L.J. An Annotated and Illustrated Catalogue of Shark Species Known to Date. In *FAO Species Catalogue*; United Nations Development Programme Food and Agriculture Organization of The United Nations: Rome, Italy, 1984; Volume 4, pp. vii + 249.
7. Compagno, L.J.V.; Dando, M.; Fowler, S. *Sharks of the World*; Princeton University Press: Princeton, NJ, USA, 2005.
8. Cortés, E. Life History Patterns and Correlations in Sharks. *Rev. Fish. Sci.* **2000**, *8*, 299–344. [CrossRef]
9. Ebert, D.; White, W.; Goldman, K.; Compagno, L.; Daly-Engel, T.; Ward, R. Resurrection and Redescription of Squalus suckleyi (Girard, 1854) from the North Pacific, with Comments on the Squalus acanthias Subgroup (Squaliformes: Squalidae). *Zootaxa* **2010**, *2612*, 22–40. [CrossRef]
10. Wourms, J.P. Reproduction and Development in Chondrichthyan Fishes. *Am. Zool.* **1977**, *17*, 379–410. [CrossRef]
11. Dell'Apa, A.; Bangley, C.W.; Rulifson, R.A. Who Let the Dogfish out? A Review of Management and Socio-Economic Aspects of Spiny Dogfish Fisheries. *Rev. Fish Biol. Fish.* **2015**, *25*, 273–295. [CrossRef]
12. Lamarca, F.; Vianna, M.; Vilasboa, A. The First Reproductive Parameters and Evidence of Multiple Paternity in One New Spiny Dogfish Species, Squalus albicaudus (Squaliformes, Squalidae). *J. Fish Biol.* **2020**, *97*, 1268–1272. [CrossRef]
13. Ketchen, K.S. Size at Maturity, Fecundity, and Embryonic Growth of the Spiny Dogfish (Squalus acanthias) in British Columbia Waters. *J. Fish. Res. Board Can.* **1972**, *29*, 1717–1723. [CrossRef]
14. Nammack, M.F.; Musick, J.A.; Colvocoresses, J.A. Life History of Spiny Dogfish off the Northeastern United States. *Trans. Am. Fish. Soc.* **1985**, *114*, 367–376. [CrossRef]
15. Jones, T.S.; Ugland, K.I. Reproduction of Female Spiny Dogfish, Squalus acanthias, in the Oslofjord. *Fish. Bull. Natl. Ocean. Atmos. Adm.* **2001**, *99*, 685–690.
16. Dulvy, N.K.; Fowler, S.L.; Musick, J.A.; Cavanagh, R.D.; Kyne, P.M.; Harrison, L.R.; Carlson, J.K.; Davidson, L.N.; Fordham, S.V.; Francis, M.P.; et al. Extinction Risk and Conservation of the World's Sharks and Rays. *eLife* **2014**, 3, e00590. [CrossRef] [PubMed]
17. List, I.R.; The IUCN Red List of Threatened Species. Version 2020-1. 2020. Available online: https://www.iucnredlist.org (accessed on 21 March 2021).
18. MMA; Bataus, Y.S.D.L.; Nogueira, C.D.C.; Marcovaldi, M.A.; Vogt, R.C.; Coutinho, M.E.; Werneck, F.D.P.; Martins, M.R.C.; Colli, G.R. *Livro Vermelho da Fauna Brasileira Ameaçada de Extinção*; ICMBio, MMA Distrito Federal; Instituto Chico Mendes de Conservação da Biodiversidade: Brasília, Brazil, 2018.
19. Geraci, M.L.; Ragonese, S.; Norrito, G.; Scannella, D.; Falsone, F.; Vitale, S. A Tale on the Demersal and Bottom Dwelling Chondrichthyes in the South of Sicily through 20 Years of Scientific Survey. In *Chondrichthyes—Multidisciplinary Approach*; IntechOpen: London, UK, 2017; pp. 13–37.

20. Last, P.R.; White, W.T.; Pogonoski, J.J.; Gledhill, D.C.; Yearsley, G.K.; Ward, R.D. *Application of a Rapid Taxonomic Approach to the Genus Squalus*; CSIRO Marine and Atmospheric Research: Hobart, Australia, 2007.
21. Veríssimo, A.; Zaera-Perez, D.; Leslie, R.; Iglésias, S.P.; Séret, B.; Grigoriou, P.; Sterioti, A.; Gubili, C.; Barría, C.; Duffy, C.; et al. Molecular Diversity and Distribution of Eastern Atlantic and Mediterranean Dogfishes Squalus Highlight Taxonomic Issues in the Genus. *Zool. Scr.* **2017**, *46*, 414–428. [CrossRef]
22. Bellodi, A.; Porcu, C.; Cau, A.; Marongiu, M.F.; Melis, R.; Mulas, A.; Pesci, P.; Follesa, M.C.; Cannas, R. Investigation on the Genus Squalus in the Sardinian Waters (Central-Western Mediterranean) with Implications on Its Management. *Mediterr. Mar. Sci.* **2018**, *19*, 256–272. [CrossRef]
23. Díaz-Jaimes, P.; Bayona-Vásquez, N.J.; Escatel-Luna, E.; Uribe-Alcocer, M.; Pecoraro, C.; Adams, D.H.; Frazier, B.S.; Glenn, T.C.; Babbucci, M. Population Genetic Divergence of Bonnethead Sharks Sphyrna tiburo in the Western North Atlantic: Implications for Conservation. *Aquat. Conserv. Mar. Freshw. Ecosyst.* **2021**, *31*, 83–98. [CrossRef]
24. Ceballos, G.; Ehrlich, P.R.; Barnosky, A.D.; García, A.; Pringle, R.M.; Palmer, T.M. Accelerated Modern Human–Induced Species Losses: Entering the Sixth Mass Extinction. *Sci. Adv.* **2015**, *1*, e1400253. [CrossRef]
25. Ziadi-Künzli, F.; Soliman, T.; Imai, H.; Sakurai, M.; Maeda, K.; Tachihara, K. Re-Evaluation of Deep-Sea Dogfishes (Genus Squalus) in Japan Using Phylogenetic Inference. *Deep. Sea Res. Part I Oceanogr. Res. Pap.* **2020**, *160*, 103261. [CrossRef]
26. Ferrari, A.; Di Crescenzo, S.; Cariani, A.; Crobe, V.; Benvenuto, A.; Piattoni, F.; Mancusi, C.; Bonnici, L.; Bonello, J.J.; Schembri, P.J.; et al. Puzzling over Spurdogs: Molecular Taxonomy Assessment of the Squalus Species in the Strait of Sicily. *Eur. Zool. J.* **2021**, *88*, 181–190. [CrossRef]
27. Naylor, G.J.P.; Caira, J.N.; Jensen, K.; Rosana, K.A.M.; White, W.T.; Last, P.R. *A DNA Sequence-Based Approach to the Identification of Shark and Ray Species and Its Implications for Global Elasmobranch Diversity and Parasitology*; Bulletin of the American Museum of Natural History, No. 367; DNA Identification of Sharks and Rays; American Museum of Natural History—Scientific Publications: New York, NY, USA, 2012.
28. Vella, A.; Vella, N.; Schembri, S. A Molecular Approach towards Taxonomic Identification of Elasmobranch Species from Maltese Fisheries Landings. *Mar. Genom.* **2017**, *36*, 17–23. [CrossRef]
29. Ward, R.D.; Holmes, B.H.; Zemlak, T.S.; Smith, P.J. Part 12—DNA Barcoding Discriminates Spurdogs of the Genus Squalus. In *Descriptions of New Dogfishes of the Genus Squalus (Squaloidea: Squalidae)*; CSIRO Marine and Atmospheric Research: Hobart, Australia, 2007; pp. 117–130.
30. Ward, R.D. DNA Barcode Divergence among Species and Genera of Birds and Fishes. *Mol. Ecol. Resour.* **2009**, *9*, 1077–1085. [CrossRef]
31. Mabragaña, E.; Díaz de Astarloa, J.M.; Hanner, R.; Zhang, J.; González Castro, M. DNA Barcoding Identifies Argentine Fishes from Marine and Brackish Waters. *PLoS ONE* **2011**, *6*, e28655. [CrossRef] [PubMed]
32. Hara, Y.; Yamaguchi, K.; Onimaru, K.; Kadota, M.; Koyanagi, M.; Keeley, S.D.; Tatsumi, K.; Tanaka, K.; Motone, F.; Kageyama, Y. Shark Genomes Provide Insights into Elasmobranch Evolution and the Origin of Vertebrates. *Nat. Ecol. Evol.* **2018**, *2*, 1761–1771. [CrossRef] [PubMed]
33. Ovenden, J.R.; Dudgeon, C.; Feutry, P.; Feldheim, K.; Maes, G.E. Genetics and Genomics for Fundamental and Applied Research on Elasmobranchs. In *Shark Research: Emerging Technologies and Applications for the Field and Laboratory*; CRC Press: Boca Raton, FL, USA, 2019; pp. 235–253.
34. Pazmiño, D.A.; van Herderden, L.; Simpfendorfer, C.A.; Junge, C.; Donnellan, S.C.; Hoyos-Padilla, E.M.; Duffy, C.A.; Huveneers, C.; Gillanders, B.M.; Butcher, P.A. Introgressive Hybridisation between Two Widespread Sharks in the East Pacific Region. *Mol. Phylogenet. Evol.* **2019**, *136*, 119–127. [CrossRef]
35. Delaval, A.; Frost, M.; Bendall, V.; Hetherington, S.J.; Stirling, D.; Hoarau, G.; Jones, C.S.; Noble, L.R. Population and Seascape Genomics of a Critically Endangered Benthic Elasmobranch, the Blue Skate Dipturus batis. *Evol. Appl.* **2022**, *15*, 78–94. [CrossRef] [PubMed]
36. Henderson, A.C.; Reeve, A.J.; Jabado, R.W.; Naylor, G.J. Taxonomic Assessment of Sharks, Rays and Guitarfishes (Chondrichthyes: Elasmobranchii) from South-Eastern Arabia, Using the NADH Dehydrogenase Subunit 2 (NADH2) Gene. *Zool. J. Linn. Soc.* **2016**, *176*, 399–442. [CrossRef]
37. Ward, R.D.; Holmes, B.H.; White, W.T.; Last, P.R. DNA Barcoding Australasian Chondrichthyans: Results and Potential Uses in Conservation. *Mar. Freshw. Res.* **2008**, *59*, 57–71. [CrossRef]
38. Hellberg, R.S.; Isaacs, R.B.; Hernandez, E.L. Identification of Shark Species in Commercial Products Using DNA Barcoding. *Fish. Res.* **2019**, *210*, 81–88. [CrossRef]
39. Puillandre, N.; Lambert, A.; Brouillet, S.; Achaz, G. ABGD, Automatic Barcode Gap Discovery for Primary Species Delimitation. *Mol. Ecol.* **2012**, *21*, 1864–1877. [CrossRef]
40. Zhang, J.; Kapli, P.; Pavlidis, P.; Stamatakis, A. A General Species Delimitation Method with Applications to Phylogenetic Placements. *Bioinformatics* **2013**, *29*, 2869–2876. [CrossRef]
41. Pons, J.; Barraclough, T.G.; Gomez-Zurita, J.; Cardoso, A.; Duran, D.P.; Hazell, S.; Kamoun, S.; Sumlin, W.D.; Vogler, A.P. Sequence-Based Species Delimitation for the DNA Taxonomy of Undescribed Insects. *Syst. Biol.* **2006**, *55*, 595–609. [CrossRef]
42. Fujisawa, T.; Barraclough, T.G. Delimiting Species Using Single-Locus Data and the Generalized Mixed Yule Coalescent Approach: A Revised Method and Evaluation on Simulated Data Sets. *Syst. Biol.* **2013**, *62*, 707–724. [CrossRef] [PubMed]

43. Melo, B.F.; Sato, Y.; Foresti, F.; Oliveira, C. The Roles of Marginal Lagoons in the Maintenance of Genetic Diversity in the Brazilian Migratory Fishes Prochilodus argenteus and *P. costatus*. *Neotrop. Ichthyol.* **2013**, *11*, 625–636. [CrossRef]
44. Kearse, M.; Moir, R.; Wilson, A.; Stones-Havas, S.; Cheung, M.; Sturrock, S.; Buxton, S.; Cooper, A.; Markowitz, S.; Duran, C.; et al. Geneious Basic: An Integrated and Extendable Desktop Software Platform for the Organization and Analysis of Sequence Data. *Bioinformatics* **2012**, *28*, 1647–1649. [CrossRef] [PubMed]
45. Edgar, R.C. MUSCLE: A Multiple Sequence Alignment Method with Reduced Time and Space Complexity. *BMC Bioinform.* **2004**, *5*, 113. [CrossRef]
46. Kousteni, V.; Kasapidis, P.; Kotoulas, G.; Megalofonou, P. Evidence of High Genetic Connectivity for the Longnose Spurdog Squalus blainville in the Mediterranean Sea. *Mediterr. Mar. Sci.* **2016**, *17*, 371–383. [CrossRef]
47. Knebelsberger, T.; Landi, M.; Neumann, H.; Kloppmann, M.; Sell, A.F.; Campbell, P.D.; Laakmann, S.; Raupach, M.J.; Carvalho, G.R.; Costa, F.O. A Reliable DNA Barcode Reference Library for the Identification of the North European Shelf Fish Fauna. *Mol. Ecol. Resour.* **2014**, *14*, 1060–1071. [CrossRef]
48. Landi, M.; Dimech, M.; Arculeo, M.; Biondo, G.; Martins, R.; Carneiro, M.; Carvalho, G.R.; Lo Brutto, S.; Costa, F.O. DNA Barcoding for Species Assignment: The Case of Mediterranean Marine Fishes. *PLoS ONE* **2014**, *9*, e106135. [CrossRef]
49. Stelbrink, B.; von Rintelen, T.; Cliff, G.; Kriwet, J. Molecular Systematics and Global Phylogeography of Angel Sharks (Genus Squatina). *Mol. Phylogenetics Evol.* **2010**, *54*, 395–404. [CrossRef]
50. Almerón-Souza, F.; Sperb, C.; Castilho, C.L.; Figueiredo, P.I.C.C.; Gonçalves, L.T.; Machado, R.; Oliveira, L.R.; Valiati, V.H.; Fagundes, N.J.R. Molecular Identification of Shark Meat from Local Markets in Southern Brazil Based on DNA Barcoding: Evidence for Mislabeling and Trade of Endangered Species. *Front. Genet.* **2018**, *9*, 138. [CrossRef]
51. McCusker, M.R.; Denti, D.; Van Guelpen, L.; Kenchington, E.; Bentzen, P. Barcoding Atlantic Canada's Commonly Encountered Marine Fishes. *Mol. Ecol. Resour.* **2013**, *13*, 177–188. [CrossRef]
52. Steinke, D.; Zemlak, T.S.; Gavin, H.; Hebert, P.D.N. DNA Barcoding Fishes of the Canadian Pacific. *Mar. Biol.* **2009**, *156*, 2641–2647. [CrossRef]
53. Gkafas, G.A.; Megalofonou, P.; Batzakas, G.; Apostolidis, A.P.; Exadactylos, A. Molecular Phylogenetic Convergence within Elasmobranchii Revealed by Cytochrome Oxidase Subunits. *Biochem. Syst. Ecol.* **2015**, *61*, 510–515. [CrossRef]
54. Kumar, S.; Stecher, G.; Li, M.; Knyaz, C.; Tamura, K. MEGA X: Molecular Evolutionary Genetics Analysis across Computing Platforms. *Mol. Biol. Evol.* **2018**, *35*, 1547–1549. [CrossRef]
55. Kimura, M. A Simple Method for Estimating Evolutionary Rates of Base Substitutions through Comparative Studies of Nucleotide Sequences. *J. Mol. Evol.* **1980**, *16*, 111–120. [CrossRef] [PubMed]
56. Felsenstein, J. Confidence Limits on Phylogenies: An Approach Using the Bootstrap. *Evolution* **1985**, *39*, 783–791. [CrossRef] [PubMed]
57. Librado, P.; Rozas, J. DnaSP v5: A Software for Comprehensive Analysis of DNA Polymorphism Data. *Bioinformatics* **2009**, *25*, 1451–1452. [CrossRef]
58. Leigh, J.W.; Bryant, D. POPART: Full-feature Software for Haplotype Network Construction. *Methods Ecol. Evol.* **2015**, *6*, 1110–1116. [CrossRef]
59. Ward, R.D.; Zemlak, T.S.; Innes, B.H.; Last, P.R.; Hebert, P.D. DNA Barcoding Australia's Fish Species. *Philos. Trans. R. Soc. B Biol. Sci.* **2005**, *360*, 1847–1857. [CrossRef]
60. Hubert, N.; Hanner, R.; Holm, E.; Mandrak, N.E.; Taylor, E.; Burridge, M.; Watkinson, D.; Dumont, P.; Curry, A.; Bentzen, P.; et al. Identifying Canadian Freshwater Fishes through DNA Barcodes. *PLoS ONE* **2008**, *3*, e2490. [CrossRef]
61. Pereira, L.H.; Hanner, R.; Foresti, F.; Oliveira, C. Can DNA Barcoding Accurately Discriminate Megadiverse Neotropical Freshwater Fish Fauna? *BMC Genet.* **2013**, *14*, 20. [CrossRef]
62. Ramirez, J.L.; Birindelli, J.L.; Carvalho, D.C.; Affonso, P.R.A.M.; Venere, P.C.; Ortega, H.; Carrillo-Avila, M.; Rodríguez-Pulido, J.A.; Galetti, P.M. Revealing Hidden Diversity of the Underestimated Neotropical Ichthyofauna: DNA Barcoding in the Recently Described Genus Megaleporinus (Characiformes: Anostomidae). *Front. Genet.* **2017**, *8*, 149. [CrossRef] [PubMed]
63. Ramirez, J.L.; Galetti, P.M. DNA Barcode and Evolutionary Relationship within Laemolyta Cope 1872 (Characiformes: Anostomidae) through Molecular Analyses. *Mol. Phylogenet. Evol.* **2015**, *93*, 77–82. [CrossRef] [PubMed]
64. Cerqueira, N.N.C.D.; Rotundo, M.M.; Marceniuk, A.P.; da Cruz, V.P.; Foresti, F.; Oliveira, C. Molecular Identification of Brachygenys and Haemulon Species (Perciformes: Haemulidae) from the Brazilian Coast. *Neotrop. Ichthyol.* **2021**, *19*. [CrossRef]
65. Ortiz, D.; Francke, O.F. Two DNA Barcodes and Morphology for Multi-Method Species Delimitation in Bonnetina tarantulas (Araneae: Theraphosidae). *Mol. Phylogenet. Evol.* **2016**, *101*, 176–193. [CrossRef] [PubMed]
66. Carstens, B.C.; Pelletier, T.A.; Reid, N.M.; Satler, J.D. How to Fail at Species Delimitation. *Mol. Ecol.* **2013**, *22*, 4369–4383. [CrossRef] [PubMed]
67. Borsa, P.; Shen, K.-N.; Arlyza, I.S.; Hoareau, T.B. Multiple Cryptic Species in the Blue-Spotted Maskray (Myliobatoidei: Dasyatidae: Neotrygon Spp.): An Update. *C. R. Biol.* **2016**, *339*, 417–426. [CrossRef]
68. Borsa, P.; Arlyza, I.S.; Hoareau, T.B.; Shen, K.-N. Diagnostic Description and Geographic Distribution of Four New Cryptic Species of the Blue-Spotted Maskray Species Complex (Myliobatoidei: Dasyatidae; Neotrygon Spp.) Based on DNA Sequences. *J. Ocean. Limnol.* **2018**, *36*, 827–841. [CrossRef]

69. Bonello, J.J.; Bonnici, L.; Ferrari, A.; Cariani, A.; Schembri, P.J. Not All That Clear Cut: Intraspecific Morphological Variability in Squalus blainville (Risso, 1827) and Implications for Identification of the Species. *J. Mar. Biol. Assoc. U. K.* **2016**, *96*, 1585–1596. [CrossRef]
70. Viana, S.T.F.; de Carvalho, M.R.; Ebert, D.A. Squalus bassi Sp. Nov., a New Long-Snouted Spurdog (Chondrichthyes: Squaliformes: Squalidae) from the Agulhas Bank. *J. Fish Biol.* **2017**, *91*, 1178–1207. [CrossRef]
71. Viana, S.T.; de Carvalho, M.R. Squalus shiraii Sp. Nov. (Squaliformes, Squalidae), a New Species of Dogfish Shark from Japan with Regional Nominal Species Revisited. *Zoosyst. Evol.* **2020**, *96*, 275. [CrossRef]
72. Schlick-Steiner, B.C.; Steiner, F.M.; Seifert, B.; Stauffer, C.; Christian, E.; Crozier, R.H. Integrative Taxonomy: A Multisource Approach to Exploring Biodiversity. *Annu. Rev. Entomol.* **2010**, *55*, 421–438. [CrossRef] [PubMed]
73. Lim, K.C.; White, W.T.; Then, A.Y.; Naylor, G.J.; Arunrugstichai, S.; Loh, K.-H. Integrated Taxonomy Revealed Genetic Differences in Morphologically Similar and Non-Sympatric Scoliodon macrorhynchos and S. laticaudus. *Animals* **2022**, *12*, 681. [CrossRef] [PubMed]
74. Petean, F.F.; Naylor, G.J.; Lima, S.M. Integrative Taxonomy Identifies a New Stingray Species of the Genus Hypanus rafinesque, 1818 (Dasyatidae, Myliobatiformes), from the Tropical Southwestern Atlantic. *J. Fish Biol.* **2020**, *97*, 1120–1142. [CrossRef] [PubMed]
75. Cerutti-Pereyra, F.; Meekan, M.G.; Wei, N.-W.V.; O'Shea, O.; Bradshaw, C.J.A.; Austin, C.M. Identification of Rays through DNA Barcoding: An Application for Ecologists. *PLoS ONE* **2012**, *7*, e36479. [CrossRef]
76. Gabbanelli, V.; Díaz de Astarloa, J.M.; Gonzalez-Castro, M.; Vazquez, D.M.; Mabragaña, E. Almost a Century of Oblivion: Integrative Taxonomy Allows the Resurrection of the Longnose Skate Zearaja brevicaudata (Marini, 1933) (Rajiformes; Rajidae). *C. R. Biol.* **2018**, *341*, 454–470. [CrossRef]
77. García-Melo, J.E.; Oliveira, C.; Silva, G.J.D.C.; Ochoa-Orrego, L.E.; Pereira, L.H.G.; Maldonado-Ocampo, J.A. Species Delimitation of Neotropical Characins (Stevardiinae): Implications for Taxonomy of Complex Groups. *PLoS ONE* **2019**, *14*, e0216786. [CrossRef]
78. Meiklejohn, K.A.; Damaso, N.; Robertson, J.M. Assessment of BOLD and GenBank—Their Accuracy and Reliability for the Identification of Biological Materials. *PLoS ONE* **2019**, *14*, e0217084. [CrossRef]
79. Pentinsaari, M.; Ratnasingham, S.; Miller, S.E.; Hebert, P.D.N. BOLD and GenBank Revisited—Do Identification Errors Arise in the Lab or in the Sequence Libraries? *PLoS ONE* **2020**, *15*, e0231814. [CrossRef]
80. Ip, Y.C.A.; Chang, J.J.M.; Lim, K.K.P.; Jaafar, Z.; Wainwright, B.J.; Huang, D. Seeing through Sedimented Waters: Environmental DNA Reduces the Phantom Diversity of Sharks and Rays in Turbid Marine Habitats. *BMC Ecol. Evol.* **2021**, *21*, 166. [CrossRef]
81. Feitosa, L.M.; Martins, A.P.B.; Giarrizzo, T.; Macedo, W.; Monteiro, I.L.; Gemaque, R.; Nunes, J.L.S.; Gomes, F.; Schneider, H.; Sampaio, I. DNA-Based Identification Reveals Illegal Trade of Threatened Shark Species in a Global Elasmobranch Conservation Hotspot. *Sci. Rep.* **2018**, *8*, 3347. [CrossRef]
82. O'Bryhim, J.R.; Parsons, E.C.M.; Lance, S.L. Forensic Species Identification of Elasmobranchs Landed in Costa Rican Artisanal Fisheries. *Fish. Res.* **2021**, *233*, 105755. [CrossRef]
83. Bellodi, A.; Benvenuto, A.; Melis, R.; Mulas, A.; Barone, M.; Barría, C.; Cariani, A.; Carugati, L.; Chatzispyrou, A.; Desrochers, M.; et al. Call Me by My Name: Unravelling the Taxonomy of the Gulper Shark Genus Centrophorus in the Mediterranean Sea through an Integrated Taxonomic Approach. *Zool. J. Linn. Soc.* **2022**, *195*, 815–840. [CrossRef]

diversity

MDPI

Article

DNA Barcoding of Invertebrates Inhabiting Olive Orchards and Vineyards Accelerates Understudied Mediterranean Biodiversity Assessment

Barbara Anđelić Dmitrović [1], Mišel Jelić [2], Emilia Rota [3] and Lucija Šerić Jelaska [1,*]

[1] Department of Biology, Faculty of Science, University of Zagreb, Rooseveltov trg 6, 10000 Zagreb, Croatia; barbara.andelic@biol.pmf.hr
[2] Varaždin City Museum, Šetalište Josipa Jurja Strossmayera 3, 42000 Varaždin, Croatia; misel.jelic@gmv.hr
[3] Department of Physics, Earth and Environmental Sciences, University of Siena, Via P.A. Mattioli 4, IT-53100 Siena, Italy; rota@unisi.it
* Correspondence: lucija.seric.jelaska@biol.pmf.hr

Abstract: The Mediterranean region has a high but unevenly studied level of invertebrate diversity. Genetic-based methods, such as DNA barcoding and metabarcoding, are proposed for biodiversity assessment; however, their application is not always straightforward. The current state of data available in genetic databases limits species identification, especially in the case of certain invertebrate groups. The aim of the study was (1) to assess the diversity of seven invertebrate groups, which are potential prey for predatory arthropods, in order to facilitate the analysis of the metabarcoding of trophic interactions and, thus, expand our knowledge on biocontrol potential, and (2) to estimate the representation of local species in BOLD and NCBI GenBank databases. The DNA barcoding results consisting of sequences for 269 specimens, collected in Zadar County, within Mediterranean part of Croatia, were used for species identification and species richness assessment through the comparison of our data with the data available in BOLD and NCBI databases, and by applying species delimitation methods. Previous barcode records enabled Lepidopteran, Hemipteran, Dipteran, and Hymenopteran species identification, while Collembolan and Oligochaetes species numbers were assessed using species delimitation. Our results showed that a high number of species were unique to the study area, especially in the case of Collembolans and Oligochaetes. We confirmed that the studied area is under-researched, which was particularly evident in taxonomically demanding groups, such as Collembolans and Oligochaetes, which are rich in rare endemic species.

Keywords: BOLD database; cytochrome c oxidase subunit; Croatia; diversity; invertebrates; Mediterranean region; NCBI database; species delimitation

Citation: Anđelić Dmitrović, B.; Jelić, M.; Rota, E.; Šerić Jelaska, L. DNA Barcoding of Invertebrates Inhabiting Olive Orchards and Vineyards Accelerates Understudied Mediterranean Biodiversity Assessment. *Diversity* **2022**, *14*, 182. https://doi.org/10.3390/d14030182

Academic Editor: Stephan Koblmüller

Received: 24 January 2022
Accepted: 22 February 2022
Published: 28 February 2022

1. Introduction

DNA barcoding is considered to be a valuable tool for rapid species identification, as it is based on molecular data and does not require specific expertise in morphological taxonomy [1,2]. The most common molecular marker used in the DNA barcoding of animals is the mitochondrial gene coding for cytochrome oxidase subunit one (*COI*). From 2003, when the DNA barcoding method was suggested [3], up to now, it has been used for taxonomic, phylogenetic, biodiversity, and monitoring purposes [2,4–8], and numerous barcode projects have been undertaken to support species identification [7–12]. In extensively studied regions, the DNA barcoding method has been successful in detecting unrecorded species, as well as in highlighting biodiversity, as shown with the example of dipteran fauna in Bavaria, Germany [13]. The use of DNA barcoding in diversity assessments has shown its usefulness in different regions and for different groups [14–16]. However, despite the extensive efforts being invested in submitting data to the National Center for Biotechnology Information's (NCBI) GenBank library and the Barcode of Life Database (BOLD), there are still huge gaps concerning some animal groups and geographical regions in these genetic databases [17,18]. This suggests that there is a difference between the pace at which new or neglected species are being identified and the pace at which they are being sequenced [19].

Information gathered from DNA barcodes can be used across many fields of biology, especially in ecology, for studying species interactions, the connection of different life stages, or trophic interactions [20]. In particular, trophic interactions can be studied through the metabarcoding of gut contents [17,21]. Many studies have been based on the sequencing of prey fragments' *COI* gene detected in the predator's gut and the subsequent identification of prey species by comparing the resulting sequences with those in reference databases [21–23]. In these and similar environmental DNA (eDNA) studies, the efficiency and accuracy of species identification strictly depend on the data available in the DNA databases [12,17]. This poses a problem when eDNA studies are conducted in geographical regions which are scarcely investigated and/or underrepresented in barcode libraries, especially if the regions are abundant in rare and endemic species, such as the Mediterranean.

Croatia's coastline is a part of the Mediterranean region in the west of the Balkan peninsula, and its invertebrate fauna is quite diversified [24,25]. This part of the country has a long tradition of olive oil and wine production, and olive orchards and vineyards are a common part of the landscape. Despite its high diversity, the region, including the agricultural land, is still under-researched, resulting in the underrepresentation of local fauna data. Oligochaetes (Lumbricidae) have a high level of diversity in the western part of the Balkan, and the Mediterranean zone, especially, is considered to be one of the biodiversity hotspots for the group [24]. However, a complete taxonomic resolution for the species present in the area has not been achieved. The Mediterranean lands harbour Europe's richest ant fauna, and half of Croatia's ant species can be found in its coastal zone [25]. Due to this diversity, the list of ant species present in Croatia is incomplete [25]. As with the species composition for the groups such as Hemipterans [26] and Lepidopterans [27], in agricultural areas, composition is related to the type of crop, given that these are mainly plant feeders and are related to certain plant species. Hemipterans especially have been extensively studied within the agricultural area due to various pests pertaining to this group [26]. Furthermore, Collembolan fauna is continuously being discovered, particularly in under-studied areas [28]. The local diversity of Collembolans, ants and all of the abovementioned groups has been negatively impacted by the agricultural land use and landscape changes [29–31].

Given that the species richness of the area is still understudied for many invertebrates, and given the land use pressures of agriculture, in the present study we implemented the DNA barcoding method to analyze species composition for Hemipterans, Dipterans, Lepidopterans, Hymenopterans, Collembolans, Oligochaetes, and Isopods, all groups important for assessing trophic interactions among invertebrates within Mediterranean agricultural lands [32–35]. Their selection was based on their potential as prey for common predatory arthropods, such as spiders and carabid beetles. Groups such as Hemipterans, Dipterans, and Lepidopterans include numerous species (e.g., *Scaphoideus titanus*, *Bactrocera oleae*, *Prays oleae*) that are frequent pests in vineyards and olive orchards. Other invertebrates, such as Oligochaetes, Collembolans, Hymenopterans, and Isopods, may serve as alternative prey to maintain predator populations during periods when pests are not present [35,36].

The aim of this research was to: (i) assess the diversity and genetic diversity of selected invertebrate groups in sampled Mediterranean vineyards, olive groves, and surrounding natural habitats; (ii) analyze the potential of the DNA barcoding method to determine species in Mediterranean vineyards, olive groves, and surrounding natural habitats in Zadar County in Croatia through available global data deposited in two databases, BOLD and NCBI; (iii) evaluate if the current state of the data available in global barcode databases is sufficient for conducting eDNA studies in Mediterranean terrestrial ecosystems and, if not, to detect the invertebrate groups that are part of the data gap.

Thus, additional data on species diversity and the genetic diversity of these groups can provide valuable information that can be implemented in subsequent ecological studies on biocontrol, diversity conservation in agricultural land, and sustainable management.

2. Materials and Methods

2.1. Sampling

Sampling was conducted in five sites within Zadar County, in the Mediterranean part of Croatia (Figure 1). A list with geographical coordinates, size of the agricultural land, and altitude is provided for the sampling sites (Supplementary Table S1). Sampling sites were located within two vineyards and two olive orchards, targeting fauna typical for such agricultural habitats in the Mediterranean part of Croatia, and within the surrounding natural habitat, with maquis and garrigue formed by the natural vegetation of the region.

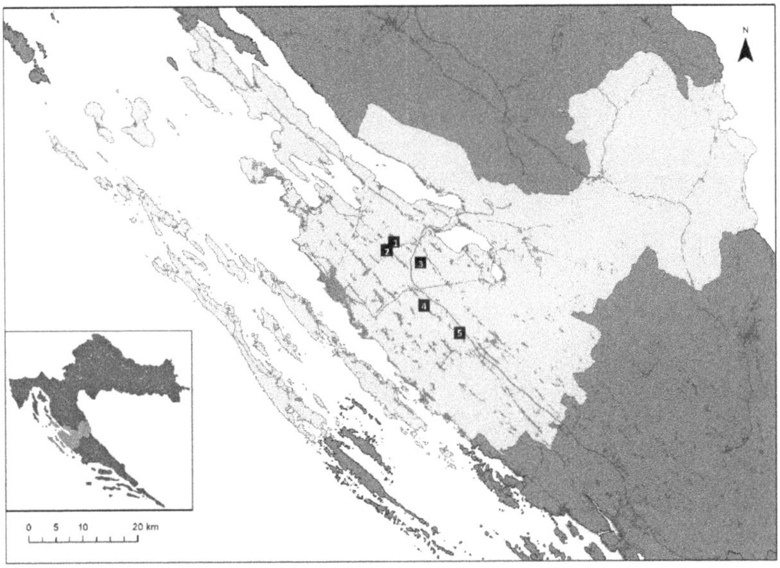

Figure 1. Map showing five sites (black squares), in Zadar County in the Mediterranean part of Croatia, selected for field research and sampling. The site within the natural habitat is located near Poličnik (C) (1), the olive orchards are located in Poličnik (OE) (2) and Škabrnja (OI) (5), the vineyards in Baštica (VI) (2) and Nadin (VE) (4).

Samples were collected during field research in 2018 and 2019. In 2018, sampling was conducted five times in the spring season, from April to July, and four times in the autumn season, from September to November. In 2018, three methods were applied: the beating method for tree canopy fauna [37], the Tullgren funnel method [38] for soil fauna, and additional sampling by hand where for anecic earthworms we used the electricity power as described below in more detailes. In 2019, sampling was performed by hand only, in the spring season from April to July, and in the autumn season from September to November. Specimens belonging to the Hemipterans, Dipterans, and Lepidopterans were collected using the beating method and handpicking, while Hymenopterans were collected with all methods. Collembolans and Isopods were sampled with the Tullgren method and by hand, as they are part of soil fauna. Oligochaetes were collected by adding water to selected soil patches (surface approximately 1 m^2) on which a current (10–100 Ah), with an adjustable frequency (20–120 Hz), was applied using a portable power generator to make them come to the surface. When on the surface, Oligochaetes were picked by hand and placed in tubes.

In the field, all collected samples were placed in absolute ethanol, except for Oligochaetes, which were, to preserve their morphological characters, firstly placed in 30% ethanol, then in 70%, to be finally stored in 100% ethanol. All the collected samples were sorted in the laboratory, divided into taxonomic groups, and processed further accordingly. They were stored at −20 °C prior to the application of molecular techniques.

2.2. DNA Extraction and PCR Reaction

DNA was extracted, depending on the invertebrate group and body size, from the entire body, abdomen, or leg in the case of arthropods, and a patch of epithelial tissue in case of earthworms. DNA was extracted using Chelex polymer (Sigma Aldrich, Merck, Darmstadt, Germany) following the protocol taken from Casquet et al. [39]. Another method of DNA extraction involved using the GenElute Mammalian Genomic DNA kit (Sigma Aldrich, Merck, Darmstadt, Germany) according to the manufacturer's instructions. Specimens' vouchers were stored at −20 °C at the research group laboratory.

The PCR analysis was performed using DreamTaq 2× (ThermoFisher Scientific, Waltham, MA, USA) or Emerald 2× (Takara, Kusatsu, Shiga, Japan) polymerase in a total volume of 10 μL using standard DNA barcode primers LCO1490/HCO2198 [40] in the final concentration of 0.2 μM. PCR analyses were optimized for all studied groups of invertebrates using gradient PCR. Annealing temperatures were adapted for all investigated groups: for Oligochaetes, Isopods, Dipterans, and Lepidopterans at 51 °C, Hymenopterans and Hemipterans at 47 °C, and Collembolans at 50 °C. Conditions for the DreamTaq polymerase were as follows: initial denaturation at 95 °C for 3 min, denaturation at 95 °C for 30 s and annealing temperatures as listed above for 30 s, extension at 72 °C for 60 s, and final extension at 72 °C for 10 min. The number of cycles for denaturation, annealing, and extension was 40. For Emerald polymerase, denaturation was at 94 °C for 30 s and extension at 72 °C for 60 s, annealing again depending on the group as listed above for 30 s, and the number of cycles was 40.

To screen for successful amplification and to measure the concentration of multiplied amplicons, we used gel electrophoresis with 1% agarose gel. To enable the sequencing of amplicons, we removed the remaining primers and dNTPs through the enzymatic purification of PCR products. The enzymes used in the reaction were exonuclease 1 (Exo1) and alkaline phosphatase (AP) (New England Biolabs, Ipswich, MA, USA) and for 10 μL reaction volume, 0.0025 μL of Exo1 and 0.005 μL of AP were added. Cycling conditions for the enzymatic PCR reaction were the following: 37 °C for 60 min, 80 °C for 20 min, and 4 °C for 10 min.

2.3. Morphological Analysis

Due to prior knowledge about the insufficiency of BOLD and NCBI GenBank data on Oligochaetes from Croatia and the need for further metabarcoding of this group, an especially careful morphological examination of the Oligochaete specimens was carried out, which revealed that they belong to the genus *Octodrilus* Omodeo, 1956. This genus has a main distribution center in the Balkan Peninsula and the identification of the species has been attempted using regional checklists and identification keys (e.g., Mršić, 1991; Szederjesi, 2017) [41,42]. Only sexually mature specimens were considered, because species diagnoses are mainly based on the position of the clitellar structures and the arrangement of internal reproductive structures. Unfortunately, a thorough examination of the taxonomic literature and comparisons with reference collection material revealed confusion and ambiguity in the nomenclature of the local *Octodrilus* taxa. Therefore, no species could be positively identified, and the morphologically based species identification provided herein remains provisional.

A key to the Collembola (Springtails) of Britain and Ireland [43] was used for Collembola specimen identification at family or genus level. For Formicidae species determination, two keys were used, Seifert (2018) [44] and Lebas et al. (2019) [45].

2.4. Sequencing and DNA Barcode Data Analysis

The sequencing of purified DNA amplicons was performed by Macrogen Inc. (Amsterdam, The Netherlands) using the LCO1490 amplification primer. Altogether, 56 sequences were obtained for Hymenopterans, 36 for Oligochaetes, 17 for Collembolans, 47 for Hemipterans, 60 for Dipterans, 16 for Lepidopterans, and 37 for Isopoda (Supplementary Table S2). Sequences were edited using BIOEDIT v.7.2. (Gene Codes Corporation, Ann Arbor, MI, USA) [46]. Chromatograms were manually checked for ambiguous nucleotides, stop codons, and indels in BIOEDIT v.7.2. Sequences were passed through databases to identify individuals at the species level, in BOLD using the BOLD Identification System (IDS) [5] and in NCBI using the Basic Local Alignment Search Tool (BLAST) [47]. We assigned Barcode Index Numbers (BINs) to our sequenced specimens in order to analyze diversity among our samples using the BOLD database algorithm [6]. For all analyzed specimens, specimen and collection data, as well as obtained sequences, were uploaded to BOLD (http://boldsystems.org, accessed on 15 February 2022) [6]. Besides our data set, for phylogenetic and species delimitation analysis, we included publicly available sequences of Collembolans (Supplementary File S1a,b) and *Octodrilus* (Oligochaeta) specimens reported for Europe (sequences available for specimens collected in France, Italy, Slovenia, and Croatia) (Supplementary File S2). Sequences of specimens from the same genus, including sequences from BOLD, were grouped and aligned in MEGA X [48] using MUSCLE [49]. Subsequently, aligned sequences were collapsed to unique haplotypes (haploid genotype) using FaBox (1.5) [50]. The DNA sequence alignments were checked for stop codons using Mesquite ver. 3.5 [51]. The Maximum Likelihood (ML) analysis was conducted using RAxML-HPC ver. 8.2.12 (https://www.phylo.org accessed on 14 February 2022) [52] on the CIPRES Gateway [53] using a GTRGAMMA model. A rapid bootstrap analysis with 1000 replicates [52] was used to search for the ML tree. The DNA barcoding and sequence alignment to current barcodes in DNA databases did not provide us with species-level identification and species numbers for Collembolans and Oligochaetes. We approached phylogenetic analyses with an awareness of the constraints of our data set, and our goal was to estimate the number of MOTUs and the diversity for these two groups, rather than solve phylogenetic relations. FigTree v1.4.4 (http://tree.bio.ed.ac.uk/software/figtree/ accessed on 14 February 2022) [54] was used for the visualization of the ML tree. Several species delimitation methods were applied to determine the number of MOTUs among studied samples. Species delimitation in the cases of *Octodrilus* (Oligochaeta, Lumbricidae), *Lepidocrytus* (Collembola, Entomobryidae), and *Orchesella* (Collembola, Entomobryidae) specimens was conducted using the Automatic Barcode Gap Discovery (ABGD) method [55], Bayesian implementation of the Poisson Tree Processes (bPTP) method [56], and multi-rate Poisson Tree Process (mPTP) method [57]. Aligned haplotype sequences served as input data for the ABGD method, which was executed with relative gap width = one and the Kimura 2 parameter (K2P) model. ML trees plotted in RAxML [52] using the above-listed parameters were input data for the bPTP and mPTP methods. P-distances for Oligochaete's haplotypes were calculated in MEGA X and compared to average p-distances between MOTU groups, identified using the mPTP method.

2.5. Rarefaction Curves

Rarefaction curves were analyzed for samples regularly collected across seasons using beating method and plotted in order to predict the expected number of BINs as a function of the sampling event's number. Plotting was performed using Microsoft Excel. If the curve reached the plateau, the number of sampled BINs was considered suitable for species richness estimation [58].

2.6. Jaccard Similarity

The Jaccard similarity index was used for comparing the absence/presence of species and morphospecies between study sites. The Jaccard indices were calculated and plotted in PAST 4.03 [59].

3. Results

3.1. DNA Barcoding Performance, Species Identification, and BIN Assignment

The alignment of obtained DNA sequences with those available in NCBI and BOLD, for the seven analyzed invertebrate groups, resulted in species-level identification for 142 out of 269 specimens (Supplementary Table S2). For Hymenopterans, the result was 83.93% (47 out of 56), for Hemipterans, 70.21% (33 out of 47), for Dipterans, 58.33% (35 out of 60), for Lepidopterans, 100% (16 out of 16), and for Isopods, 29.73% (11 out of 37). Such identification was not possible for Oligochaete and Collembolan specimens. Out of 269 obtained sequences, 244 sequences were arranged in 119 BINs, where 74 BINs (62.18%) were previously recorded in BOLD, and 45 BINs (37.82%) were unique BINs that were, until now, not reported in BOLD (Figure 2). Among the previously reported BINs, twenty contained up to ten sequences, including sequences from this research, nine of them pertaining to Hemipterans. This is a relatively low sequence number compared to other BINs in this database. For example, in the case of the Lepidoptera species *Prays oleae* (Praydidae), the formed BIN contained 173 reported sequences, and for *Yponomeuta evonymellus* (Yponomeutidae), the BIN contained 395 reported sequences. A higher number was detected for the Diptera species *Psilopa obscuripes* (Ephydridae) and *Scaptomyza pallida* (Drosophilidae), where both BIN clusters contained 2043 sequences. The highest number of BINs was observed for Hemipterans (35), Dipterans (27), and Hymenopterans (19). On the other hand, the lowest number of unique BINs was observed in earthworms (six) and Collembolans (five). Most of the recorded species were assigned at least one specific BIN. The exceptions were *Tetramorium semilaevae* (Hymenoptera, Formicidae), *Armadillidium vulgare* (Isopoda, Armadillidiidae), and *Prays oleae* (Lepidoptera, Praydidae), which were assigned two or more BINs.

(**a**)

(**b**)

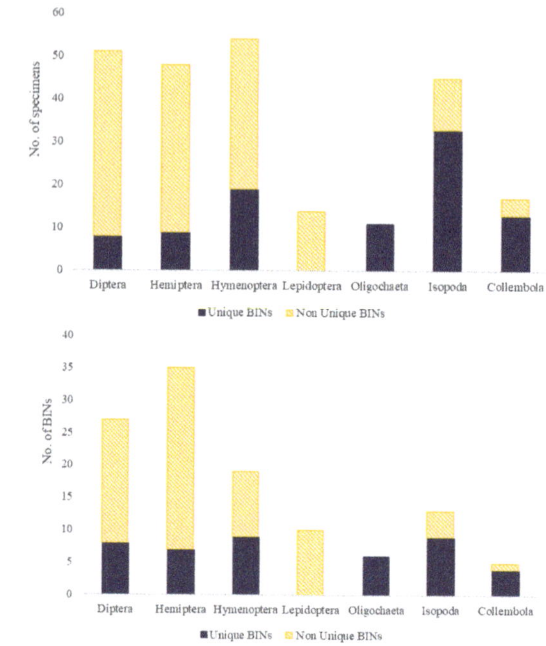

Figure 2. Observed BINs, shown as a ratio between Non-Unique BINs (yellow pattern) and Unique BINs (blue) within different groups: Diptera, Hemiptera, Lepidoptera, Hymenoptera, Isopoda, Collembola, and Oligochaeta, represented as a number of specimens belonging to each BIN type (**a**) or as a number of each BIN type (**b**).

3.2. *Rarefaction Curves*

Rarefaction curves showed that for some groups the number of recorded BINs did not reach the plateau and that further sampling efforts would provide us with a higher BIN (Figure 3).

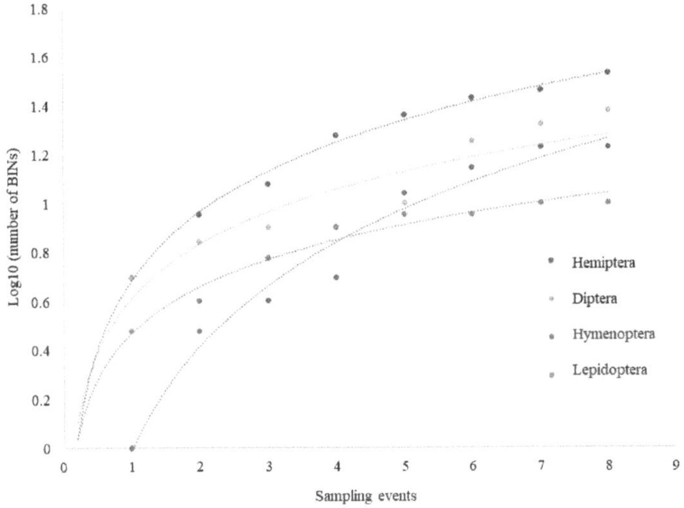

Figure 3. Rarefaction curves of the number of different BINs for selected groups shown as a function of the number of sampling events.

3.3. *Species Delimitation for Oligochaetes and Collembolans*

Oligochaeta sequences obtained from our study were grouped in six unique BINs and were all reported for the first time in BOLD. The obtained 36 sequences collapsed to 21 unique haplotypes, which were grouped in three MOTUs using species delimitation methods (ABGD, bPTP, and mPTP). The first MOTU aggregated the majority of BINs (four out of six) and haplotypes (17 out of 21). BINs clustered to the first MOTU were the following: BOLD:AEH4576, BOLD:AEJ1183, BOLD:AEJ1182, and BOLD:AEH3543. The remaining BINs were split into two MOTUs. The second MOTU was formed by the BIN BOLD:AEH3542 (MOTU 2). The third MOTU was the BIN BOLD:AEI1132 (MOTU3). The first two MOTUs were the most closely related to the species *Octodrilus complanatus* (Dugès, 1828), whereas MOTU3 was more genetically distant. The phylogenetic ML tree result is shown in Figure 4. The first and the second MOTU were present in the two vineyards and the integrated olive orchard from the studied area. The third MOTU was found in one of the studied vineyards. The calculated p-distance between MOTU groups 1 and 2 was 6.68% (0.0668 +/− SD 0.004283), between MOTU groups 1 and 3 was 11.59% (0.1159 +/− SD 0.098806), and between groups 2 and 3 was 19.28% (0.1928 +/− SD 0.00133) (Supplementary Table S3).

Morphologically, the single *Octodrilus* individual (MOTU3) belongs to a group of small-sized, red-pigmented *Octodrilus* species, which includes *O. croaticus* (Rosa, 1895), *O. juvyi* Zicsi & Cuendet, 2005, *O. bretscheri* (Zicsi, 1969), *O. lissaensis* (Michaelsen, 1891), and *O. argoviensis* (Bretscher, 1899) [60], and possibly a few other nominal species (cf. Mršić, 1991) [41]. Several members of this group have been recorded as present in the Balkan peninsula [41,42]. Our finding locality is very close to the Croatian locality given for *O. croaticus* by Mršić (1991), whereas *O. bretscheri* (like *O. juvyi* in France) seems to be confined to mountain habitats. Therefore, we identified this specimen provisionally as *O. croaticus*.

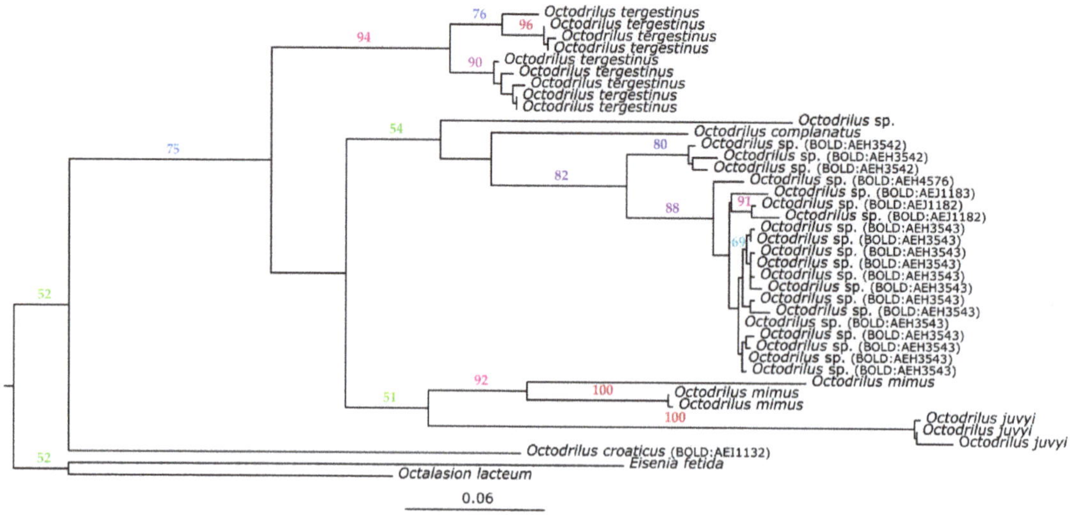

Figure 4. Maximum likelihood phylogram inferred for specimens of the genus *Octodrilus*. Numbers on the branches represent the nonparametric bootstrap support (≥50). The tree is rooted on *Octolasion lacteum* (BOLD:ACF5848) and *Eisenia fetida* (BOLD:AAB2558) as outgroups.

Other examined specimens had an abundant overlap in many aspects of morphology and anatomy, without any separation in classical diagnostic characters. Accordingly, they were assorted in the same morpho group. Their general morphology would suggest a relationship with *O. complanatus*. This relation was also observed in the genetic analyses.

Collembolan specimens from our sites were divided into five MOTUs using the BIN assignment method. Two of the recorded BINs (BOLD:AEH5984 and BOLD:AEE1997) were morphologically identified as genus *Orchesella* Templeton, 1835 (Entomobryomorpha, Entomobryidae). The BOLD:AEH5615 BIN was morphologically identified as belonging to genus *Lepidocyrtus* Bourlet, 1839 (Entomobryomorpha, Entomobryidae), and BIN BOLD:AEH9402 as genus *Heteromurus* Wankel, 1860 (Entomobryomorpha, Entomobryidae). The remaining BIN (BOLD:AEJ0895) morphologically belongs to the family Sminthuridae (Symphypleona).

Our *Lepidocyrtus* specimens were shown to be most closely related to the species *Lepidocyrtus pallidus*, according to the phylogenetic analysis (Figure 5a). In Europe, the genus *Lepidocyrtus* has an increasing number of species, with the *Lepidocyrtus pallidus* species group consisting of six species [61]. By using the mPTP species delimitation approach, both BINs (BOLD:AEH5984 and BOLD:AEE1997) belonging to the genus *Orchesella* were clustered into one MOTU. *Orchesella cincta* and the aforementioned MOTU are genetically and morphologically related (Figure 5b). The remaining BINS (BOLD:AEH9402 and BOLD:AEJ0895) were not subjected to maximum likelihood analysis, since they were only represented by one specimen.

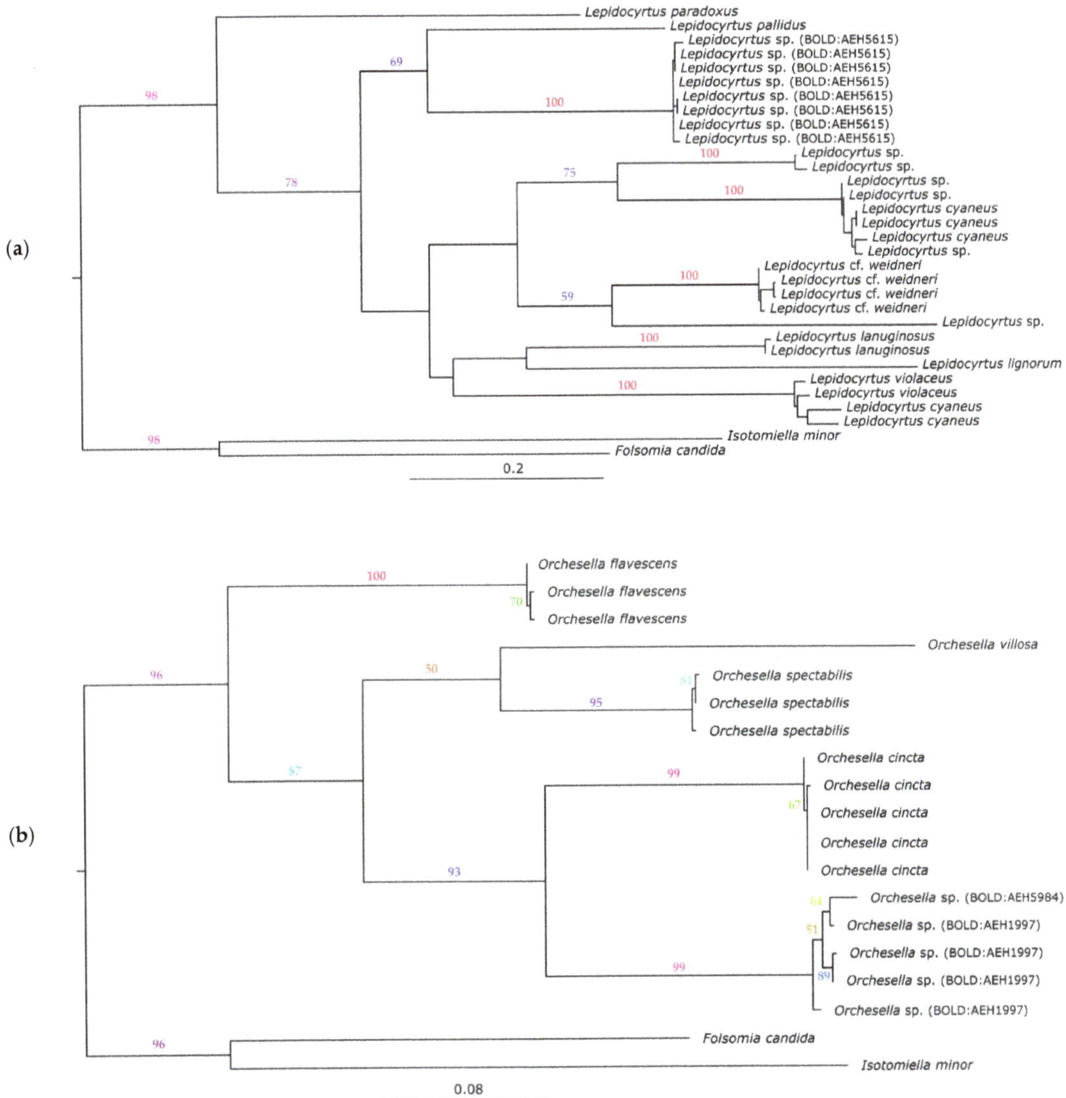

Figure 5. Maximum likelihood phylogram inferred for specimens of the genus *Lepidocyrtus* (**a**) and the genus *Orchesella* (**b**). Numbers on the branches represent the nonparametric bootstrap support (≥50). The trees are rooted on *Folsomia candida* (BOLD:AAB6463) and *Isotomiella minor* (BOLD:ACQ1061) as outgroups.

3.4. Similarity Measures

The Jaccard similarity index was compared between the five sites. The lowest values were found between all of the sites in the case of Lepidopteran species, where there was no overlap in species present between the sites. The result was probably affected by the low species number. The highest index values were found in the case of Isopoda morphospecies between integrated olive orchard and natural habitat, as well as in the ecological olive orchard. Between the integrated olive orchard and the integrated vineyard, as well as the ecological vineyard and natural habitat, high index values were recorded for the Hymenopteran species. High index values were also observed for the Collembolan morphospecies in the case of the two vineyards and the integrated olive orchard. On the other hand, low index values were observed between all sites for the Hemipteran as well as Dipteran species. The results are shown on Figure 6. Oligochaeta were not included in the analysis due to low species number (two).

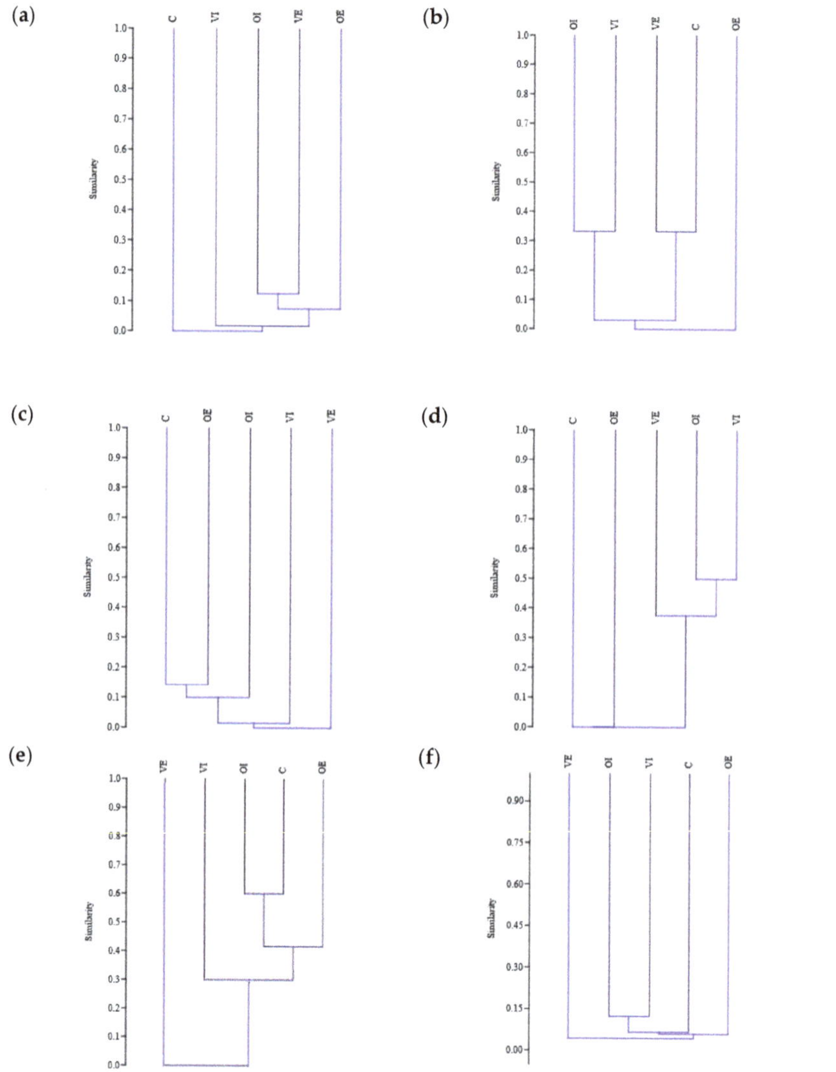

Figure 6. Jaccard similarity between the study sites for Hemipteran (**a**), Hymenopteran (**b**), Dipteran (**c**), Collembolan (**d**), and Isopods (**e**) species, and for all the groups combined (**f**). Study sites are annotated with abbreviations as follows: OE, olive orchard with ecological pest management; OI, olive orchard with integrated pest management; VE, vineyard with ecological pest management; VI, vineyard with integrated pest management, C, natural habitat.

4. Discussion

Our results revealed a lower coverage of sequences in databases for Collembolans and Oligochaetes, and a better coverage for Lepidopterans, Hemipterans, Dipterans, and Hymenopterans. As expected, the highest number of species and morphospecies was observed for Hemipterans and Dipterans, due to their high abundance in agricultural areas. As mentioned above, most species were grouped to one specific BIN, with the exception of three species, *Tetramorium semilaevae* (Hymenoptera, Formicidae), *Armadillidium vulgare* (Isopoda, Armadillidiidae), and *Prays oleae* (Lepidoptera, Praydidae), which were sorted into two or more BINs. This can point to an intraspecific genetic divergence higher than 2.2% [6,62], leading to the conclusion that *COI* marker is not ideal for the genetic distinction of these species. The application of species delimitation methods used to estimate species number for Collembolans and Oligochaetes, as well as the overall results obtained for these two groups, suggest that there is an underestimated species richness in the Mediterranean area.

The analysis of the species composition for the recorded species/morphospecies suggested that there was more similarity in the overall species composition for the soil-dwelling organisms compared to the canopy fauna. The potential explanation can be found in the strong effect of the surrounding agricultural landscape on the canopy fauna. It is worth mentioning that Formicidae, making up the majority of our Hymenoptera species, have similar compositions depending on the management type, indicating that they are strong bio-indicators of land use effects, which is in accordance with the earlier research [63].

As was expected, the majority of Hemipterans were identified at the species level, seeing as this group was studied in numerous DNA barcoding efforts [64–67]. From this group, the highest number of unidentified species belonged to the family Cicadellidae, which is widely present at agricultural sites, and the species from this family are often vectors of various plant diseases. Some members of the family are invasive species, such as *Scaphoideus titanus* (Hemiptera, Cicadellidae), a species that causes agronomical damage throughout Europe as a "*Flavescence dorée*" vector [68]. While this insect family is highly diverse with more than 20,000 described species, only 2000 species are present with sequence records in the BOLD database. This makes the identification of Cicadellidae species through DNA barcodes challenging.

The identification of Oligochaete specimens also presented a challenge due to the low sequence coverage in the databases. This group is genetically understudied in Croatia, with only 10 available sequences (out a total of 7532 available sequences representing 218 species) according to the BOLD database. Moreover, high species diversity in Mediterranean and the existence of many cryptic species also make species identification, morphologically as well as genetically, more demanding [69,70]. We did not identify any of the 36 specimens at species level through pre-existing sequences, but we confirmed that all specimens belong to the genus *Octodrilus*. The Balkans are among the main centers of diversification of the genus *Octodrilus*, and the phenomena of morphological convergence between unrelated species is frequent, possibly caused by ecological adaptations [41,71]. In addition, the morphological diagnostic characters of many species of *Octodrilus* do not consider the variations between populations (within a population there is more generally uniformity). Therefore, the definition of taxonomic boundaries (and synonymies) is quite problematic, and discrimination using morphological characters is highly complex. The DNA barcoding of specimens appears to be useful for assessing the biodiversity of this genus in the Balkans, including the Mediterranean part. Species delimitation methods revealed that specimens from our sample could be divided into three MOTUs. Considering the results of Huang et al. [72], where interspecific distance was in some cases greater than 15% and intraspecific difference was up to 7.8%, in analyzing genetic distances for our dataset, earthworms were grouped into two MOTUs. That was consistent with the morphological results that placed our specimens into two morphospecies, *Octodrilus croaticus* and an unidentified species. The unidentified species was most closely related to *O. complanatus*, which was also confirmed by the examination of morphological characters. A sexually mature specimen identified as

O. croaticus was genetically distant compared to other specimens from our sample and from the database. The distribution of *O. croaticus* encompasses the Balkan Peninsula, including the Mediterranean part of Croatia [24]. This complex group is not only underrepresented in two main databases, BOLD and NCBI, but also contains a high number of endemic species. Importantly, for better understanding their taxonomy, more systematic sampling and barcoding efforts are necessary.

The number of described Collembolan species is only a part of the total number, as earlier research shows, due to the lack of data for many geographical regions, including parts of southern Europe [73]. Due to the abovementioned, and the fact that described species are not present with a sequence in databases, difficulties with identification at species level were expected. Our results showed that Collembolans are underrepresented in the databases. Specimens included in our research were most likely endemic species, and their species identification could not be carried out with global sequence data. Similar results have been reported in Shaw and Benefer [74], where they have found that 25 out of 48 species do not correspond to a known BIN. The comparison of morphological with genetic data and the species delimitation placement of our specimens grouped them into four MOTUs. The highest number of specimens belonged to the genus *Lepidocyrtus* which among the Collembola genera comprises the highest number of species [75]. Because of the prevalence of cryptic species and species groups that cannot be resolved using solely morphological criteria, some authors report that the real number of *Lepidocyrtus* species could be significantly higher than the species number currently recognizes [76–79], which is probably also the case with the *Lepidocyrtus* specimens collected at our research sites. Specimens of the genus *Orchesella* were grouped into two BINs that finally formed one MOTU. A morphological difference between the two BINs was also evident in the different pigmentation levels, but this intraspecific variability, which affects pigmentation, is sometimes linked to their life cycle [80,81]. The *Orchesella* species from the Mediterranean area have a high genetic divergence, which is probably the reason why our specimens differ compared to the other *Orchesella* species [80].

The species delimitation conducted on our dataset showed discordance among the methods used for species assignment, suggesting a different number of potential species, in the case of both Oligochaetes and Collembolans. Likewise, even though species delimitation and BIN assignment proved to be a valuable tool in various species identification [81,82], a number of papers reported taxa incongruence among implemented methods [83].

Contrary to earthworms and springtails, all of the Lepidopteran specimens were identified at the species level due to the high number of available sequences in DNA barcode databases [84–86]. Our results indicated that the DNA barcoding method could be used as a reliable tool for Lepidopteran species identification because there is a DNA sequence available in BOLD for 73% of European Lepidopterans [8].

With regard to Formicidae (Hymenoptera), there are DNA barcode records for more than half of the described species. Earlier studies also report good DNA sequence coverage in databases [87], with many research projects providing numerous DNA barcode sequences for the Mediterranean area [88]. However, the high divergence of ant species in the Mediterranean area, from the mainland to the islands [89], complicates species identification. Namely, species identification through DNA barcodes can be affected by high intraspecific divergence, as in the case of *Pheidole pallidula*. Different species exhibit different levels of intraspecific divergence, and it is important to keep this in mind while analyzing ant species composition through DNA barcode data.

In the case of Isopods, only 1300 species were presented with DNA barcodes. These were mainly marine species. We observed that some species could be identified only to the genus level or lower, mostly because there are not enough DNA barcode data on this group.

There are not many papers focusing on the DNA barcoding of the whole Diptera fauna for a specific geographical region [90]; however, most studies have been focused on specific families or genera from this insect order. As noticed in Morinière et al. [91], many of so called "dark taxa" of Dipteran, or species without records in databases, are small in size.

Families with the smallest body size often have a higher number of overlooked species, either taxonomically or genetically. We observed a similar pattern, where small-sized specimens could not be identified at the species level. In numerous studies, such specimens have been overlooked and, thus, their identification through DNA barcoding is made more difficult. Authors in Morinière et al. [91] have concluded that sequences which were not joined with species names, but classified only as OTUs, are equally important, as they can be used to compare the diversity of samples collected at different sites. It is also important to add that such DNA sequences are indicators of what we have missed in our barcoding efforts, and that many species are still waiting not just to be barcoded, but also discovered.

Through our experiment, we identified, using DNA barcode tools, an invasive species *Drosophila suzukii* (Diptera, Drosophilidae) in the olive orchard in Škabrnja. This species has already been recorded for the Zadar County in different plantations [92], but not in olive orchards. However, our sampling site was surrounded by cherry orchards, suggesting that *Drosophila suzukii* spread in the area even between different cultivars. This species, native of eastern and south-eastern Asia, represents an economical problem in Europe, given that it is easily distributed. This only reinforces the fact that a rapid biodiversity assessment using DNA barcoding can be valuable in detecting non-native invasive species, and among native species, especially rare and locally distributed ones. It is necessary to gain more biodiversity information on natural and also agricultural habitats to be able to solely rely on the DNA barcode ID tool.

The efficiency of DNA barcoding in Mediterranean agricultural areas of Croatia differs between groups of organisms depending on several factors: firstly, on the availability of the data for the different groups of organisms in the two largest databases (NCBI and BOLD), and secondly, on the level of taxonomy relationship resolution and the association of molecular data with the morphospecies concept for the particular group [93]. Some groups, such as Collembola and Oligochaeta, were more difficult to identify through previous data, as their fauna seemed to be more endemic to the area. On the other hand, the identified species of Dipterans, Hemipterans, and Lepidopterans showed geographically wide distribution and presence in agricultural habitats. Consequently, their identification was facilitated through a great amount of data available in the databases. Ant fauna in the studied area showed a similar species composition with other Mediterranean areas previously barcoded and available in the BOLD database [88,89], which facilitated their identification using the DNA barcoding method. This method proved suitable for biodiversity analysis in this part of the Mediterranean. However, it is important to keep in mind that, especially for some groups, such as Collembolans and Oligochaetes, DNA barcode sequencing is not sufficient for species identification, and it should be supplemented with classical morphological analysis. Nevertheless, this does not reduce the value of DNA barcode data, as databases are continuously appended with different species' sequences. It is, thus, expected that, at some point, DNA barcodes will be able to make species-level determinations for the vast majority of groups.

Supplementary Materials: The following supporting information can be downloaded at: https://www.mdpi.com/article/10.3390/d14030182/s1, Table S1. For each sampling site, a list of geographical coordinates, altitude, and surface area is provided. Table S2. List of specimens included in the research. Table S3. Calculated p-distances between *Octodrilus* MOTU groups. Minimum and maximum values and calculated standard deviation and standard error are listed. Supplementary File S1a. Supplementary File S1b. Supplementary File S2.

Author Contributions: Conceptualization, L.Š.J., B.A.D. and M.J.; methodology, L.Š.J., B.A.D. and M.J.; validation, L.Š.J., M.J. and E.R.; formal analysis, B.A.D.; resources, L.Š.J.; data curation, B.A.D. and E.R.; writing—original draft preparation, L.Š.J. and B.A.D.; writing—review and editing, L.Š.J., M.J. and E.R.; visualization, B.A.D.; supervision, L.Š.J.; project administration, L.Š.J.; funding acquisition, L.Š.J. All authors have read and agreed to the published version of the manuscript.

Funding: This research was funded by The Croatian Science Foundation under the MEDITERATRI Project (UIP-2017-05-1046), granted to Lucija Šerić Jelaska, and co-funded by the Department of Biology, Faculty of Science at the University of Zagreb and the Croatian Academy of Sciences and Art.

Institutional Review Board Statement: Not applicable.

Data Availability Statement: Data supporting reported results can be found in Barcode of Life Database under the DOI: dx.doi.org/10.5883/DS-070222, and Mendeley Data public depository under the DOI: 10.17632/3gpfzz9rzv.1.

Acknowledgments: We are thankful to Lara Ivanković Tatalović and Tomislav Kos for their help in the field and sample photographing, Petra Furčić for creating the map of the study sites, L'ubomír Kováč for help with Collembolan species identification, Josip Primorac for sorting ant samples in the laboratory. We are also thankful to the Croatian Academy of Sciences and Art for co-funding the research within the Croatian CryoEntoArk project.

Conflicts of Interest: The authors declare no conflict of interest.

References

1. Hebert, P.D.N.; Gregory, T.R. The promise of DNA barcoding for taxonomy. *Syst. Biol.* **2005**, *54*, 852–859. [CrossRef]
2. Jung, S.; Duwal, R.K.; Lee, S. COI barcoding of true bugs (*Insecta, Heteroptera*). *Mol. Ecol. Resour.* **2011**, *11*, 266–270. [CrossRef]
3. Hebert, P.D.N.; Cywinska, A.; Ball, S.L.; de Waard, J.R. Biological identifications through DNA barcodes. *Proc. R. Soc. B Biol. Sci.* **2003**, *270*, 313–321. [CrossRef]
4. Greenstone, M.H.; Rowley, D.L.; Heimbach, U.; Lundgren, J.G.; Pfannenstiel, R.S.; Rehner, S.A. Barcoding generalist predators by polymerase chain reaction: Carabids and spiders. *Mol. Ecol.* **2005**, *14*, 3247–3266. [CrossRef]
5. Ratnasingham, S.; Hebert, P.D.N. BOLD: The Barcode of Life Data System. *Mol. Ecol. Notes* **2007**, *7*, 355–364. [CrossRef]
6. Ratnasingham, S.; Hebert, P.D.N. A DNA-based registry for all animal species: The barcode index number (BIN) system. *PLoS ONE* **2013**, *8*, e66213. [CrossRef]
7. Pentinsaari, M.; Blagoev, G.A.; Hogg, I.D.; Levesque-Beaudin, V.; Perez, K.; Sobel, C.N.; Vandenbrink, B.; Borisenko, A. A DNA Barcoding Survey of an Arctic Arthropod Community: Implications for Future Monitoring. *Insects* **2020**, *11*, 46. [CrossRef] [PubMed]
8. Lopez-Vaamonde, C.; Kirichenko, N.; Cama, A.; Doorenweerd, C.; Godfray, H.C.J.; Guiguet, A.; Gomboc, S.; Huemer, P.; Landry, J.F.; Laštůvka, A.; et al. Evaluating DNA Barcoding for Species Identification and Discovery in European Gracillariid Moths. *Front. Ecol. Evol.* **2021**, *9*, 66. [CrossRef]
9. Gregory, T. DNA barcoding does not compete with taxonomy. *Nature* **2005**, *434*, 1067. [CrossRef] [PubMed]
10. Costa, F.; Dewaard, J.R.; Boutillier, J.; Ratnasingham, S.; Dooh, R.T.; Hajibabaei, M.; Hebert, P. Biological identifications through DNA barcodes: The case of the Crustacea. *Can. J. Fish. Aquat. Sci.* **2007**, *64*, 272–295. [CrossRef]
11. Hendrich, L.; Morinière, J.; Haszprunar, G.; Hebert, P.D.N.; Hausmann, A.; Köhler, F.; Balke, M. A comprehensive DNA barcode database for Central European beetles with a focus on Germany: Adding more than 3500 identified species to BOLD. *Mol. Ecol. Resour.* **2015**, *15*, 795–818. [CrossRef] [PubMed]
12. Villalta, I.; Ledet, R.; Baude, M.; Genoud, D.; Bouget, C.; Cornillon, M.; Moreau, S.; Courtial, B.; Lopez-Vaamonde, C. A DNA barcode-based survey of wild urban bees in the Loire Valley, France. *Sci. Rep.* **2021**, *11*, 4770. [CrossRef] [PubMed]
13. Chimeno, C.; Hausmann, A.; Schmidt, S.; Raupach, M.J.; Doczkal, D.; Baranov, V.; Hübner, J.; Höcherl, A.; Albrecht, R.; Jaschhof, M.; et al. Peering into the Darkness: DNA Barcoding Reveals Surprisingly High Diversity of Unknown Species of Diptera (*Insecta*) in Germany. *Insects* **2022**, *13*, 82. [CrossRef] [PubMed]
14. Smith, M.A.; Fisher, B.L.; Hebert, P.D.N. DNA barcoding for effective biodiversity assessment of a hyperdiverse arthropod group: The ants of Madagascar. *Philos. Trans. R. Soc. Lond. Ser. B* **2005**, *360*, 1825–1834. [CrossRef]
15. De Leon, L.F.; Cornejo, A.; Gavilan, R.G.; Aguilar, C. Hidden biodiversity in Neotropical streams: DNA barcoding uncovers high endemicity of freshwater macroinvertebrates at small spatial scales. *PLoS ONE* **2020**, *15*, e0231683. [CrossRef]
16. Nneji, L.M.; Adeola, A.C.; Wang, Y.Y.; Ajao, A.M.; Anyaele, O.; Malann, Y.; Olatunde, O.; Nneji, I.C.; Ayoola, A.O.; Rahman, M.M.; et al. Testing the Effectiveness of DNA Barcoding for Biodiversity Assessment of Moths from Nigeria. *Diversity* **2020**, *12*, 85. [CrossRef]
17. Kennedy, S.R.; Prost, S.; Overcast, I.; Rominger, A.J.; Gillespie, R.G.; Krehenwinkel, H. High-throughput sequencing for community analysis: The promise of DNA barcoding to uncover diversity, relatedness, abundances and interactions in spider communities. *Dev. Genes Evol.* **2020**, *230*, 185–201. [CrossRef]
18. Meiklejohn, K.A.; Damaso, N.; Robertson, J.M. Assessment of BOLD and GenBank—Their accuracy and reliability for the identification of biological materials. *PLoS ONE* **2019**, *14*, e0217084. [CrossRef]
19. Page, R.D. DNA barcoding and taxonomy: Dark taxa and dark texts. *Philos. Trans. R. Soc. Lond. Ser. B* **2016**, *371*, 20150334. [CrossRef]
20. Jinbo, U.; Kato, T.; Ito, M. Current progress in DNA barcoding and future implications for entomology. *Entomol. Sci.* **2011**, *14*, 107–124. [CrossRef]

21. Cuff, J.P.; Drake, L.E.; Tercel, M.P.; Stockdale, J.E.; Orozco-ter Wengel, P.; Bell, J.R.; Vaughan, I.P.; Müller, C.T.; Symondson, W.O. Money spider dietary choice in pre- and post-harvest cereal crops using metabarcoding. *Ecol. Entomol.* **2021**, *46*, 249–261. [CrossRef]
22. Lafage, D.; Elbrecht, V.; Cuff, J.P.; Steinke, D.; Hambäck, P.A.; Erlandsson, A. A new primer for metabarcoding of spider gut contents. *Environ. DNA* **2020**, *2*, 234–243. [CrossRef]
23. O'Rourke, D.R.; Mangan, M.T.; Mangan, K.E.; Bokulich, N.A.; MacManes, M.D.; Foster, J.T. Lord of the Diptera (and Moths and a Spider): Molecular Diet Analyses and Foraging Ecology of Indiana Bats in Illinois. *Front. Ecol. Evol.* **2021**, *9*, 12. [CrossRef]
24. Kutuzović, D.H.; Kutuzović, B.H. Checklist of the earthworm fauna of Croatia (*Oligochaeta: Lumbricidae*). *Zootaxa* **2013**, *3710*, 1–30. [CrossRef]
25. Bračko, G. Review of the ant fauna (*Hymenoptera: Formicidae*) of Croatia. *Acta Entomol. Slov.* **2006**, *14*, 131–156.
26. Franin, K.; Franin, G.K.; Maričić, B.; Marcelić, Š.; Pavlović, M.; Kos, T.; Laznik, Ž. True bugs (*Heteroptera*) assemblage and diversity in the ecological infrastructures around the Mediterranean vineyards. *Bull. Insectol.* **2021**, *74*, 65–78.
27. Botham, M.S.; Fernandez-Ploquin, E.C.; Brereton, T.; Harrower, C.A.; Roy, D.B.; Heard, M.S. Lepidoptera communities across an agricultural gradient: How important are habitat area and habitat diversity in supporting high diversity? *J. Insect Conserv.* **2015**, *19*, 403–420. [CrossRef]
28. Porco, D.; Bedos, A.; Greenslade, P.; Janion, C.; Skarżyński, D.; Stevens, M.I.; Deharveng, L. Challenging species delimitation in Collembola: Cryptic diversity among common springtails unveiled by DNA barcoding. *Invertebr. Syst.* **2012**, *26*, 470–477. [CrossRef]
29. Ponge, J.F.; Gillet, S.; Dubs, F.; Fedoroff, E.; Haese, L.; Sousa, J.P.; Lavelle, P. Collembolan communities as bioindicators of land use intensification. *Soil Biol. Biochem.* **2003**, *35*, 813–826. [CrossRef]
30. Hendrickx, F.; Maelfait, J.P.; Van Wingerden, W.; Schweiger, O.; Speelmans, M.; Aviron, S.; Augenstein, I.; Billeter, R.; Bailey, D.; Bukacek, R.; et al. How landscape structure, land-use intensity and habitat diversity affect components of total arthropod diversity in agricultural landscapes. *J. Appl. Ecol.* **2007**, *44*, 340–351. [CrossRef]
31. Hevia, V.; Ortega, J.; Azcárate, F.M.; López, C.A.; González, J.A. Exploring the effect of soil management intensity on taxonomic and functional diversity of ants in Mediterranean olive groves. *Agric. For. Entomol.* **2019**, *21*, 109–118. [CrossRef]
32. McNabb, D.M.; Halaj, J.; Wise, D.H. Inferring trophic positions of generalist predators and their linkage to the detrital food web in agroecosystems: A stable isotope analysis. *Pedobiologia* **2001**, *45*, 289–297. [CrossRef]
33. King, R.A.; Vaughan, I.P.; Bell, J.R.; Bohan, D.A.; Symondson, W.O. Prey choice by carabid beetles feeding on an earthworm community analysed using species- and lineage-specific PCR primers. *Mol. Ecol.* **2010**, *19*, 1721–1732. [CrossRef] [PubMed]
34. Šerić Jelaska, L.; Franjević, D.; Jelaska, S.D.; Symondson, W.O.C. Prey detection in carabid beetles (*Coleoptera: Carabidae*) in woodland ecosystems by PCR analysis of gut contents. *Eurpean J. Entomol.* **2014**, *111*, 631–638. [CrossRef]
35. Šerić Jelaska, L.; Symondson, O.C.W. Predation on epigeic, endogeic and anecic earthworms by carabids active in spring and autumn. *Period. Biol.* **2016**, *118*, 281–289. [CrossRef]
36. Sanders, D.; Platner, C. Intraguild interactions between spiders and ants and top-down control in a grassland food web. *Oecologia* **2007**, *150*, 611–624. [CrossRef]
37. Schowalter, T.; Chao, J.T. Canopy Insect Sampling. In *Measuring Arthropod Biodiversity*; Santos, J.C., Fernandes, G.W., Eds.; Springer: Berlin/Heidelberg, Germany, 2021. [CrossRef]
38. Macfadyen, A. Notes on methods for the extraction of small arthropods. *J. Anim. Ecol.* **1953**, *21*, 65–77. [CrossRef]
39. Casquet, J.; Thebaud, C.; Gillespie, R.G. Chelex without boiling, a rapid and easy technique to obtain stable amplifiable DNA from small amounts of ethanol-stored spiders. *Mol. Ecol. Resour.* **2012**, *12*, 136–141. [CrossRef]
40. Folmer, O.; Black, M.; Hoeh, W.; Lutz, R.; Vrijenhoek, R. DNA primers for amplification of mitochondrial cytochrome c oxidase subunit I from diverse metazoan invertebrates. *Mol. Mar. Biol. Biotechnol.* **1994**, *3*, 294–299.
41. Mršić, N. Monograph on earthworms (*Lumbricidae*) of the Balkans. *Acad. Sci. Et Artium Slov. Cl. IV Hist. Nat.* **1991**, *30*, 1–757.
42. Szederjesi, T. The first combined checklist of earthworms of the Northeastern Mediterranean region (Clitellata: Megadrili). *Opusc. Zool. Bp.* **2017**, *48*, 77–116. [CrossRef]
43. Hopkin, S.P. *A key to the Collembola (springtails) of Britain and Ireland*; Field Studies Council Publ.: Telford, Shropshire, UK, 2007.
44. Seifert, B. *The ants of Central and North Europe*; Lutra Verlags und Vertriebsgesellschaft: Tauer, Germany, 2018.
45. Lebas, C.; Galkowski, C.; Blatrix, R.; Wegnet, P. *Ants of Britain and Europe*; Bloomsbury Publishing PLC.: London, UK, 2019.
46. Hall, T.A. BioEdit: A User-Friendly Biological Sequence Alignment Editor and Analysis Program for Windows 95/98/NT. *Nucleic Acids Symp. Ser.* **1999**, *41*, 95–98.
47. Altschul, S.F.; Gish, W.; Miller, W.; Myers, E.W.; Lipman, D.J. Basic local alignment search tool. *J. Mol. Biol.* **1990**, *215*, 403–410. [CrossRef]
48. Kumar, S.; Stecher, G.; Li, M.; Knyaz, C.; Tamura, K. MEGA X: Molecular Evolutionary Genetics Analysis across computing platforms. *Mol. Biol. Evol.* **2018**, *35*, 1547–1549. [CrossRef]
49. Edgar, R.C. MUSCLE: A multiple sequence alignment method with reduced time and space complexity. *BMC Bioinform.* **2004**, *5*, 113. [CrossRef]
50. Villesen, P. FaBox: An online toolbox for fasta sequences. *Mol. Ecol. Notes* **2007**, *7*, 965–968. [CrossRef]
51. Maddison, W.P.; Maddison, D.R. Mesquite: A Modular System for Evolutionary Analysis. Version 3.50. 2019. Available online: http://www.mesquiteproject.org (accessed on 9 February 2022).

52. Stamatakis, A. RAxML version 8: A tool for phylogenetic analysis and post-analysis of large phylogenies. *Bioinformatics* **2014**, *30*, 1312–1313. [CrossRef]
53. Miller, M.A.; Pfeiffer, W.; Schwartz, T. Creating the CIPRES Science Gateway for inference of large phylogenetic trees. In *2010 Gateway Computing Environments Workshop (GCE), Proceedings of the Gateway Computing Environments Workshop (GCE), New Orleans, LA, USA, 14 November 2010*; Institute of Electrical and Electronics Engineers (IEEE): Manhattan, NY, USA, 2010; pp. 1–8.
54. Rambaut, A. *FigTree*, Version 1.4.4. (Computer Program). 2014. Available online: http://tree.bio.ed.ac.uk/software/figtree/ (accessed on 14 February 2022).
55. Puillandre, N.; Lambert, A.; Brouillet, S.; Achaz, G. ABGD, Automatic Barcode Gap Discovery for primary species delimitation. *Mol. Ecol.* **2011**, *21*, 1864–1877. [CrossRef]
56. Zhang, J.; Kapli, P.; Pavlidis, P.; Stamatakis, A.A. General Species Delimitation Method with Applications to Phylogenetic Placements. *Bioinformatics* **2013**, *29*, 2869–2876. [CrossRef]
57. Kapli, P.; Lutteropp, S.; Zhang, J.; Kobert, K.; Pavlidis, P.; Stamatakis, A.; Flouri, T. Multi-rate Poisson tree processes for single-locus species delimitation under maximum likelihood and Markov chain Monte Carlo. *Bioinformatics* **2017**, *33*, 1630–1638. [CrossRef]
58. Albertini, A.; Pizzolotto, R.; Petacchi, R. Carabid patterns in olive orchards and woody semi-natural habitats: First implications for conservation biological control against Bactrocera oleae. *BioControl* **2017**, *62*, 71–83. [CrossRef]
59. Hammer, Ø.; Harper, D.A.T.; Ryan, P.D. PAST: Paleontological statistics software package for education and data analysis. *Palaeontol. Electron.* **2001**, *4*, 9.
60. Zicsi, A.; Cuendet, G. Über eine neue Octodrilus-Art aus Frankreich (*Oligochaeta*: *Lumbricidae*). *Rev. Suisse Zool.* **2005**, *112*, 329–331. [CrossRef]
61. Winkler, D.; Traser, G. Explanation of the European Lepidocyrtus pallidus–serbicus group (Collembola, Entomobryidae), with description of new species from Hungary. *Zootaxa* **2012**, *3394*, 35–47. [CrossRef]
62. Raupach, M.J.; Hannig, K.; Morinière, J.; Hendrich, L. A DNA barcode library for ground beetles (*Insecta, Coleoptera, Carabidae*) of Germany: The genus Bembidion Latreille, 1802 and allied taxa. *ZooKeys* **2016**, *592*, 121–141. [CrossRef]
63. Andersen, A.N.; Hoffmann, B.D.; Müller, W.J.; Griffiths, A.D. Using ants as bioindicators in land management: Simplifying assessment of ant community responses. *J. Appl. Ecol.* **2002**, *39*, 8–17. [CrossRef]
64. Coeur d'acier, A.; Cruaud, A.; Artige, E.; Genson, G.; Clamens, A.L.; Pierre, E.; Hudaverdian, S.; Simon, J.-C.; Jousselin, E.; Rasplus, J.-Y. DNA Barcoding and the Associated PhylAphidB@se Website for the Identification of European Aphids (*Insecta*: *Hemiptera*: *Aphididae*). *PLoS ONE* **2014**, *9*, e97620. [CrossRef]
65. Foottit, R.G.; Maw, H.E.L.; Von Dohlen, C.D.; Hebert, P.D.N. Species identification of aphids (*Insecta*: *Hemiptera*: *Aphididae*) through DNA barcodes. *Mol. Ecol. Resour.* **2008**, *8*, 1189–1201. [CrossRef]
66. Gwiazdowski, R.A.; Foottit, R.G.; Maw, H.E.L.; Hebert, P.D.N. The Hemiptera (*Insecta*) of Canada: Constructing a Reference Library of DNABarcodes. *PLoS ONE* **2015**, *4*, e0125635. [CrossRef]
67. Park, D.; Suh, S.; Hebert, P.; Oh, H.; Hong, K. DNA barcodes for two scale insect families, mealybugs (*Hemiptera*: *Pseudococcidae*) and armored scales (*Hemiptera*: *Diaspididae*). *Bull. Entomol. Res.* **2011**, *101*, 429–434. [CrossRef]
68. Chuche, J.; Thiéry, D. Biology and ecology of the Flavescence dorée vector Scaphoideus titanus: A review. *Agron. Sustain. Dev.* **2014**, *34*, 381–403. [CrossRef]
69. Csuzdi, C.; Szederjesi, T.; Marchán, D.F.; Sosa, I.D.; Gavinelli, F.; Dorigo, L.; Pamio, A.; Dreon, A.L.; Fusaro, S.; Moretto, E.; et al. DNA barcoding of the Italian anecic Octodrilus species in rural (vineyard) and forested areas with description of Octodrilus zicsiniello sp. nov.(*Clitellata, Megadrili*). *Zootaxa* **2018**, *4496*, 43–64. [CrossRef] [PubMed]
70. Novo, M.; Almodóvar, A.; Fernández, R.; Trigo, D.; Cosín, D.J.D. Cryptic speciation of hormogastrid earthworms revealed by mitochondrial and nuclear data. *Mol. Phylogenetics Evol.* **2010**, *56*, 507–512. [CrossRef] [PubMed]
71. Mršić, N. Allolobophora oculata (Hoffmeister 1845) new Earthworms species (*Lumbricidae*) in the Fauna of Yugoslavia. *Biolški Vestn.* **1987**, *35*, 67–77.
72. Huang, J.; Xu, Q.; Sun, Z.J.; Tang, G.L.; Su, Z.Y. Identifying earthworms through DNA barcodes. *Pedobiologia* **2007**, *51*, 301–309. [CrossRef]
73. Rusek, J. Biodiversity of Collembola and their functional role in the ecosystem. *Biodivers. Conserv.* **1998**, *7*, 1207–1219. [CrossRef]
74. Shaw, P.; Benefer, C.M. Development of a barcoding database for the UK Collembola: Early results. *Soil Org.* **2015**, *87*, 197–202.
75. Mateos, E.; Lukić, M. New European Lepidocyrtus Bourlet, 1839 (*Collembola, Entomobryidae*) with the first description of feeding-related dancing behaviour in Collembola. *Zootaxa* **2019**, *4550*, 221–235. [CrossRef]
76. Mateos, E.; Escuer, P.; Buşmachiu, G.; Riutort, M.; Álvarez-Presas, M. Untangling Lepidocyrtus (*Collembola, Entomobryidae*): New molecular data shed light on the relationships of the European groups. *Invertebr. Syst.* **2018**, *32*, 639–651. [CrossRef]
77. Soto-Adames, F.N. Phylogeny of Neotropical Lepidocyrtus (*Collembola*: *Entomobryidae*): First assessment of patterns of speciation in Puerto Rico and phylogenetic relevance of some subgeneric diagnostic characters. *Syst. Entomol.* **2020**, *25*, 485–502. [CrossRef]
78. Cicconardi, F.; Nardi, F.; Emerson, B.C.; Frati, F.; Fanciulli, P.P. Deep phylogeographic divisions and long-term persistence of forest invertebrates (*Hexapoda*: *Collembola*) in the north-western Mediterranean basin. *Mol. Ecol.* **2010**, *19*, 386–400. [CrossRef] [PubMed]
79. Frati, F.; Dell'Ampio, E.; Casasanta, S.; Carapelli, A.; Paolo Fanciulli, P. Large Amounts of Genetic Divergence among Italian Species of the Genus Orchesella (*Insecta, Collembola*) and the Relationships of Two New Species. *Mol. Phylogenetics Evol.* **2020**, *17*, 456–461. [CrossRef] [PubMed]

80. Parisi, V. Ricerche sui Collemboli: Dati morfometrici sullo sviluppo della livrea della *Orchesella villosa* (L.) e descrizione di una nuova specie. *Boll. Di Zool. Agrar. E Di Bachic.* **1960**, *3*, 117–126.
81. Galimberti, A.; Spada, M.; Russo, D.; Mucedda, M.; Agnelli, P.; Crottini, A.; Ferri, E.; Martinoli, A.; Casiraghi, M. Integrated Operational Taxonomic Units (IOTUs) in Echolocating Bats: A Bridge between Molecular and Traditional Taxonomy. *PLoS ONE* **2012**, *7*, e40122. [CrossRef]
82. Shen, Y.Y.; Chen, X.; Murphy, R.W. Assessing DNA Barcoding as a Tool for Species Identification and Data Quality Control. *PLoS ONE* **2013**, *8*, e57125. [CrossRef]
83. Hartop, E.; Srivathsan, A.; Ronquist, F.; Meier, R. Large-scale Integrative Taxonomy (LIT): Resolving the data conundrum for dark taxa. *bioRxiv* **2021**. [CrossRef]
84. Ball, S.L.; Armstrong, K.F. DNA barcodes for insect pest identification: A test case with tussock moths (Lepidoptera: Lymantriidae). *Can. J. For. Res.* **2006**, *36*, 337–350. [CrossRef]
85. Huemer, P.; Mutanen, M.; Sefc, K.M.; Hebert, P.D.N. Testing DNA barcode performance in 1000 species of European Lepidoptera: Large geographic distances have small genetic impacts. *PLoS ONE* **2014**, *9*, e115774. [CrossRef]
86. Wilson, J.J. Assessing the value of DNA barcodes and other priority gene regions for molecular phylogenetics of Lepidoptera. *PLoS ONE* **2010**, *5*, e10525. [CrossRef]
87. Surya, P.; Sebastian, C.D. Revisiting DNA barcoding of ants (Formicidae: Hymenoptera) from India. *Int. J. Entomol. Res.* **2020**, *5*, 161–169.
88. Blatrix, R.; Aubert, C.; Decaens, T.; Berquier, C.; Andrei-Ruiz, M.C.; Galkowski, C. Contribution of a DNA barcode to an assessment of the specificity of ant taxa (*Hymenoptera: Formicidae*) on Corsica. *Eur. J. Entomol.* **2020**, *117*, 420–429. [CrossRef]
89. Schär, S.; Menchetti, M.; Schifani, E.; Hinojosa, J.C.; Platania, L.; Dapporto, L.; Vila, R. Integrative biodiversity inventory of ants from a Sicilian archipelago reveals high diversity on young volcanic islands (*Hymenoptera: Formicidae*). *Org. Divers. Evol.* **2020**, *20*, 405–416. [CrossRef]
90. Ferreira, S.A.; Andrade, R.; Gonçalves, A.R.; Sousa, P.; Paupério, J.; Fonseca, N.A.; Beja, P. The InBIO barcoding initiative database: DNA barcodes of Portuguese Diptera 01. *Biodivers. Data J.* **2020**, *8*, e49985. [CrossRef] [PubMed]
91. Morinière, J.; Balke, M.; Doczkal, D.; Geiger, M.F.; Hardulak, L.A.; Haszprunar, G.; Hausmann, A.; Hendrich, L.; Regalado, L.; Rulik, B.; et al. A DNA barcode library for 5200 German flies and midges (*Insecta: Diptera*) and its implications for metabarcoding-based biomonitoring. *Mol. Ecol. Resour.* **2019**, *19*, 900–928. [CrossRef]
92. Bjeliš, M.; Buljubašić, I.; Popović, L.; Masten Milek, T. Spread of the spotted wing drosophila—Drosophila suzukii (*Diptera, Drosophlidae*) and new distribution records in Dalmatia region of Croatia. *EPPO Bull.* **2015**, *45*, 214–217. [CrossRef]
93. DeSalle, R.; Goldstein, P. Review and interpretation of trends in DNA barcoding. *Front. Ecol. Evol.* **2019**, *7*, 302. [CrossRef]

diversity

MDPI

Article

Molecular Species Delimitation Using COI Barcodes of Mealybugs (Hemiptera: Pseudococcidae) from Coffee Plants in Espírito Santo, Brazil

Pablo Viana Oliveira [1,*], Alexandre Rosa dos Santos [2], Emily Lopes Olive [3], Karolinni Bianchi Britto [1], Francine Alves Nogueira de Almeida [4], Vitor Cezar Pacheco da Silva [5], Carolina Barros Machado [6], Maurício José Fornazier [7], José Aires Ventura [7], Mark Paul Culik [7] and Greiciane Gaburro Paneto [1,8]

1 Programa de Pós-Graduação em Biotecnologia, Universidade Federal do Espírito Santo, Vitoria 29047-105, ES, Brazil
2 Departamento de Engenharia Rural, Universidade Federal do Espírito Santo, Alegre 29500-000, ES, Brazil
3 Programa de Pós-Graduação em Agronomia, Universidade Federal do Espírito Santo, Alegre 29500-000, ES, Brazil
4 Programa de Pós-Graduação em Genética e Melhoramento, Universidade Federal do Espírito Santo, Alegre 29500-000, ES, Brazil
5 Sección Entomología, Facultad de Ciencias, Unidad de Entomología, Facultad de Agronomía, Universidad de la República, Montevideo 11400, Uruguay
6 Laboratório de Biodiversidade Molecular e Conservação, Departamento de Genética e Evolução, Universidade Federal de São Carlos, São Carlos 13565-905, SP, Brazil
7 Instituto Capixaba de Pesquisa, Assistência Técnica e Extensão Rural (INCAPER), Vitoria 29052-010, ES, Brazil
8 Departamento de Farmácia e Nutrição, Universidade Federal do Espírito Santo, Alegre 29500-000, ES, Brazil
* Correspondence: pablo.viana27@gmail.com

Citation: Oliveira, P.V.; dos Santos, A.R.; Olive, E.L.; Britto, K.B.; de Almeida, F.A.N.; Pacheco da Silva, V.C.; Machado, C.B.; Fornazier, M.J.; Ventura, J.A.; Culik, M.P.; et al. Molecular Species Delimitation Using COI Barcodes of Mealybugs (Hemiptera: Pseudococcidae) from Coffee Plants in Espírito Santo, Brazil. *Diversity* **2023**, *15*, 305. https://doi.org/10.3390/d15020305

Academic Editors: Stephan Koblmüller and Michael Wink

Received: 2 August 2022
Accepted: 13 November 2022
Published: 20 February 2023

Abstract: Mealybugs are insects belonging to the family Pseudococcidae. This family includes many plant-pest species with similar morphologies, which may lead to errors in mealybug identification and delimitation. In the present study, we employed molecular-species-delimitation approaches based on distance (ASAP) and coalescence (GMYC and mPTP) methods to identify mealybugs collected from coffee and other plant hosts in the states of Espírito Santo, Bahia, Minas Gerais, and Pernambuco, Brazil. We obtained 171 new COI sequences, and 565 from the BOLD Systems database, representing 26 candidate species of Pseudococcidae. The MOTUs estimated were not congruent across different methods (ASAP-25; GMYC-30; mPTP-22). Misidentifications were revealed in the sequences from the BOLD Systems database involving *Phenacoccus solani* × *Ph. solenopsis*, *Ph. tucumanus* × *Ph. baccharidis*, and *Planacoccus citri* × *Pl. minor* species. Ten mealybug species were collected from coffee plants in Espírito Santo. Due to the incorrect labeling of the species sequences, the COI barcode library of the dataset from the database needs to be carefully analyzed to avoid the misidentification of species. The systematics and taxonomy of mealybugs may be improved by integrative taxonomy which may facilitate the integrated pest management of these pests.

Keywords: biodiversity; agricultural pests; DNA barcoding; species identification; coffee plant; phylogenetic analysis; entomology; systematics; integrative taxonomy

1. Introduction

Coffee (*Coffea* spp.) is the second-most valuable commodity in the world and a crop that drives a multi-billion-dollar worldwide market [1,2]. From coffee planting and harvest to the consumer's table, the production chain faces numerous challenges, such as climate change, price volatility, post-harvest storage, and mycotoxins [3–6], as well as insect pests and diseases [7–10]. Mealybugs (Hemiptera: Pseudococcidae) are phytophagous insects with a global distribution [11–14]. They are vectors of plant pathogens that harm agricultural crops, and they are significant pests of coffee [15–17].

Because mealybugs are typically small and cryptic, it is challenging to identify and control them. They also secrete wax, which creates a coating that protects them from pesticides [18–21]. The effective prevention and management of pests such as mealybugs depends on the accurate identification of species, which is often difficult using traditional methods based on morphological characteristics [22,23]. Mealybug identification based on morphology is a challenge because it requires the microscopic examination of mature females; it is especially difficult to differentiate closely related species using this method [24]. The other limitations of using morphological methods for species identification include immature specimens, specimens with physical deterioration, and multiple species with overlapping morphological characteristics that inhibit accurate identification [25].

These taxonomic issues have recently been investigated using molecular methods [26–31]. DNA barcoding is a genetic technique that uses a portion of the mitochondrial gene cytochrome c oxidase I (COI), which has approximately 658 base pairs, to identify species [32–34]. DNA barcoding has contributed to the discovery of new species, assessment of biodiversity, clarification of taxonomic ambiguities, and correction of misidentifications in studies of diversity [35–37]. Taxonomic and systematic investigations have also examined the delimitation of species based on the COI gene, with DNA barcoding used with genetic distance or tree-based methods for species delimitation to increase the accuracy of identifications [38–44].

Mealybugs are a group of insects that have been subjected to relatively little research in Brazil, despite significant recent advancements in the study of species delimitation. A search in the BOLD Systems database [45] in April 2022 using the term "Pseudococcidae" produced 3417 published records and 276 BINs, whereas the phrase "Pseudococcidae Brazil" produced 69 records and 11 BINs. In other words, just 2% of the world records are from Brazil, despite the fact that Brazil is a country with high species richness [37]. This percentage demonstrates the lack of awareness about the biodiversity of mealybugs in Brazil.

Because of the economic importance of coffee, mealybugs, as coffee pests, and the challenges involved in their identification based on morphology, information on the species of mealybugs present in the coffee crop in state of Espírito Santo is essential. Thus, the aims of this study were to develop a DNA barcode library for mealybug species collected in Espírito Santo, Brazil, based on morphological analysis, to assess the accuracy of COI barcodes, and to perform phylogenetic analyses for species delimitation in the family in Brazil.

2. Materials and Methods

2.1. Sample Collection and Morphological Identification

Adult female mealybugs were collected from coffee plants in the Brazilian states of Espírito Santo and Bahia (Figure 1), with three or more specimens collected at each sample location when possible. Specimens were preserved in 95% ethanol and stored at −30 °C after collection. Mealybug specimens were also collected from other host plants to investigate mealybug species composition in the state.

Collected mealybugs, including DNA voucher specimens, were slide-mounted and identified by light microscopy using methods described by Sirisena et al. (2013) [46]. Mealybugs were examined with a phase-contrast compound microscope (Ernst Leitz GmbH Biomed, Wetzlar, Germany) and identified using taxonomic keys [47–52].

Planococcus minor (Maskell, 1897) from Minas Gerais, *Ferrisia dasylirii* (Cockerell, 1896), *Phenacoccus solenopsis* (Tinsley, 1898), and *Planococcus citri* (Risso, 1813) from Pernambuco, Brazil, were included in the research as reference samples. This research did not include any endangered or protected insect species and consent from landowners was acquired for collection of the mealybugs. A total of 226 mealybug samples from four states were used in this study. Information on host plants from which specimens were collected and geographic coordinates of collection locations are available in Table S1 (Supplementary Materials).

2.2. DNA Extraction, Amplification, and Sequencing

The methodology described by Arseneau, Steeves, and Laflamme (2017) [53] was used to extract genomic DNA from each individual mealybug specimen separately, without crushing. The original technique was adjusted to include a minimum of 8 h of incubation, with no RNase A in this stage, and an elution step using 40 µL of DNAse-free water. A NanoDropTM 2000 Spectrophotometer (Thermo Scientific, Waltham, MA, USA) was used to verify DNA yield and purity. DNA extracts were preserved in a freezer at −30 °C.

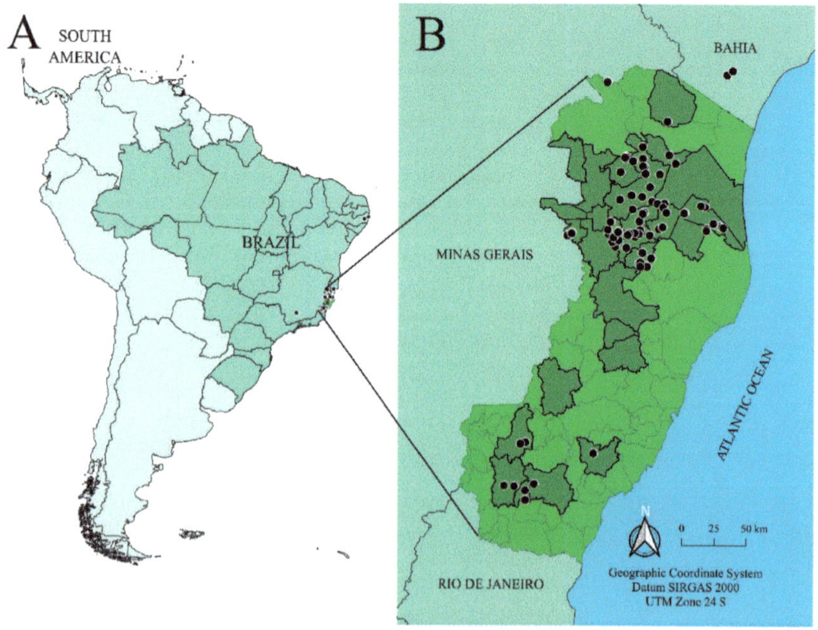

Figure 1. Sampling locations in Brazil for mealybugs (Hemiptera: Pseudococcidae) collected in the present study. (**A**) Map of South America indicating the geographical location of samples collected in the states of Minas Gerais (MG) and Pernambuco (PE). (**B**) Map of Espírito Santo indicating the geographical location of sampling sites (black dots) and municipalities sampled without specific geographical location (dark green with black borders). Specific collection locations and additional information for samples are available in Table S1 (Supplementary Materials). The map was created using QGIS Geographic Information System v3.16 (https://qgis.org accessed on 12 November 2022) by P.V. Oliveira.

COI gene amplicons (~670 bp) were amplified and sequenced using PCO-F1 and LEP-R1 primers [24]. The reaction master mix comprised 1X PCR Buffer, 1.5 mM MgCl2, 0.2 mM each dNTP, 0.2 µM forward and reverse primers, 1 unit of Platinum™ Taq DNA Polymerase (Invitrogen), 20 ng of DNA and ultrapure DNase/RNase-Free distilled water up to a final volume 12.5 µL. The following PCR procedures were modified from Park et al. (2010) [24]: A 2-min initial denaturation at 94 °C, followed by 5 cycles of 40 s at 94 °C, 40 s at 45 °C, and 1 min at 72 °C. This preceded 35 cycles of 40 s at 94 °C, 40 s at 51 °C, and 1 min at 72 °C, followed by a final extension of 10 min at 72 °C, which ended at 4 °C.

Successful PCR products were confirmed using electrophoresis on a 1.5% agarose gel stained with SYBRTM Safe DNA Gel Stain (Invitrogen). Primer residues and unincorporated nucleotides were eliminated using ExoSAP enzymes. Sequencing reactions were carried out using the BigDye® Terminator v3.1 Cycle Sequencing Kit in accordance with the manufacturer's instructions. Samples were precipitated with EDTA/NaOAc/ethanol following the sequencing procedure, dried at 50 °C for 30 min, and then sequenced using an ABI 3500 Genetic Analyzer (Applied Biosystems, Carlsbad, CA, USA).

2.3. Data Assembly and Analysis

MEGA X was used to analyze, edit, and align raw sequences [54]. The MUSCLE algorithm was used for multiple sequence alignment [55]. All sequences translated into amino acids were examined for the presence of stop-codons and indels to ensure there were no nuclear pseudogenes [56].

Our mealybug COI-sequence dataset was created by combining 171 new COI sequences obtained in the present study, all sequences from other countries classified to mealybug species level by BOLD methods, and all publicly available records of mealybugs restricted to Brazil. To prevent missing data, sequences of less than 580 bp were not included in the analysis. The 736 sequences that made up the final dataset are provided in SM1 (Supplementary Materials). DNA sequences generated in this study are available in the GenBank database under accession numbers OP381504-OP381598, OP391569-OP391593, OP391594-OP391608, OP425673-OP425695, and OP450828-OP450839.

2.4. Phylogenetic Reconstruction and Species-Delimitation Methods

This study employed a haplotype-based species delimitation method, with each haplotype represented by a single sequence (except for *Phenacoccus solani* × *Ph. solenopsis* and *Planococcus citri* × *Pl. minor* species) to simplify computing requirements. The haplotype sequences used to infer species delimitation (n = 110) were determined using DnaSP 6 software [57]. The aligned COI haplotypes used for the species-delimitation study are available in SM2 (Supplementary Materials). The numbers of sequences used for *Phenacoccus solani* × *Ph. solenopsis* and *Planococcus citri* × *Pl. minor* species were different because these closely related species shared haplotypes, with three sequences selected for *Phenacoccus solani* × *Ph. solenopsis* species and two for *Planococcus citri* × *Pl. minor* species.

Maximum likelihood (ML) and Bayesian inference (BI) were used to evaluate the phylogenetic connections among the studied specimens. The Bayesian Information Criterion (BIC) was used to infer the substitution model for phylogenetic-tree reconstruction using jModelTest v.2.1.10 [58]. GTR+I+G was used as the best-fit model in ML analysis with RAxML v 8.2.12 [59] and 1000 bootstrap replicates. *Ripersiella emarai* and *R. multiporifera* sequences were employed as outgroups [60]. The San Diego Supercomputer Center's CIPRES Science Gateway (https://www.phylo.org/index.php accessed on 12 November 2022) was used for the ML analysis.

To derive an ultrametric tree, BI analysis was carried out in BEAST v.1.8.4 [61] utilizing the GTR+I+G as a substitution model, the speciation birth–death model as a tree prior, and the lognormal relaxed clock model. Three independent runs of 10 million generations, sampled every 10,000 generations, and 25% burn-in were conducted. Convergence was evaluated with Tracer v.1.7.1 [62], the trees were summarized in TreeAnnotator v.2.4.8 [63], and the resultant tree was constructed using FigTree v1.4.3 (http://tree.bio.ed.ac.uk/software/figtree/ accessed on 12 November 2022)

To estimate the molecular operational taxonomic units (MOTUs), species-delimitation techniques using distance- and tree-based methodologies were used. Assemble Species by Automatic Partitioning (ASAP), a webserver for species delimitation (https://bioinfo.mnhn.fr/abi/public/asap/ accessed on 12 November 2022) was used with the K2P model and default settings for the distance-based method [64]. Bayesian Poisson Tree Processes (bPTP; https://species.h-its.org/ptp/ accessed on 12 November 2022) [65] and multi-rate Poisson Tree Processes (mPTP; https://mptp.h-its.org accessed on 12 November 2022) with default settings were used for the tree-based method [66].

A non-ultrametric tree estimated by RAxML version 8.2.12 was used to calculate MOTUs for both methods. The Generalized Mixed Yule Coalescent model (GMYC; https://species.h-its.org/gmyc/ accessed on 12 November 2022) made use of an ultrametric tree that BEAST v.1.8.4 had reconstructed using a single-threshold approach with default settings [67]. When at least two of the three methods were in agreement, a consensus species delimitation was chosen. The Kimura-2-parameter (K2P) model was used to compute overall mean distance, intraspecific distances, and interspecific distances among MOTUs [68]. MEGA X was used to determine mean pairwise distances using pairwise deletion [54].

3. Results and Discussion

A total of 171 novel COI barcode sequences from 7 genera and 10 species were recovered, with the identifications confirmed morphologically. Furthermore, 565 publically available BOLD sequences were added to our dataset of 736 sequences without outgroups, including 219 from Brazil and 517 from other countries, representing the mealybug species recorded in Brazil (SM1, Supplementary Materials). This study included 10 morphologically identified candidate species and 16 candidate species from the database. There were 328 variable sites (53.33%), of which 257 (78.35%) were parsimony-informative and 71 (21.65%) were singletons.

In the public databases, conflicting findings were observed for two genera: three sequences of "*Phenacoccus solani*" and three of *Phenacoccus solenopsis*; and two sequences of "*Planococcus citri*" and two of *Planococcus minor*, which shared the same haplotype. In the cases in which species identification was questionable, the name of the species was noted with quotation marks. The final haplotype alignment included 118 sequences (without outgroups), representing 110 different haplotypes (Supplementary Materials). The estimated molecular operational taxonomic units were not consistent among the methods: 25 for ASAP, 30 for GMYC, and 22 for mPTP (Figure 2).

For species delimitation, the use of multiple approaches may be more reliable than a single method [69]. In the present study, three approaches were studied and ASAP was found to be the best species-delimitation model for mealybugs. *Planococcus citri* and *Pl. minor* were recognized as a single species, suggesting an inability to delimit these two closely related species. The weaker genetic divergence between them can explain this clustering [70–72]. This method identified 25 candidate species, near to the number used initially (26 species). Thus, ASAP has better biological significance (delimiting species concordantly) and a lower number of singletons than GMYC.

With a strong theoretical basis, GMYC was created to delimit species using single-locus data. However, compared to other approaches, it often produces more OTUs. [73]. *Dysmicoccus brevipes*, *M. hirsutus*, and *Ps. longispinus*, for example, were each separately split into two, two, and three candidate species by GMYC, respectively. This could mean cryptic species, but it is necessary to be cautious. More gene sequences are necessary to perform a reliable species delimitation in these cases. For this to be definitive, an integrated taxonomic strategy and more thorough sampling are required [71,74].

Although mPTP has been regarded as a useful technique for species delimitation, in the present study, mPTP was combined with well-known taxa such as *Ferrisia virgata* (Cockerell), *F. dasylirii*, *Pseudoccoccus jackbeardsleyi* (Gimpel & Miller), and *Ps. annonae* (Pacheco da Silva & Kaydan) [38,75]. The bPTP approach was eliminated from our study because it estimated too many singletons, which might have led to an incorrect interpretation of the data [73]. As input, bPTP does not require an ultrametric tree or a sequence-similarity threshold [65]. Instead, it adds Bayesian support values to the input tree to delimit species. The greatest Bayesian support value at a node suggests that all of that node's descendants are most likely species [42,76,77].

Due to character loss or poor specimen quality, it can be challenging to distinguish closely related mealybug species based on morphology. In the present study, sequences from the BOLD Systems database that were deposited as *Ph. solani* were noted. However, our results strongly suggest that the sequences are from *Ph. solenopsis*. All 14 specimens collected and identified as *Ph. solenopsis* based on their morphology were sequenced and matched to *Ph. solenopsis* in BOLD to support this conclusion.

Similar circumstances apply to the *Pl. citri* sequences from BOLD, which apparently belonged to the *Pl. minor* species. These findings may help to explain why *Ph. solani* x *Ph. solenopsis* and *Pl. citri* × *Pl. minor* have inconsistent species identifications based on morphology, since they share geographic regions and have similar morphologies [78]. In addition, *Phenacoccus tucumanus* (Granara de Willink, 1983) sequences were matched to *Ph. baccharidis* (Williams; 99.77% likelihood of placement).

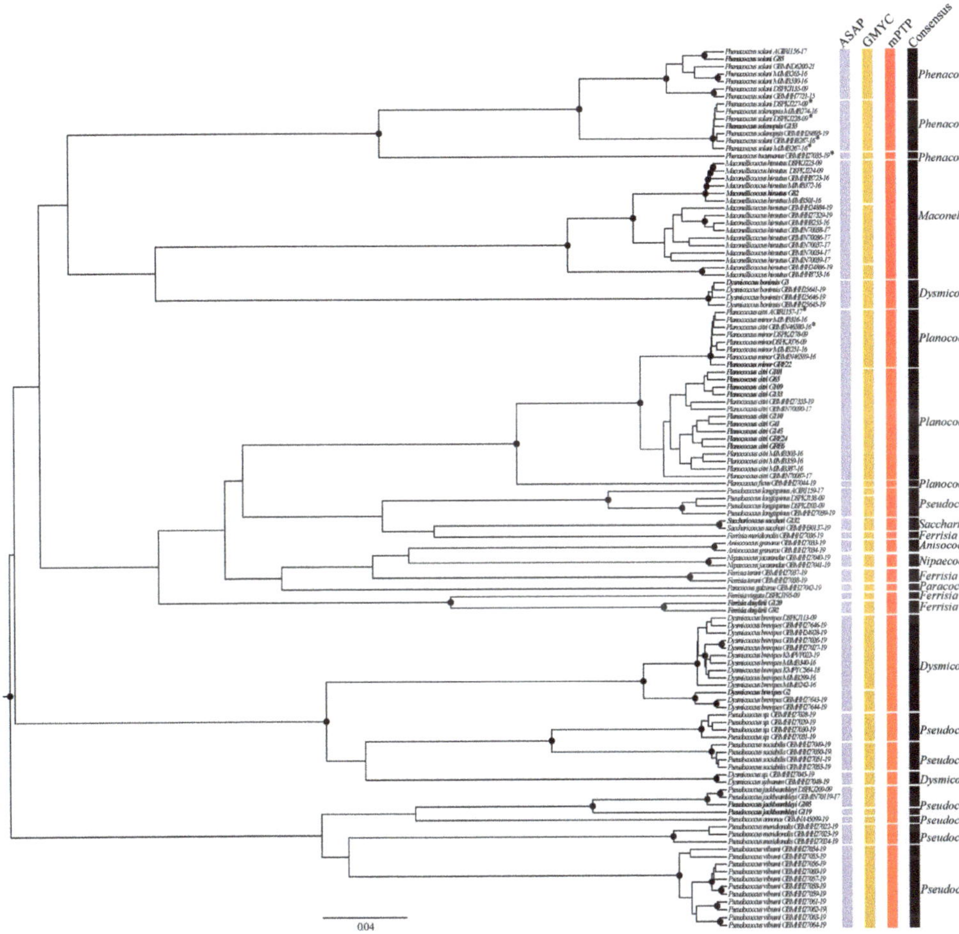

Figure 2. Phylogenetic analysis of mealybugs (Hemiptera: Pseudococcidae) based on COI-gene haplotypes and results of species-delimitation approaches. This tree was estimated using Bayesian inference. Black dots in the nodes indicate statistical support of posterior probability (>0.9). Nominal species in bold indicate new sequences obtained in the present study and asterisks (*) indicate errors in the species identification using the BOLD Systems database. Colored bars on the right show the putative species inferred by ASAP, GMYC, and mPTP. The last bar (far right) indicates species consensus through the approaches.

The use of genetic/molecular databases has been shown to increase the accuracy of identification [79]. Nevertheless, many sequences obtained from databases may result in incorrect identifications and create taxonomic confusion. To prevent problems with accuracy in biodiversity data, care must be taken when taxonomic information is obtained from repositories [80,81].

The genera *Maconellicoccus*, *Phenacoccus*, and *Planococcus* were sampled with a significant number of haplotypes, according to the BI phylogeny (Figure 2). These monophyletic groups have PP values over 0.9. Although only one or two haplotypes were examined, we also found monophyletism for the genera *Anisococcus*, *Nipaecoccus*, *Paracoccus*, and

Saccharicoccus. However, the delimitation of putative species may be biased because of this underrepresentation of sequences [82].

The paraphyletic groups for the genera *Dysmicoccus* and *Pseudococcus* are shown. These taxonomic groups had comparable phylogenetic relationships in the ML phylogeny, with high support values (Figure S1, Supplementary Materials). Despite the fact that only one marker was utilized for defining the species, our results are comparable to those of other publications that used multilocus approaches [60].

In the present study, we discovered 25 candidate species for Brazil using the consensus reached by the methodologies, which included the ASAP, GMYC, and mPTP approaches. In the state of Espírito Santo, 10 species of mealybugs were identified. With the exception of *Saccharicoccus sacchari* (Cockerell), the other nine mealybug species were previously identified on *Coffea* spp. However, in this study, one specimen of *Dysmicoccus boninsis* (Kuwana) and two of *S. sacchari* (Cockerell) were the only species collected from sugarcane, and one specimen of *Ph. solenopsis* was collected from okra.

The rapid screening and identification of mealybugs of commercial significance and quarantine concern is made possible by COI barcodes [83]. Our findings suggest a successful identification strategy that can facilitate pest management in the Brazilian coffee industry. The closely related species *Pl. citri* and *Pl. minor*, which are both common and abundant in Brazil, were identified. By using known natural enemies of the identified species, coffee growers may also gain from species identification. For instance, the citrus mealybug (*Pl. citri*) is attacked by a variety of natural enemies that have been identified around the world, including a large number of hymenopteran parasitoids, predatory fly larvae, lacewings, and ladybug beetles [84–88].

4. Conclusions

Because mealybugs are common and serious pests of many agriculture crops, the accurate identification of mealybug species is essential to enable the integrated pest management (IPM) of these insects. In addition, the accurate identification of mealybugs is necessary for quarantine interceptions to prevent the spread of these pests to uninfested regions. In the present study, a DNA barcode library for mealybug species was produced, using mealybug species obtained mainly in Espírito Santo, Brazil, a major coffee-producing region.

Ten mealybug species were collected and identified from coffee plants in Espírito Santo, and our results demonstrate that the molecular delimitation of species is useful to ascertain the diversity of the species in this crop. Despite possible limitations in the mealybug database, our results shows that two of the three delimitation strategies examined were reliable.

Accurate species identification is essential for integrated pest management. The Pseudococcidae COI barcode library for the state of Espírito Santo obtained in the present study provides an advantageous reference for the reliable and rapid identification of mealybug species, particularly for closely related and cryptic species. In addition, coffee farming will benefit from the ability to use recognized natural enemies to manage mealybugs in coffee plantations identified by integrative taxonomy.

Supplementary Materials: The following supporting information can be downloaded at: https://www.mdpi.com/article/10.3390/d15020305/s1, Figure S1: Phylogenetic relationships of mealybugs (Hemiptera: Pseudococcidae) based on COI-gene haplotypes. This tree was estimated using Maximum Likelihood with RaxML. Black dots in the nodes indicate bootstrap values (>50%). Nominal species in bold indicate new sequences obtained in the present study and asterisks (*) indicate errors in the species identification using the BOLD Systems database; Table S1: List of specimen samples used in this study, geographic coordinates and Best ID in the BOLD Systems database; The alignments of COI sequences of all specimens and haplotypes are available in fasta file.

Author Contributions: Conceptualization, P.V.O., M.J.F., J.A.V., M.P.C. and G.G.P.; methodology, P.V.O., E.L.O., K.B.B., F.A.N.d.A. and G.G.P.; validation, C.B.M.; formal analysis, P.V.O. and C.B.M.; investigation, P.V.O., E.L.O., K.B.B., F.A.N.d.A., V.C.P.d.S. and C.B.M.; resources, A.R.d.S., V.C.P.d.S., M.J.F., J.A.V. and M.P.C.; data curation, P.V.O., E.L.O., K.B.B., F.A.N.d.A., V.C.P.d.S. and C.B.M.; writing—original draft preparation, P.V.O.; writing—review and editing, K.B.B., V.C.P.d.S., C.B.M.,

M.P.C. and G.G.P.; visualization, P.V.O., C.B.M. and G.G.P.; supervision, A.R.d.S. and G.G.P.; project administration, A.R.d.S., J.A.V. and G.G.P.; funding acquisition, G.G.P. All authors have read and agreed to the published version of the manuscript.

Funding: Coordenação de Aperfeicoamento de Pessoal de Nível Superior: 001; Fundação de Amparo à Pesquisa do Espírito Santo: PROCAP 02/2018 and T.O. number 024/2023

Institutional Review Board Statement: Not applicable.

Data Availability Statement: Not applicable.

Acknowledgments: We thank Perseu Perdoná, Manager of the Technical Department of COOABRIEL, the coffee producers and technicians of INCAPER, and Pablo Oliveira's family and friends, who assisted in the collection of mealybugs for this study. Thanks also to GAGEN and Taís Moreira for their review.

Conflicts of Interest: The authors declare no conflict of interest.

References

1. Cure, J.R.; Rodríguez, D.; Gutierrez, A.P.; Ponti, L. The Coffee Agroecosystem: Bio-Economic Analysis of Coffee Berry Borer Control (*Hypothenemus hampei*). *Sci. Rep.* **2020**, *10*, 12262. [CrossRef] [PubMed]
2. Davis, A.P.; Mieulet, D.; Moat, J.; Sarmu, D.; Haggar, J. Arabica-like Flavour in a Heat-Tolerant Wild Coffee Species. *Nat. Plants* **2021**, *7*, 413–418. [CrossRef] [PubMed]
3. Jaramillo, J.; Muchugu, E.; Vega, F.E.; Davis, A.; Borgemeister, C.; Chabi-Olaye, A. Some Like It Hot: The Influence and Implications of Climate Change on Coffee Berry Borer (*Hypothenemus hampei*) and Coffee Production in East Africa. *PLoS ONE* **2011**, *6*, e24528. [CrossRef] [PubMed]
4. Jayakumar, M.; Rajavel, M.; Surendran, U.; Gopinath, G.; Ramamoorthy, K. Impact of Climate Variability on Coffee Yield in India—With a Micro-Level Case Study Using Long-Term Coffee Yield Data of Humid Tropical Kerala. *Clim. Chang.* **2017**, *145*, 335–349. [CrossRef]
5. Martins, M.Q.; Fortunato, A.S.; Rodrigues, W.P.; Partelli, F.L.; Campostrini, E.; Lidon, F.C.; DaMatta, F.M.; Ramalho, J.C.; Ribeiro-Barros, A.I. Selection and Validation of Reference Genes for Accurate RT-QPCR Data Normalization in *Coffea* Spp. under a Climate Changes Context of Interacting Elevated [CO_2] and Temperature. *Front. Plant Sci.* **2017**, *8*, 307. [CrossRef]
6. Vega, F.E.; Rosenquist, E.; Collins, W. Global Project Needed to Tackle Coffee Crisis. *Nature* **2003**, *425*, 343. [CrossRef]
7. Correa, L.R.; Souza, B.; Santa-Cecília, L.V.C.; Prado, E. Estudos biológicos de cochonilhas do gênero *Planococcus* (Hemiptera: Pseudococcidae) em diferentes hospedeiros. *Arq. Inst. Biol.* **2011**, *78*, 233–240. [CrossRef]
8. Läderach, P.; Ramirez–Villegas, J.; Navarro-Racines, C.; Zelaya, C.; Martinez–Valle, A.; Jarvis, A. Climate Change Adaptation of Coffee Production in Space and Time. *Clim. Chang.* **2017**, *141*, 47–62. [CrossRef]
9. Pham, Y.; Reardon-Smith, K.; Mushtaq, S.; Cockfield, G. The Impact of Climate Change and Variability on Coffee Production: A Systematic Review. *Clim. Chang.* **2019**, *156*, 609–630. [CrossRef]
10. Chemura, A.; Mudereri, B.T.; Yalew, A.W.; Gornott, C. Climate Change and Specialty Coffee Potential in Ethiopia. *Sci. Rep.* **2021**, *11*, 8097. [CrossRef]
11. Culik, M.P.; Martins, D.D.S.; Gullan, P.J. First Records of Two Mealybug Species in Brazil and New Potential Pests of Papaya and Coffee. *J. Insect Sci.* **2006**, *6*, 23. [CrossRef] [PubMed]
12. Santa-Cecília, L.V.C.; Souza, B.; de Souza, J.C.; Prado, E.; Junior, A.M.; Fornazier, M.J.; Carvalho, G.A. *Cochonilhas-Farinhentas em Cafeeiros: Bioecologia, Danos e Métodos de Controle*; Epamig: Belo Horizonte, Brazil, 2007.
13. Costa, M.B.; Souza, B.; Santa-Cecília, L.V.C.; Prado, E. Tabela de Vida de Fertilidade de *Planococcus citri* (Risso) e *Planococcus minor* (Maskell) (Hemiptera: Pseudococcidae) Em Cafeeiro. *Coffee Sci.* **2016**, *11*, 204–210.
14. Rondelli, V.M.; Peronti, A.L.B.G.; Dias, J.R.M.; Fogaça, I.; Dos Santos, I.L.V.; Nery, A.G. New Records of Mealybugs (Hemiptera: Pseudococcidae) Infesting Rosettes of Conilon Coffee Plants in the State of Rondônia, South-Western Amazon, Brazil. *Fla. Entomol.* **2018**, *101*, 705. [CrossRef]
15. He, Y.B.; Wan, X.W.; Liu, Y.H.; Sun, G.M.; Zhan, R.L. Mitochondrial COI from *Dysmicoccus brevipes* (Hemiptera: Pseudococcidae) Suggests Cryptic Lineage and Pinpoints the Source of the Introduction to China. *Fla. Entomol.* **2012**, *95*, 183–191. [CrossRef]
16. Pacheco da Silva, V.C.; Bertin, A.; Blin, A.; Germain, J.-F.; Bernardi, D.; Rignol, G.; Botton, M.; Malausa, T. Molecular and Morphological Identification of Mealybug Species (Hemiptera: Pseudococcidae) in Brazilian Vineyards. *PLoS ONE* **2014**, *9*, e103267. [CrossRef]
17. Daane, K.M.; Middleton, M.C.; Sforza, R.F.H.; Kamps-Hughes, N.; Watson, G.W.; Almeida, R.P.P.; Correa, M.C.G.; Downie, D.A.; Walton, V.M. Determining the Geographic Origin of Invasive Populations of the Mealybug *Planococcus ficus* Based on Molecular Genetic Analysis. *PLoS ONE* **2018**, *13*, e0193852. [CrossRef] [PubMed]
18. Cid, M.; Fereres, A. Characterization of the Probing and Feeding Behavior of *Planococcus citri* (Hemiptera: Pseudococcidae) on Grapevine. *Ann. Entomol. Soc. Am.* **2010**, *103*, 404–417. [CrossRef]
19. Correa, M.C.G.; Zaviezo, T.; Le Maguet, J.; Herrbach, E.; Malausa, T. Characterization of Microsatellite DNA Libraries from Three Mealybug Species and Development of Microsatellite Markers for *Pseudococcus viburni* (Hemiptera: Pseudococcidae). *Bull. Entomol. Res.* **2014**, *104*, 213–220. [CrossRef]

20. Correa, M.C.G.; Lombaert, E.; Malausa, T.; Crochard, D.; Alvear, A.; Zaviezo, T.; Palero, F. Mealybug Species from Chilean Agricultural Landscapes and Main Factors Influencing the Genetic Structure of *Pseudococcus viburni*. *Sci. Rep.* **2015**, *5*, 16483. [CrossRef]
21. Brahmachari, V.; Kohli, S.; Gulati, P. In Praise of Mealybugs. *J. Genet.* **2018**, *97*, 379–389. [CrossRef]
22. Heya, H.M.; Khamis, F.M.; Onyambu, G.K.; Akutse, K.S.; Mohamed, S.A.; Kimathi, E.K.; Ombura, F.L.O.; Ekesi, S.; Dubois, T.; Subramanian, S.; et al. Characterization and Risk Assessment of the Invasive Papaya Mealybug, *Paracoccus marginatus*, in Kenya under Changing Climate. *J. Appl. Entomol.* **2020**, *144*, 442–458. [CrossRef]
23. Andrews, K.R.; Gerritsen, A.; Rashed, A.; Crowder, D.W.; Rondon, S.I.; van Herk, W.G.; Vernon, R.; Wanner, K.W.; Wilson, C.M.; New, D.D.; et al. Wireworm (Coleoptera: Elateridae) Genomic Analysis Reveals Putative Cryptic Species, Population Structure, and Adaptation to Pest Control. *Commun. Biol.* **2020**, *3*, 489. [CrossRef] [PubMed]
24. Park, D.-S.; Leem, Y.J.; Hahn, K.-W.; Suh, S.-J.; Hong, K.-J.; Oh, H.-W. Molecular Identification of Mealybugs (Hemiptera: Pseudococcidae) Found on Korean Pears. *J. Econ. Entomol.* **2010**, *103*, 25–33. [CrossRef] [PubMed]
25. Wang, F.; Wang, D.; Guo, G.; Hu, Y.; Wei, J.; Liu, J. Species Delimitation of the Dermacentor Ticks Based on Phylogenetic Clustering and Niche Modeling. *PeerJ* **2019**, *7*, e6911. [CrossRef]
26. Malausa, T.; Fenis, A.; Warot, S.; Germain, J.; Ris, N.; Prado, E.; Botton, M.; Vanlerberghe-Masutti, F.; Sforza, R.; Cruaud, C.; et al. DNA Markers to Disentangle Complexes of Cryptic Taxa in Mealybugs (Hemiptera: Pseudococcidae). *J. Appl. Entomol.* **2011**, *135*, 142–155. [CrossRef]
27. Park, D.S.; Suh, S.J.; Hebert, P.D.N.; Oh, H.W.; Hong, K.J. DNA Barcodes for Two Scale Insect Families, Mealybugs (Hemiptera: Pseudococcidae) and Armored Scales (Hemiptera: Diaspididae). *Bull. Entomol. Res.* **2011**, *101*, 429–434. [CrossRef]
28. Amouroux, P.; Crochard, D.; Germain, J.F.; Correa, M.; Ampuero, J.; Groussier, G.; Kreiter, P.; Malausa, T.; Zaviezo, T. Genetic Diversity of Armored Scales (Hemiptera: Diaspididae) and Soft Scales (Hemiptera: Coccidae) in Chile. *Sci. Rep.* **2017**, *7*, 2014. [CrossRef]
29. Oliveira, P.V.; Matos, N.S.; Klippel, A.H.; Oliveira-Costa, J.; Careta, F.D.P.; Paneto, G.G. Using DNA Barcodes to Identify Forensically Important Species of Diptera in Espírito Santo State, Brazil. *Braz. Arch. Biol. Technol.* **2017**, *60*, e17160106. [CrossRef]
30. Pacheco da Silva, V.C.; Kaydan, M.B.; Malausa, T.; Germain, J.-F.; Palero, F.; Botton, M. Integrative Taxonomy Methods Reveal High Mealybug (Hemiptera: Pseudococcidae) Diversity in Southern Brazilian Fruit Crops. *Sci. Rep.* **2017**, *7*, 15741. [CrossRef]
31. Dewer, Y.; Abdel-Fattah, R.S.; Schneider, S.A. Molecular and Morphological Identification of the Mealybug, *Phenacoccus solani* Ferris (Hemiptera: Pseudococcidae): First Report in Egypt. *EPPO Bull.* **2018**, *48*, 155–159. [CrossRef]
32. Hebert, P.D.N.; Ratnasingham, S.; de Waard, J.R. Barcoding Animal Life: Cytochrome c Oxidase Subunit 1 Divergences among Closely Related Species Barcoding Animal Life: Cytochrome c Oxidase Subunit 1 Divergences among Closely Related Species. *Proc. R. Soc. Lond. B* **2003**, *270*, S96–S99. [CrossRef] [PubMed]
33. Klippel, A.H.; Oliveira, P.V.; Britto, K.B.; Freire, B.F.; Moreno, M.R.; Dos Santos, A.R.; Banhos, A.; Paneto, G.G. Using DNA Barcodes to Identify Road-Killed Animals in Two Atlantic Forest Nature Reserves, Brazil. *PLoS ONE* **2015**, *10*, e0134877. [CrossRef] [PubMed]
34. Oliveira, P.V.; de Almeida, F.A.N.; Lugon, M.D.; Britto, K.B.; Oliveira-Costa, J.; Santos, A.R.; Paneto, G.G. Using High-Resolution Melting to Identify Calliphoridae (Blowflies) Species from Brazil. *PeerJ* **2020**, *8*, e9680. [CrossRef] [PubMed]
35. Zhou, Z.; Guo, H.; Han, L.; Chai, J.; Che, X.; Shi, F. Singleton Molecular Species Delimitation Based on COI-5P Barcode Sequences Revealed High Cryptic/Undescribed Diversity for Chinese Katydids (Orthoptera: Tettigoniidae). *BMC Evol. Biol.* **2019**, *19*, 79. [CrossRef] [PubMed]
36. Lopez-Vaamonde, C.; Kirichenko, N.; Cama, A.; Doorenweerd, C.; Godfray, H.C.J.; Guiguet, A.; Gomboc, S.; Huemer, P.; Landry, J.-F.; Laštůvka, A.; et al. Evaluating DNA Barcoding for Species Identification and Discovery in European Gracillariid Moths. *Front. Ecol. Evol.* **2021**, *9*, 66. [CrossRef]
37. Bukowski, B.; Ratnasingham, S.; Hanisch, P.E.; Hebert, P.D.N.; Perez, K.; deWaard, J.; Tubaro, P.L.; Lijtmaer, D.A. DNA Barcodes Reveal Striking Arthropod Diversity and Unveil Seasonal Patterns of Variation in the Southern Atlantic Forest. *PLoS ONE* **2022**, *17*, e0267390. [CrossRef]
38. Martoni, F.; Bulman, S.; Pitman, A.; Taylor, G.; Armstrong, K. DNA Barcoding Highlights Cryptic Diversity in the New Zealand Psylloidea (Hemiptera: Sternorrhyncha). *Diversity* **2018**, *10*, 50. [CrossRef]
39. Dalstein, V.; Eberle, J.; Fabrizi, S.; Etzbauer, C.; Ahrens, D. COI-Based Species Delimitation in Indochinese *Tetraserica chafers* Reveal Hybridisation despite Strong Divergence in Male Copulation Organs. *Org. Divers. Evol.* **2019**, *19*, 277–286. [CrossRef]
40. Sabadini, C.P.; Machado, C.B.; Vilhena, P.D.S.; Garófalo, C.A.; Del Lama, M.A. Species Delimitation and Phylogenetic Relationships in the Genus Trypoxylon (Hymenoptera: Crabronidae) Using Molecular Markers: An Alternative to Taxonomic Impediment. *Syst. Biodivers.* **2020**, *18*, 315–327. [CrossRef]
41. Martínez-Arce, A.; De Jesús-Navarrete, A.; Leasi, F. DNA Barcoding for Delimitation of Putative Mexican Marine Nematodes Species. *Diversity* **2020**, *12*, 107. [CrossRef]
42. Zhang, H.; Ning, X.; Yu, X.; Bu, W.J. Integrative Species Delimitation Based on COI, ITS, and Morphological Evidence Illustrates a Unique Evolutionary History of the Genus *Paracercion* (Odonata: Coenagrionidae). *PeerJ* **2021**, *9*, e11459. [CrossRef] [PubMed]
43. Collado, G.A.; Torres-Díaz, C.; Valladares, M.A. Phylogeography and Molecular Species Delimitation Reveal Cryptic Diversity in *Potamolithus* (Caenogastropoda: Tateidae) of the Southwest Basin of the Andes. *Sci. Rep.* **2021**, *11*, 15735. [CrossRef] [PubMed]

44. Koroiva, R.; Gomes, V.G.N.; Vilela, D.S. DNA Barcoding and New Records of Odonates (Insecta: Odonata) from Paraíba State, Brazil. *Diversity* **2022**, *14*, 203. [CrossRef]
45. Ratnasingham, S.; Hebert, P.D.N. BOLD: The Barcode of Life Data System (www.Barcodinglife.Org). *Mol. Ecol. Notes* **2007**, *7*, 355–364. [CrossRef] [PubMed]
46. Sirisena, U.G.A.I.; Watson, G.W.; Hemachandra, K.S.; Wijayagunasekar, H.N.P. A Modified Technique for the Preparation of Specimens of Sternorryncha for Taxonomic Studies. *Trop. Agric. Res.* **2013**, *24*, 139–149.
47. Cox, J.M.; Ben-Dov, Y. Planococcine Mealybugs of Economic Importance from the Mediterranean Basin and Their Distinction from a New African Genus (Hemiptera: Pseudococcidae). *Bull. Entomol. Res.* **1986**, *76*, 481–489. [CrossRef]
48. Miller, D.R.; Gimpel, W.F. *Systematic Analysis of the Mealybugs in the Pseudococcus Maritimus Complex (Homoptera: Pseudococcidae)*; Associated Publishers: Gainesville, FL, USA, 1996; Volume 2.
49. Granara de Willink, M.C.; Szumik, C. Phenacoccinae de Centro y Sudamérica (Hemiptera:Coccoidea:Pseudococcidae): Sistemática y Filogenia. *Rev. Soc. Entomológica Argent.* **2007**, *66*, 29–129.
50. Granara de Willink, M.C. *Dysmicoccus* de La Región Neotropical (Hemiptera: Pseudococcidae). *Rev. Soc. Entomológica Argent.* **2009**, *68*, 11–95.
51. Granara de Willink, M.C.; González, P. Revisión Taxonómica de *Pseudococcus* Westwood (Hemiptera: Pseudococcidae) de Centro y Sud América Con Descripciones de Especies Nuevas. *Insecta Mundi* **2018**, *1775*, 1–117.
52. Kaydan, M.B.; Gullan, P.J. A Taxonomic Revision of the Mealybug Genus *Ferrisia* Fullaway (Hemiptera: Pseudococcidae), with Descriptions of Eight New Species and a New Genus. *Zootaxa* **2012**, *3543*. [CrossRef]
53. Arseneau, J.R.; Steeves, R.; Laflamme, M. Modified Low-Salt CTAB Extraction of High-Quality DNA from Contaminant-Rich Tissues. *Mol. Ecol. Resour.* **2017**, *17*, 686–693. [CrossRef] [PubMed]
54. Kumar, S.; Stecher, G.; Li, M.; Knyaz, C.; Tamura, K. MEGA X: Molecular Evolutionary Genetics Analysis across Computing Platforms. *Mol. Biol. Evol.* **2018**, *35*, 1547–1549. [CrossRef] [PubMed]
55. Edgar, R.C. MUSCLE: A Multiple Sequence Alignment Method with Reduced Time and Space Complexity. *BMC Bioinform.* **2004**, *5*, 113. [CrossRef] [PubMed]
56. Song, H.; Buhay, J.E.; Whiting, M.F.; Crandall, K.A. Many Species in One: DNA Barcoding Overestimates the Number of Species When Nuclear Mitochondrial Pseudogenes Are Coamplified. *Proc. Natl. Acad. Sci. USA* **2008**, *105*, 13486–13491. [CrossRef]
57. Rozas, J.; Ferrer-Mata, A.; Sánchez-DelBarrio, J.C.; Guirao-Rico, S.; Librado, P.; Ramos-Onsins, S.E.; Sánchez-Gracia, A. DnaSP 6: DNA Sequence Polymorphism Analysis of Large Data Sets. *Mol. Biol. Evol.* **2017**, *34*, 3299–3302. [CrossRef]
58. Darriba, D.; Taboada, G.L.; Doallo, R.; Posada, D. JModelTest 2: More Models, New Heuristics and Parallel Computing. *Nat. Methods* **2012**, *9*, 772. [CrossRef]
59. Stamatakis, A. RAxML Version 8: A Tool for Phylogenetic Analysis and Post-Analysis of Large Phylogenies. *Bioinformatics* **2014**, *30*, 1312–1313. [CrossRef]
60. Choi, J.; Lee, S. Higher Classification of Mealybugs (Hemiptera: Coccomorpha) Inferred from Molecular Phylogeny and Their Endosymbionts. *Syst. Entomol.* **2022**, *47*, 354–370. [CrossRef]
61. Drummond, A.J.; Suchard, M.A.; Xie, D.; Rambaut, A. Bayesian Phylogenetics with BEAUti and the BEAST 1.7. *Mol. Biol. Evol.* **2012**, *29*, 1969–1973. [CrossRef]
62. Rambaut, A.; Drummond, A.J.; Xie, D.; Baele, G.; Suchard, M.A. Posterior Summarization in Bayesian Phylogenetics Using Tracer 1.7. *Syst. Biol.* **2018**, *67*, 901–904. [CrossRef]
63. Bouckaert, R.; Heled, J.; Kühnert, D.; Vaughan, T.; Wu, C.H.; Xie, D.; Suchard, M.A.; Rambaut, A.; Drummond, A.J. BEAST 2: A Software Platform for Bayesian Evolutionary Analysis. *PLOS Comput. Biol.* **2014**, *10*, e1003537. [CrossRef] [PubMed]
64. Puillandre, N.; Brouillet, S.; Achaz, G. ASAP: Assemble Species by Automatic Partitioning. *Mol. Ecol. Resour.* **2021**, *21*, 609–620. [CrossRef] [PubMed]
65. Zhang, J.; Kapli, P.; Pavlidis, P.; Stamatakis, A. A General Species Delimitation Method with Applications to Phylogenetic Placements. *Bioinformatics* **2013**, *29*, 2869–2876. [CrossRef]
66. Kapli, P.; Lutteropp, S.; Zhang, J.; Kobert, K.; Pavlidis, P.; Stamatakis, A.; Flouri, T. Multi-Rate Poisson Tree Processes for Single-Locus Species Delimitation under Maximum Likelihood and Markov Chain Monte Carlo. *Bioinformatics* **2017**, *33*, 1630–1638. [CrossRef]
67. Fujisawa, T.; Barraclough, T.G. Delimiting Species Using Single-Locus Data and the Generalized Mixed Yule Coalescent Approach: A Revised Method and Evaluation on Simulated Data Sets. *Syst. Biol.* **2013**, *62*, 707–724. [CrossRef]
68. Kimura, M. A Simple Method for Estimating Evolutionary Rates of Base Substitutions through Comparative Studies of Nucleotide Sequences. *J. Mol. Evol.* **1980**, *16*, 111–120. [CrossRef]
69. Carstens, B.C.; Pelletier, T.A.; Reid, N.M.; Satler, J.D. How to Fail at Species Delimitation. *Mol. Ecol.* **2013**, *22*, 4369–4383. [CrossRef] [PubMed]
70. Kol-Maimon, H.; Ghanim, M.; Franco, J.C.; Mendel, Z. Evidence for Gene Flow between Two Sympatric Mealybug Species (Insecta; Coccoidea; Pseudococcidae). *PLoS ONE* **2014**, *9*, e88433. [CrossRef]
71. Wang, X.B.; Zhang, J.T.; Deng, J.; Zhou, Q.S.; Zhang, Y.Z.; Wu, S.A. DNA Barcoding of Mealybugs (Hemiptera: Coccoidea: Pseudococcidae) from Mainland China. *Ann. Entomol. Soc. Am.* **2016**, *109*, 438–446. [CrossRef]
72. Rung, A.; Scheffer, S.J.; Evans, G.; Miller, D. Molecular Identification of Two Closely Related Species of Mealybugs of the Genus *Planococcus* (Homoptera: Pseudococcidae). *Entomol. Soc. Am* **2008**, *101*, 525–532. [CrossRef]

73. Talavera, G.; Dincă, V.; Vila, R. Factors Affecting Species Delimitations with the GMYC Model: Insights from a Butterfly Survey. *Methods Ecol. Evol.* **2013**, *4*, 1101–1110. [CrossRef]
74. Tan, D.S.H.; Ang, Y.; Lim, G.S.; Bin Ismail, M.R.; Meier, R. From "cryptic Species" to Integrative Taxonomy: An Iterative Process Involving DNA Sequences, Morphology, and Behaviour Leads to the Resurrection of *Sepsis Pyrrhosoma* (Sepsidae: Diptera). *Zool. Scr.* **2010**, *39*, 51–61. [CrossRef]
75. Delrieu-Trottin, E.; Durand, J.; Limmon, G.; Sukmono, T.; Kadarusman; Sugeha, H.Y.; Chen, W.; Busson, F.; Borsa, P.; Dahruddin, H.; et al. Biodiversity Inventory of the Grey Mullets (Actinopterygii: Mugilidae) of the Indo-Australian Archipelago through the Iterative Use of DNA-based Species Delimitation and Specimen Assignment Methods. *Evol. Appl.* **2020**, *13*, 1451–1467. [CrossRef] [PubMed]
76. Huang, W.; Xie, X.; Huo, L.; Liang, X.; Wang, X.; Chen, X. An Integrative DNA Barcoding Framework of Ladybird Beetles (Coleoptera: Coccinellidae). *Sci. Rep.* **2020**, *10*, 10063. [CrossRef]
77. Song, C.; Lin, X.L.; Wang, Q.; Wang, X.H. DNA Barcodes Successfully Delimit Morphospecies in a Superdiverse Insect Genus. *Zool. Scr.* **2018**, *47*, 311–324. [CrossRef]
78. Koroiva, R.; Rodrigues, L.R.R.; Santana, D.J. DNA Barcoding for Identification of Anuran Species in the Central Region of South America. *PeerJ* **2020**, *8*, e10189. [CrossRef]
79. Bergsten, J.; Bilton, D.T.; Fujisawa, T.; Elliott, M.; Monaghan, M.T.; Balke, M.; Hendrich, L.; Geijer, J.; Herrmann, J.; Foster, G.N.; et al. The Effect of Geographical Scale of Sampling on DNA Barcoding. *Syst. Biol.* **2012**, *61*, 851–869. [CrossRef]
80. Meiklejohn, K.A.; Damaso, N.; Robertson, J.M. Assessment of BOLD and GenBank—Their Accuracy and Reliability for the Identification of Biological Materials. *PLoS ONE* **2019**, *14*, e0217084. [CrossRef]
81. Guimarães, K.L.A.; Lima, M.P.; Santana, D.J.; de Souza, M.F.B.; Barbosa, R.S.; Rodrigues, L.R.R. DNA Barcoding and Phylogeography of the *Hoplias Malabaricus* Species Complex. *Sci. Rep.* **2022**, *12*, 5288. [CrossRef]
82. Puillandre, N.; Lambert, A.; Brouillet, S.; Achaz, G. ABGD, Automatic Barcode Gap Discovery for Primary Species Delimitation. *Mol. Ecol.* **2012**, *21*, 1864–1877. [CrossRef]
83. Ren, J.M.; Ashfaq, M.; Hu, X.N.; Ma, J.; Liang, F.; Hebert, P.D.N.; Lin, L.; Germain, J.F.; Ahmed, M.Z. Barcode Index Numbers Expedite Quarantine Inspections and Aid the Interception of Nonindigenous Mealybugs (Pseudococcidae). *Biol. Invasions* **2018**, *20*, 449–460. [CrossRef]
84. Wakgari, W.M.; Giliomee, J.H. Natural Enemies of Three Mealybug Species (Hemiptera: Pseudococcidae) Found on Citrus and Effects of Some Insecticides on the Mealybug Parasitoid *Coccidoxenoides peregrinus* (Hymenoptera: Encyrtidae) in South Africa. *Bull. Entomol. Res.* **2003**, *93*, 243–254. [CrossRef]
85. Lopes, F.S.C.; de Oliveira, J.V.; de Morais Oliveira, J.E.; de Oliveira, M.D.; de Souza, A.M. Host Plants for Mealybugs (Hemiptera: Pseudococcidae) in Grapevine Crops. *Pesq. Agropec. Trop.* **2019**, *49*, e54421. [CrossRef]
86. Ahmed, N.; Abd-Rabou, S. Host Plants, Geographical Distribution, Natural Enemies and Biological Studies of the Citrus Mealybug, *Planococcus citri* (Risso) (Hemiptera: Pseudococcidae). *Egypt. Acad. J. Biol. Sci. A Entomol.* **2010**, *3*, 39–47. [CrossRef]
87. Fernandes, M.H.D.A.; Oliveira, J.E.D.M.; Costa, V.A.; de Menezes, K.O. *Coccidoxenoides perminutus* Parasitizing *Planococcus citri* on Vine in Brazil. *Ciência Rural* **2016**, *46*, 1130–1133. [CrossRef]
88. Mansour, R.; Belzunces, L.P.; Suma, P.; Zappalà, L.; Mazzeo, G.; Grissa-Lebdi, K.; Russo, A.; Biondi, A. Vine and Citrus Mealybug Pest Control Based on Synthetic Chemicals. A Review. *Agron. Sustain. Dev.* **2018**, *38*, 37. [CrossRef]

diversity

MDPI

Article

An Incomplete European Barcode Library Has a Strong Impact on the Identification Success of Lepidoptera from Greece

Peter Huemer [1,*] and Marko Mutanen [2]

[1] Naturwissenschaftliche Sammlungen, Sammlungs- und Forschungszentrum, Tiroler Landesmuseen Betriebsges.m.b.H., 6060 Hall in Tirol, Austria
[2] Ecology and Genetics Research Unit, University of Oulu, 90014 Oulu, Finland; marko.mutanen@oulu.fi
* Correspondence: p.huemer@tiroler-landesmuseen.at; Tel.: +43-51259489-721

Abstract: Species identification by means of DNA barcodes depends essentially on the scope and quality of a relevant reference library. The first analysis of a large number (about 600 morphospecies) of southern European Lepidoptera (Greece: Peloponnese) shows both the advantages and disadvantages with regard to a reliable identification of Mediterranean species. We determined 946 DNA barcode sequences from 47 families, of which 929 sequences from 46 families were successfully assigned to a Barcode Index Number (BIN) in the global Barcode of Life Data Systems (BOLD) database. A species level identification for 485 BINs representing 477 Linnaean names was successful. These taxa include 34 new records for Greece. However, 128 BINs (c. 20% of the inventory) could not be attached to a Linnaean name from referenced sequences available in BOLD. Of these BINs, 99 are new and hence represent unique records for BOLD. Intra- and inter-BIN divergences are presented and discussed. An initial and preliminary in-depth analysis of randomly selected species indicates an incomplete DNA barcode library in terms of Linnean taxa, in addition to a considerable number of probably undescribed species. It is therefore strongly recommended that the already advanced European barcode library of Lepidoptera should be supplemented with not-yet-sequenced taxa from the Mediterranean.

Keywords: Barcode Index Number (BIN); DNA barcoding; faunistics; Mediterranean; new species; unidentified species

Citation: Huemer, P.; Mutanen, M. An Incomplete European Barcode Library Has a Strong Impact on the Identification Success of Lepidoptera from Greece. *Diversity* **2022**, *14*, 118. https://doi.org/10.3390/d14020118

Academic Editor: Stephan Koblmüller

Received: 29 December 2021
Accepted: 4 February 2022
Published: 7 February 2022

1. Introduction

During the last decade, considerable efforts have been made in several European countries to establish mostly national DNA barcode libraries of regional faunas, with only Finland so far having published a comprehensive reference library for the arthropod fauna [1]. Lepidoptera are furthermore almost completely covered in Norway, The Netherlands, Germany and Austria, although published results only deal with parts of the fauna [2–5], leaving numerous sequences for private use only. In addition to national initiatives, only a few studies have dealt with larger biogeographical regions, the project on the Lepidoptera of the Alps by the Tyrolean State Museums, Innsbruck, Austria being an exception [6]. Furthermore, selected larger families or superfamilies have been genetically studied on a continental scale by means of DNA barcodes, in particular the Gelechiidae [7,8], Gracillariidae [9], large parts of the Geometridae [10] and Papilionoidea [11]. With about 40 K barcodes, the largest dataset on Lepidoptera has been released by Mutanen et al. [12]. Apart from a few individually funded research projects, the vast majority of the sequences have been, and are being, incorporated into the global, publicly accessible Barcode of Life Data Systems (BOLD; [13]). About three-quarters of around 10,700 species of Lepidoptera known from Europe [14] are represented in BOLD by at least one reference sequence, while around 2600 species, typically very rare, do not yet have a reference DNA barcode. Due to the national initiatives mentioned above, DNA barcodes of almost all species in Northern and Central Europe can be assigned to Linnaean names or already known genetic clusters.

The extent to which these favorable conditions also affect the identification success of southern European samples was tested here on a larger scale for the first time using barcode data from more than 600 morphospecies from Greece.

2. Materials and Methods

A representative portion of Lepidoptera species from Greece was collected during two excursions to the Peloponnese peninsula, from 9 to 25 May 2019 and 9 to 29 September 2020. Collecting efforts covered all taxonomic groups except for butterflies, which have already been studied on a continental level [11]. Various survey methods, in particular light capture, were used to sample a representative species spectrum in the selected study areas on the Peloponnese peninsula. A priority goal was the sampling of about two individuals on average from the most possible number of species and thus from all lepidopteran families, including the so-called microlepidoptera. Of those taxa that are known to be difficult to distinguish from congeners, up to five specimens were collected. Species differentiations were provisionally carried out in the field according to morphological criteria. The material was pinned directly on site and immediately dried for further investigations. The final sample selection was later carried out in the laboratory according to external characteristics such as wing markings and color, head characteristics and occasionally also after preliminary examination of the genital morphology.

Finally, tissue samples (dried legs) of 1056 specimens of provisionally identified morphospecies were prepared according to the prescribed standards to obtain DNA barcode sequences of the mitochondrial COI gene (cytochrome c oxidase 1) [15]. Samples covered 47 families, namely, the Gelechiidae (n = 166), Geometridae (n = 136), Noctuidae (n = 123), Pyralidae (n = 101), Crambidae (n = 63), Erebidae (n = 57), Tortricidae (n = 43) and another 40 families (n = 367).

Material was processed at the Canadian Centre for DNA Barcoding (CCDB, Biodiversity Institute of Ontario, University of Guelph, Guelph, ON, Canada) using the standard high-throughput protocol described in deWaard et al. [15]. Details including complete voucher data and images can be accessed in the public dataset "Lepidoptera Barcoding Greece" (https://dx.doi.org/10.5883/DS-LEPGREEC in BOLD, accessed on 28 December 2021). Sequences were then submitted to GenBank.

All sequences were assigned to the Barcode Index Numbers (BINs), algorithm-based operational taxonomic units that provide an accurate proxy for the true species [16]. BINs were automatically calculated for records in BOLD that are compliant with the DNA Barcode standard [17]. As BOLD presently does not have a functionality of calculating intra- and inter-BIN divergences, those values were provided for us by BOLD support. Follow-up species identification strictly followed available reference sequences in BOLD with a cross-check control of external morphology. In the case of BINs covering more than one taxon in BOLD (BIN-sharing, misidentifications, contaminations), identification was based on external morphology and, in critical cases, also on genital morphology. BINs attributed to a single Linnean name were accepted as correct although potential misidentifications cannot be fully ruled out.

Degrees of intra- and interspecific variation in DNA barcode fragments were calculated under the Kimura 2 parameter model of nucleotide substitution using analytical tools of BOLD systems v. 4.0. Finally, a Neighbor-Joining tree was constructed from these data in Newick format and a tree was drawn using FigTree v1.4.4 (http://tree.bio.ed.ac.uk/software/figtree/, accessed on 28 December 2021).

3. Results

3.1. General Overview

Sequencing of 1056 tissue samples resulted in 946 DNA barcode sequences (c. 90% success rate). Full barcodes of 658 bp were recovered for 733 specimens, and for a further 167 specimens a sequence ranging between 600 and 657 bp was recovered. Sequences less than 600 bp were obtained for only 46 specimens. Of all records with sequence data recovered, 834 sequences were considered to be barcode compliant following BOLD standards.

The vast majority of 929 sequences was assigned to a total of 614 different BINs in BOLD, leaving only 17 sequences without a BIN (see Figure S1). Of all BINs, 535 belong to only 15 families, the remaining 79 BINs representing 31 usually smaller families (Figure 1, Table S1). For one family (Adelidae) only a short sequence without a BIN was recovered. Average intra-BIN variability ranged from 0% to a maximum of 1.98% (mean 0.44%), and maximum intra-BIN distances ranged from 0% to 5.41% (mean 1.18%). Distances to nearest neighbor BINs ranged from a minimum of 1.04% to a maximum of 13.14% (mean 4.16%) (Table S1).

3.2. Unidentified BINs—Incomplete Barcode Library

A total of 128 BINs could not be attached to a Linnaean name from referenced sequences available in BOLD. Of these BINs, 99 are new and hence represent unique records for BOLD, whereas the remaining 29 BINs have been recorded from elsewhere (Table S1).

Judging from morphology, the unidentified BINs probably represent separate taxa at species level belong to 27 families. The Autostichidae with 17 and Gelechiidae with 14 unidentified BINs are the most important families in this respect, followed by Noctuidae, Pyralidae and Tineidae, each with nine, and Coleophoridae and Erebidae, each with seven unidentified BINs. Cosmopterigidae, Crambidae, Geometridae and Tortricidae have six unidentified BINs each, and the remaining 16 families have between one and four unidentified BINs (Figure 1).

An unknown portion of the so-far unidentified BINs is based on an incomplete barcode library of European Lepidoptera. The lack of reference sequences in BOLD thus prevents unambiguous identification. Time-consuming morphological investigations are therefore necessary in order to ultimately assign BINs to Linnaean names. Through such research, several taxa have already been identified in the context of the present study that did not have relevant reference sequences in BOLD, e.g., *Nukusa cinerella* (Rebel, 1891), *Aglossa signicostalis* Staudinger, 1870, and *Paidia cinerascens* (Herrich-Schäffer, 1847).

3.3. Potential Cryptic Diversity

The BINs that have not yet been assigned to species level certainly also include previously overlooked cryptic species, but the fraction of nameless taxa cannot be estimated as they require meticulous taxonomic scrutiny on a case-by-case basis. Examples of cryptic diversity can be found in the genera *Ypsolopha*, *Aristotelia* and *Coleophora*, where preliminary analysis of morphology indicates new taxa. However, comprehensive taxonomic investigations are required for further clarification.

3.4. BINs Attributed to Linnaean Names

BOLD identification analytics resulted in a species level identification for 485 BINs representing 477 Linnaean names (Table S1). Eight species with more than one sequenced specimen from Greece showed large genetic variation and were consequently clustered in two BINs. Such cases require in-depth analysis for possible cryptic diversity. Furthermore, two species, viz. *Cryphia ochsi* (Boursin, 1940) and *Trichoplusia circumscripta* (Freyer, 1831) shared their BINs with other taxa in BOLD, for which reason identifications were made based on external and genital morphology.

The Noctuidae with 87 identified species (18% of identified BINs) are followed by Geometridae (71 spp.), Gelechiidae (57 spp.), Pyralidae (46 spp.), Erebidae (36 spp.), Crambidae (32 spp.), Tortricidae (27 spp.) and another 35 families (Figure 1).

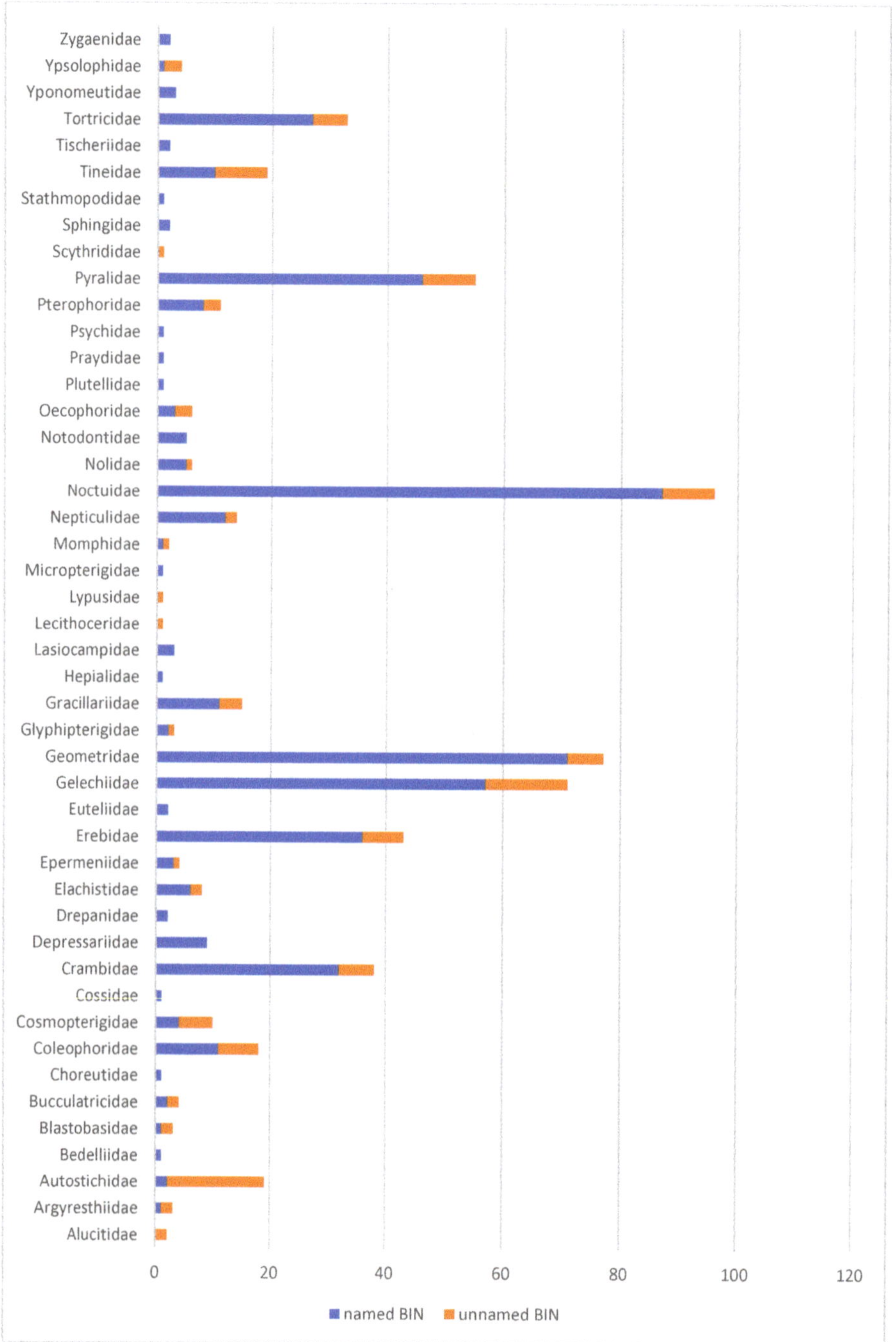

Figure 1. Named and unnamed BINs in a sample of Lepidoptera from Greece.

3.5. New Faunistic Records

Following refs. [18,19] 34 species attributed to a Linnaean taxon had not been recorded from Greece (Table 1). Further, three species, viz. *Helcystogramma lamprostoma* (Zeller, 1847), *Ornativalva heluanensis* (Debski, 1913) and *Caradrina levantina* Hacker, 2004 are new records for the Greek mainland. With the exception of *Eupithecia ultimaria* Boisduval, 1840 and *Bryophila felina* (Eversmann, 1852) all the remaining species belong to 11 families of so-called microlepidoptera, which are obviously under-represented in faunistic papers from Greece.

Table 1. New faunistic records for Greece.

Taxon	Family
Coleophora helgada (Anikin, 2005)	Coleophoridae
Coleophora gardesanella Toll, 1953	Coleophoridae
Eteobalea siciliae (Riedl, 1966)	Cosmopterigidae
Agriphila brioniellus (Zerny, 1914)	Crambidae
Friedlanderia cicatricella (Hübner, 1824)	Crambidae
Spoladea recurvalis (Fabricius, 1775)	Crambidae
Elachista argentella (Clerck, 1759)	Elachistidae
Elachista atricomella Stainton, 1849	Elachistidae
Elachista biatomella (Stainton, 1848)	Elachistidae
Elachista chrysodesmella Zeller, 1850	Elachistidae
Elachista obliquella Stainton, 1854	Elachistidae
Anarsia leberonella Réal, 1994	Gelechiidae
Aproaerema sangiella (Stainton, 1863)	Gelechiidae
Aristotelia subdecurtella (Stainton, 1859)	Gelechiidae
Ivanauskiella occitanica (Nel & Varenne, 2013)	Gelechiidae
Mesophleps oxycedrella (Millière, 1871)	Gelechiidae
Scrobipalpa superstes Povolný, 1977	Gelechiidae
Thiotricha subocellea (Stephens, 1834)	Gelechiidae
Eupithecia ultimaria Boisduval, 1840	Geometridae
Stigmella obliquella (Heinemann, 1862)	Nepticulidae
Bryophila felina (Eversmann, 1852)	Noctuidae
Batia inexpectella Jäckh, 1972	Oecophoridae
Acrobasis bithynella Zeller, 1848	Pyralidae
Acrobasis fallouella (Ragonot, 1871)	Pyralidae
Assara concicolella (Constant, 1884)	Pyralidae
Ceutolopha isidis (Zeller, 1867)	Pyralidae
Phycita torrenti Agenjo, 1962	Pyralidae
Tischeria decidua Wocke, 1876	Tischeriidae
Clepsis burgasiensis (Rebel, 1916)	Tortricidae
Cydia rymarczyki Varenne & Nel, 2013	Tortricidae
Lobesia bicinctana (Duponchel, 1844)	Tortricidae
Neocochylis dubitana (Hübner, 1799)	Tortricidae
Pammene herrichiana (Heinemann, 1854)	Tortricidae
Ypsolopha alpella (Denis & Schiffermüller, 1775)	Ypsolophidae

4. Discussion

Due to extensive recent achievements in the implementation of a DNA barcode reference library, the Lepidoptera in Central and Northern Europe can now be largely identified to the species level [12].

According to our study, however, the situation in southern Europe is very different. Although 485 of 614 BINs could be assigned to a Linnaean name, about one-fifth could not, as 128 BINs were not assignable to species level. Unidentified species are present in practically all families, but the majority of these can be found in some groups of the so-called "microlepidoptera" (Figures 1 and S1, Table S1). Increased fractions are found, e.g., in the Autostichidae, Gelechiidae and Coleophoridae, and are on the one hand caused by the lack of reference sequences, and on the other hand probably due to additional cryptic diversity. However, the intraspecific variation of many taxa seems to be incompletely covered due to under-represented geographical coverage of samples in BOLD. Gaytán et al. [20] found in a small group of *Quercus* herbivore moths that even a single sequence from one of the Mediterranean Peninsulas (Iberia and Italy) increased genetic divergence. Therefore, to verify that a new BIN represents an undescribed species, additional, independent evidence, e.g., from the nuclear genome or morphology, is needed.

Although many previous studies have focused on intra- and interspecific variability of DNA barcodes, we are not aware of previous works discussing intra- and inter-BIN variability. This is likely largely due to the fact that presently BOLD does not have this functionality. We observed intra-BIN variability to be on average 1.18%, while that between the closest BINs to be 4.16% on average, i.e., the inter-BIN divergences being on average about 3.5 times as large as intra-BIN divergences (see Table S1). Interestingly, within two BINs, variability exceeded 5%, which we find remarkable given that the BIN algorithm has an initial proxy for species at 2.2% [17]. This highlights that BINs are not merely defined by a fixed cut-off value but are heavily affected by the shapes of distributions of genetic variability within and between the clusters. For the same reason, with distributions being particularly narrow and sampling intensity being high, a gap of just over 1% is sufficient to assign clusters into two different BINs in our data. Compared to the previous studies on intra- and interspecific variability in lepidopteran barcodes (e.g., [12]), the BINs appear to be characterized by somewhat lower distances to nearest neighbors. The reasons for this should be studied in detail but it is likely affected by undetected cryptic diversity and other operational factors when basing the comparison on Linnean names.

Unfortunately, barcoding failed for about 10% of the material. Particularly samples of Micropterigidae did not work well with standard protocols. Based on our previous experience, DNA barcoding of species of this family generally shows a low rate of success, probably due to partial primer incompatibility. The same may be true for some other groups and species, but to our knowledge this has not been systematically studied in Lepidoptera. Furthermore, a considerable number of specimens from taxa with small body size, such as Nepticulidae and Elachistidae, likely failed because of low sample quality or quantity. Barcoding in general is unproblematic in these families, suggesting that the primer issues are not common. In our experience, the recovered 90% success rate is normal with pinned fresh samples of Lepidoptera.

The unexpectedly high number of hitherto unpublished species for the fauna of Greece largely reflects the lack of publicly available faunistic data rather than an absence of samples in private and institutional collections. We are personally aware that several of the new national records published here have been observed in the country but remained unreported. Furthermore, there is always a risk that already published faunistic data have remained unnoticed. Taxa not yet identified to species level very likely include additional records of faunistic or taxonomic interest.

The high rate (c. 20% of the inventory) of undetermined and possibly undescribed species in our sample is a significant limiting factor for faunistic-ecological and nature conservation-oriented studies. These gaps are particularly significant for possible future monitoring programs, especially for automated methods such as the use of malaise traps.

It is therefore strongly recommended to supplement the already advanced European barcode library of Lepidoptera specifically with Mediterranean taxa that have not yet been sequenced. A (near) complete DNA barcode reference library for this area would probably support taxonomic research and accelerate descriptions of new species, hence bringing us one step closer to the full bio-literacy of European lepidopteran diversity.

Supplementary Materials: The following supporting information can be downloaded at: https://www.mdpi.com/article/10.3390/d14020118/s1, Figure S1: Neighbor joining tree based on COI of specimens used in this study; Table S1: Specimens used in this study and inter- and intra-BIN distances from BOLD.

Author Contributions: Conceptualization, P.H.; Data curation, P.H.; Formal analysis, P.H. and M.M.; Methodology, P.H. and M.M.; Writing—original draft, P.H. and M.M. All authors have read and agreed to the published version of the manuscript.

Funding: This research received no external funding.

Institutional Review Board Statement: Not applicable.

Informed Consent Statement: Not applicable.

Data Availability Statement: All 946 COI sequences are available in DS-LEPGREEC on BOLD (https://dx.doi.org/10.5883/DS-LEPGREEC) at https://www.boldsystems.org/ (accessed on 28 December 2021).

Acknowledgments: P.H. is particularly grateful to Maria Kamilari for tremendous help and support with applications. Furthermore, the BOLD staff under the leadership of Paul D.N. Hebert are thanked for their continuous support. We specifically thank Dina Soliman and Brianne St. Jacques for help with conducting calculations of BIN divergences. The Greece Ministry of Environment is acknowledged for issuing necessary permits to P.H. Robert (Bob) Heckford kindly helped with linguistic improvement of the manuscript.

Conflicts of Interest: The authors declare no conflict of interest.

References

1. Roslin, T.; Somervuo, P.; Pentinsaari, M.; Hebert, P.D.N.; Agda, J.; Ahlroth, P.; Anttonen, P.; Aspi, J.; Blagoev, G.; Blanco, S.; et al. A molecular-based identification resource for the arthropods of Finland. *Mol. Ecol. Res.* **2021**, *22*, 803–822. [CrossRef] [PubMed]
2. Hausmann, A.; Haszprunar, G.; Segerer, A.H.; Speidel, W.; Behounek, G.; Hebert, P.D.N. Now DNA barcoded: The butterflies and larger moths of Germany (Lepidoptera: Rhopalocera, Macroheterocera). *Spixiana* **2011**, *34*, 47–58.
3. Huemer, P.; Mutanen, M.; Sefc, K.M.; Hebert, P.D.N. Testing DNA Barcode Performance in 1000 Species of European Lepidoptera: Large Geographic Distances Have Small Genetic Impacts. *PLoS ONE* **2014**, *9*, e115774. [CrossRef] [PubMed]
4. Huemer, P.; Wiesmair, B. DNA-Barcoding der Tagfalter (Lepidoptera, Papilionoidea) Österreichs—unbekannte genetische Vielfalt im Zentrum Europas. *Wiss. Jahrb. Tirol. Landesmuseen* **2017**, *10*, 8–33.
5. Huemer, P.; Wieser, C.; Stark, W.; Hebert, P.D.N.; Wiesmair, B. DNA barcode library of megadiverse Austrian Noctuoidea (Lepidoptera)—a nearly perfect match of Linnean taxonomy. *Biodivers. Data J.* **2019**, *7*, e37734. [CrossRef] [PubMed]
6. Huemer, P.; Hebert, P.D.N. DNA Barcode Bibliothek der Schmetterlinge Südtirols und Tirols (Italien, Österreich)—Impetus für integrative Artdifferenzierung im 21. Jahrhundert. *Gredleriana* **2016**, *16*, 141–164.
7. Huemer, P.; Karsholt, O. Commented checklist of European Gelechiidae (Lepidoptera). *ZooKeys* **2020**, *921*, 65–140. [CrossRef] [PubMed]
8. Huemer, P.; Karsholt, O.; Aarvik, L.; Berggren, K.; Bidzilya, O.; Junnilainen, J.; Landry, J.-F.; Mutanen, M.; Nupponen, K.; Segerer, A.; et al. DNA barcode library for European Gelechiidae (Lepidoptera) suggests greatly underestimated species diversity. *ZooKeys* **2020**, *921*, 141–157. [CrossRef] [PubMed]
9. Lopez-Vaamonde, C.; Kirichenko, N.; Cama, A.; Doorenweerd, C.; Godfray, H.C.J.; Guiguet, A.; Gomboc, S.; Huemer, P.; Landry, J.-F.; Laštůvka, A.; et al. Evaluating DNA Barcoding for Species Identification and Discovery in European Gracillariid Moths. *Front. Ecol. Evol.* **2021**, *9*, 626752. [CrossRef]
10. Hausmann, A.; Godfray, H.C.J.; Huemer, P.; Mutanen, M.; Rougerie, R.; van Nieukerken, E.J.; Ratnasingham, S.; Hebert, P.D.N. Genetic Patterns in European Geometrid Moths Revealed by the Barcode Index Number (BIN) System. *PLoS ONE* **2013**, *8*, e84518. [CrossRef]
11. Dinca, V.; Dapporto, L.; Somervuo, P.; Vodă, R.; Cuvelier, S.; Gascoigne-Pees, M.; Huemer, P.; Mutanen, M.; Hebert, P.D.N.; Vila, R. High resolution DNA barcode library for European butterflies reveals continental patterns of mitochondrial genetic diversity. *Comm. Biol.* **2021**, *4*, 1–11. [CrossRef] [PubMed]
12. Mutanen, M.; Kivelä, S.M.; Vos, R.A.; Doorenweerd, C.; Ratnasingham, S.; Hausmann, A.; Huemer, P.; Dincă, V.; van Nieukerken, E.J.; Lopez-Vaamonde, C.; et al. Species-Level Para- and Polyphyly in DNA Barcode Gene Trees: Strong Operational Bias in European Lepidoptera. *Syst. Biol.* **2016**, *65*, 1024–1040. [CrossRef] [PubMed]
13. Ratnasingham, S.; Hebert, P.D.N. BOLD: The Barcode of Life Data System. *Mol. Ecol. Notes* **2007**, *7*, 355–364. [CrossRef] [PubMed]
14. Lepiforum: Website zur Bestimmung von Schmetterlingen (Lepidoptera) und Ihren Präimaginalstadien. Available online: http://www.lepiforum.eu/bh/downloads/2021-iii-20/Lepiforums-Europaliste_Schmetterlinge_Version_8.3_Stand_2021_03_20 (accessed on 28 December 2021).

15. De Waard, J.R.; Ivanova, N.V.; Hajibabaei, M.; Hebert, P.D.N. Assembling DNA barcodes: Analytical protocols. In *Methods in Molecular Biology: Environmental Genomics*; Martin, C.C., Ed.; Humana Press Inc.: Totowa, NJ, USA, 2008; pp. 275–293.
16. Ortiz, A.S.; Rubio, R.M.; Guerrero, J.J.; Garre, M.J.; Serrano, J.; Hebert, P.D.N.; Hausmann, A. Close congruence between Barcode Index Numbers (Bins) and species boundaries in the Erebidae (Lepidoptera: Noctuoidea) of the Iberian Peninsula. *Biodiv. Data J.* **2017**, *5*, e19840. [CrossRef] [PubMed]
17. Ratnasingham, S.; Hebert, P.D.N. A DNA-based registry for all animal species: The Barcode Index Number (BIN) System. *PLoS ONE* **2013**, *8*, e66213. [CrossRef] [PubMed]
18. Gozmány, L. *The Lepidoptera of Greece and Cyprus. Fauna Graecia IX*; Hellenic Zoological Society: Athens, Greece, 2012; Volume 1.
19. Karsholt, O.; van Nieukerken, E.J. Lepidoptera, Moths. Fauna Europaea 2013, Version 2021.12. Available online: https://fauna-eu.org (accessed on 1 December 2021).
20. Gaytán, Á.; Bergsten, J.; Canelo, T.; Pérez-Izquierdo, C.; Santoro, M.; Bonal, R. DNA Barcoding and geographical scale effect: The problems of undersampling genetic diversity hotspots. *Ecol. Evol.* **2020**, *10*, 10754–10772. [CrossRef] [PubMed]

diversity

MDPI

Article

DNA Barcoding Data Reveal Important Overlooked Diversity of *Cortinarius* sensu lato (*Agaricales, Basidiomycota*) in the Romanian Carpathians

Emerencia Szabó [1,2,3,†], Bálint Dima [4,*,†], Avar L. Dénes [1], Viktor Papp [5] and Lujza Keresztes [1,6]

1 Centre of Systems Biology, Biodiversity and Bioresources, Faculty of Biology and Geology, University of Babeș-Bolyai Cluj-Napoca, Clinicilor 5-7, 400006 Cluj-Napoca, Romania; emerenciaszabo@gmail.com (E.S.); avar.lehel@gmail.com (A.L.D.); keresztes2012@gmail.com (L.K.)
2 Doctoral School of Integrative Biology, Babeş-Bolyai University, Republicii 44, 400015 Cluj-Napoca, Romania
3 Institute of Interdisciplinary Research in Bio-Nano-Sciences, Babeș-Bolyai University, Treboniu Laurian 42, 400271 Cluj-Napoca, Romania
4 Department of Plant Anatomy, Institute of Biology, Eötvös Loránd University, Pázmány Péter sétány 1/C, H-1117 Budapest, Hungary
5 Department of Botany, Institute of Agronomy, Hungarian University of Agriculture and Life Sciences, Villányi út 29-43, H-1118 Budapest, Hungary; papp.viktor@uni-mate.hu or agaricum@gmail.com
6 Hungarian Department of Biology and Ecology, Faculty of Biology and Geology, University of Babeș-Bolyai Cluj-Napoca, Clinicilor 5-7, 400006 Cluj-Napoca, Romania
* Correspondence: cortinarius1@gmail.com
† These authors contributed equally to this work.

Abstract: *Cortinarius* s.l. is a globally distributed agaricoid genus that has been well studied in Europe with over 1000 described species. However, the information about their taxonomy and diversity in eastern Central Europe is still limited. Only 124 species have been reported so far from Romania, based solely on morphological observations. The aim of this study was to re-examine the diversity of the genus *Cortinarius* s.l. in the Romanian Carpathian area, employing molecular phylogenetic and morphological methods. During intensive field work in the period 2017–2020, a total of 234 *Cortinarius* s.l. specimens were collected and studied with integrative taxonomic methods. For all the samples, we amplified and sequenced the nrDNA ITS region, which is the widely used official barcode marker of fungi. These sequences were compared to the data found in public databases (GenBank, UNITE, BOLD). Based on phylogenetic analyses, we identified 109 *Cortinarius* s.l. species, which represent 40 sections and 3 clades. Out of these species, 43 have previously been documented from Romania based on morphological identification methods, while 66 species are reported as new to the country.

Keywords: Eastern Europe; MrBayes; nrDNA ITS; RAxML; phylogeny; revision; taxonomy

Citation: Szabó, E.; Dima, B.; Dénes, A.L.; Papp, V.; Keresztes, L. DNA Barcoding Data Reveal Important Overlooked Diversity of *Cortinarius* sensu lato (*Agaricales, Basidiomycota*) in the Romanian Carpathians. *Diversity* **2023**, *15*, 553. https://doi.org/10.3390/d15040553

Academic Editor: Stephan Koblmüller

Received: 14 March 2023
Revised: 3 April 2023
Accepted: 9 April 2023
Published: 13 April 2023

1. Introduction

Cortinarius (Pers.) Gray s.l. (sensu lato) has been considered to be the biggest genus of the order *Agaricales* [1], with a cosmopolitan distribution of over 3000 described species [2]. *Cortinarius* species are important ectomycorrhizal fungi (EcM) associated with different trees and shrubs, belonging to the order *Fagales*, families *Caesalpiniaceae*, *Cistaceae*, *Dipterocarpaceae*, *Myrtaceae*, *Pinaceae*, *Polygonaceae*, *Rhamnaceae*, *Rosaceae* and *Salicaceae* as well as a few herbaceous plants in the *Cyperaceae*. Because of their EcM nutritional mode, they also play a key role in carbon cycling, especially in boreal forests. Several groups have narrow ecological preferences, and they are sensitive to environmental changes; therefore, some species have been used as indicators for valuable natural environments [3]. Species belonging to *Cortinarius* have a highly variable appearance, from mycenoid to tricholomatoid basidiomata. Their color can be uniformly brown or colorful, and the surface of the pilei is dry, fibrillose, silky, squamose or viscid [4]. Another common feature of *Cortinarius* is a cobweb-like partial veil, the usually rusty-brown spore print and the lack of a germ pore and perisporium [5]. In the beginning of the 21st century, molecular phylogenetic studies showed that the genus also contains several species (mainly extra-European) with sequestrate fruiting body forms [6]. According to IndexFungorum (http://www.indexfungorum.org, accessed on 7 March 2023), 5819 *Cortinarius* names have been published worldwide (accessed 2 March 2023). However, this number includes all infrageneric taxa, and still the number of existing species names are estimated to be over 5000. Many of these names were inconsistently used in the literature, and there is also a high number of synonyms. One of the most challenging parts of *Cortinarius* taxonomy is to examine which species have already been described [3].

Since the early 19th century, many researchers have focused on the genus *Cortinarius* in Europe, e.g., [7–9]. From the 20th century onwards, several fundamental morphological works were published from Europe, e.g., [10–18]. From the late 1980s, two taxonomic schools have spread in Europe, representing the South European (especially French, Italian and Spanish) and the Scandinavian directions. During this period, the Atlas des Cortinaires [19–45] described approximately 1500 new species, while the more conservative species concept of the Scandinavian school, the *Cortinarius*, Flora Photographica [46–49], dealt with only 300 species. The introduction of molecular tools in addition to the macro- and micromorphological character-based identification led to a more precise and reliable classification and helped avoid misunderstandings in the taxonomy and nomenclature of the genus. Northern Europe became one of the most investigated and well-studied regions [4,50–63], but in the last few decades, several other studies focused also on the Central European region [64–70]; however, the information from this region is still limited. Similarly, there is a lack of information regarding the southern parts of Europe [3], with only a few papers [71–73] published from the Mediterranean area so far. Depending on the identification and classification concepts, the genus *Cortinarius* s.l. was separated into several subgenera by different authors. Moser [15] named seven subgenera, while Moënne-Loccoz et al. [19,20] and Bidaud et al. [25,26] mentioned six subgenera in their works. Scandinavian researchers [46–49] separated the genus into five subgenera. In the molecular era, phylogenetic analyses showed that these subgenera are mostly artificial, e.g., [3,70]. New infrageneric classifications of *Cortinarius* based on multigene phylogenetic analyses were proposed by Garnica et al. [74] and Soop et al. [75]. Recently, Liimatainen et al. [76] has split the genus *Cortinarius* into 10 genera based on genomic and multi-gene sequence data. This new classification, however, is not adopted in the current study, due to practical reasons.

Fungal research in Romania has received increasing interest since 1950, leading to 8727 reported fungal species [77,78], including both macro- and microfungi. Although there are a number of publications that mention the genus *Cortinarius* and provide scarce species lists from few areas in the Carpathians [79–89], systematic study that only focuses on this genus is still lacking. Based on an extensive literature search, we know about 124 reported *Cortinarius* species from Romania, identified based only on their morphological characteristics. This is a relatively low number when compared to the more than 1000 known European *Cortinarius* species [90].

Using solely morphological characteristics to identify *Cortinarius* species is very challenging and needs a lot of experience and solid knowledge. During the development of the basidiocarps, several characteristics may change significantly and overlap with other species, i.e., their intraspecific variability is high. Micromorphological characteristics, such as the size, shape or ornamentation of the basidiospores, also play an important role in the identification process, e.g., [4]. Nonetheless, DNA-sequence-based analyses (e.g., DNA barcoding) are the most reliable identification methods to date, allowing the investigation of the differences between species and varieties too. The nrDNA ITS (Internal Transcribed Spacer) region has been proposed and started to be widely used in molecular taxonomy as the universal and official barcoding region for fungi [91]. ITS is composed of three subregions: ITS1, 5.8S and ITS2, of which the ITS1 and ITS2 spacers show higher evolutionary rates than the 5.8S; therefore, these are the best suited for studying interspecific level differences [92]. This region evolved relatively rapidly compared to the mitochondrial genes; therefore, in the case of *Cortinarius* too, it is useful and widely used for species identification purposes [3,56,74,93,94].

To study fresh material, we based our study mainly on sampling in the Apuseni Mts, which is a prominent forested area of Romania [95] and located in northern part of the Western Romanian Carpathians. The mountain range is regarded as an important biogeographical area between the Pannonian Plain and the Transylvanian Plateau as well as in the Southern and Eastern Carpathians. Due to its location, several biogeographical regions are present in the area (e.g., alpine, arctic, Mediterranean, etc.), which has resulted in it being one of the regions in Europe with high biodiversity [96]. Several suitable forest types and hot spots for *Cortinarius* are represented in the area, both on calcareous and siliceous bedrocks. The forest vegetation is dominated by oak (*Quercus petraea* agg.), mixed hornbeam–beech (*Carpinus betulus, Fagus sylvatica*), beech (*F. syvatica*), mixed beech and spruce (*F. sylvatica, Picea abies*) and spruce (*P. abies*) forests [97]. Additionally, we also extended our sampling to some localities in the Eastern Carpathians (Ținutul Sării, Baraolt and Liban).

Due to the limited knowledge and the lack of molecular genetic data of the genus *Cortinarius* in Romania, the aims of this study were to make a pilot revision of the diversity, taxonomy and distribution of *Cortinarius* s.l. species, using DNA barcoding combined with morphological methods, and update the list of species of this important ectomycorrhizal genus.

2. Materials and Methods

2.1. Taxon Sampling

For our pilot survey, samples were collected from sites including the main habitat types, characteristic for the Transylvanian part of Romania, such as broad-leaved forests (40%), coniferous forests (35%), mixed forests (11%) and other habitat types (14%; e.g., transitional woodland–shrub, pastures, natural grasslands). The collections were made from 2017 to the autumn of 2020, with the professional collaboration of K. Babos, L. Bartha, G. Bélfenyéri, A. Dénes, R. Erös, K. Fehér, L. Gál, B. Jancsó, K. Macalik and Cs. Szabó.

A total of 234 *Cortinarius* samples were collected (Table 1). Every sample has a unique code (e.g., C001, where the C means the first letter of the genus name), and in every case the collection coordinates were recorded. The coordinates were registered in the decimal degrees (DD) format, with six decimals (Table 1). The majority of the samples were photographed in the field (Figure 1), and gross morphology was noted. All samples were dried with a dehydrator at 45 °C. Samples were deposited at the Faculty of Biology and Geology, Babeș-Bolyai University, Cluj-Napoca, Romania.

Table 1. Metadata of *Cortinarius* samples studied and identified during this work. All sequences are newly generated.

Species	DNA-Codes	Collection Date	Location	Latitude	Longitude	GenBank acc. no.
C. alboviolaceus	C213	04.10.2020	Sovata	46.662003	25.215961	OP099671
C. anfractoides	C361	07.11.2020	Făgetul Clujului	46.6993848	23.5488839	OP099771
C. anfractoides	C363	07.11.2020	Făgetul Clujului	46.6993848	23.5488839	OP099773
C. anomalus	C145	30.08.2020	Liban	46.552611	25.525509	OP099625
C. anomalus	C160	10.10.2020	Statiunea Stana de Vale	46.697647	22.626014	OP099635
C. anomalus	C324	07.11.2020	Cheile Someșului Cald	46.67071058	22.81810123	OP099745

Table 1. *Cont.*

Species	DNA-Codes	Collection Date	Location	Latitude	Longitude	GenBank acc. no.
C. anomalus	C159 *	10.10.2020	Statiunea Stana de Vale	46.697647	22.626014	OP132853
C. aureopulverulentus	C338	07.11.2020	Cheile Someșului Cald	46.62928894	22.7806518	OP099756
C. balteatocumatilis	C290	01.11.2020	Chinteni	46.895154	23.520302	OP099723
C. balteatocumatilis	C291	01.11.2020	Chinteni	46.895154	23.520302	OP099724
C. balteatus	C103	10.07.2020	Zetea	46.4733373	25.3454161	OP099607
C. bergeronii	C223	24.10.2020	Cheile Vârghişului	46.183191	25.590322	OP099680
C. brunneus	C246A	24.10.2020	Statiunea Muntele Baisoara	46.50627098	23.26640838	OP099695
C. bulliardii	C359	07.11.2020	Făgetul Clujului	46.6993848	23.5488839	OP099769
C. caesiophylloides	C009	27.07.2018	Statiunea Muntele Baisoara	46.5382	23.3071099	OP099571
C. calochrous	C221	24.10.2020	Cheile Vârghişului	46.205054	25.558222	OP099678
C. calochrous	C233	24.10.2020	Cheile Vârghişului	46.203207	25.559007	OP099685
C. camphoratus	C097	19.08.2017	Statiunea Muntele Baisoara	46.520139	23.271811	OP099603
C. camphoratus	C153	10.10.2020	Statiunea Stana de Vale	46.697647	22.626014	OP099630
C. camphoratus	C202	04.10.2020	Chiheru de Jos	46.679428	25.035528	OP099661
C. caninus	C040	22.09.2018	Valea Ierii	46.524239	23.275145	OP099579
C. caninus	C146	30.08.2020	Liban	46.552103	25.524967	OP099626
C. caninus	C149	03.09.2020	Mărtiniș	46.2647	25.355	OP099627
C. caninus	C189	04.10.2020	Chiheru de Jos	46.686903	25.040178	OP099648
C. caninus	C192	04.10.2020	Pădurea Buneților	46.081617	25.034936	OP099651
C. caninus	C193	04.10.2020	Pădurea Buneților	46.081617	25.034936	OP099652
C. caninus	C194	04.10.2020	Pădurea Buneților	46.081617	25.034936	OP099653
C. caninus	C199	04.10.2020	Chiheru de Jos	46.679428	25.035528	OP099658
C. caninus	C200	04.10.2020	Chiheru de Jos	46.679428	25.035528	OP099659
C. caninus	C206	04.10.2020	Chiheru de Jos	46.681697	25.035186	OP099665
C. caninus	C207	04.10.2020	Chiheru de Jos	46.681697	25.035186	OP099666
C. caninus	C255	01.11.2020	Harghita	46.538506	25.612982	OP099701
C. caperatus	C074	13.10.2019	Statiunea Muntele Baisoara	46.522375	23.273463	OP099594
C. caperatus	C092	19.08.2017	Statiunea Muntele Baisoara	46.520139	23.271811	OP099598
C. caperatus	C093	26.08.2017	Statiunea Muntele Baisoara	46.520139	23.271811	OP099599
C. caperatus	C120	28.07.2020	Cheile Somesului Cald	46.626862	22.788439	OP099617
C. caperatus	C143 *	08.08.2020	Statiunea Muntele Baisoara	46.53262527	23.28001233	OP132852
C. catharinae	C089	02.12.2019	Făgetul Clujului	46.735133	23.539162	OP099596
C. cinereobrunneolus	C343	07.11.2020	Făgetul Clujului	46.6993848	23.5488839	OP099759
C. cinnamomeus	C105	10.07.2020	Zetea	46.4733373	25.34546	OP099608

Table 1. *Cont.*

Species	DNA-Codes	Collection Date	Location	Latitude	Longitude	GenBank acc. no.
C. claricolor	C094	26.08.2017	Statiunea Muntele Baisoara	46.520139	23.271811	OP099600
C. collinitus	C003	27.07.2018	Statiunea Muntele Baisoara	46.5382	23.3071099	OP099565
C. collinitus	C011	27.07.2018	Statiunea Muntele Baisoara	46.5382	23.3071099	OP099573
C. collinitus	C045	22.09.2018	Valea Ierii	46.525571	23.274678	OP099580
C. collinitus	C048	22.09.2018	Valea Ierii	46.524456	23.274586	OP099583
C. collocandoides	C349	07.11.2020	Făgetul Clujului	46.6993848	23.5488839	OP099760
C. colymbadinus	C114	23.06.2020	Liban	46.553983	25.520218	OP099613
C. corrosus	C263	01.11.2020	Ghimes	46.528277	26.033167	OP099704
C. corrosus	C273	01.11.2020	Ghimes	46.528277	26.033167	OP099710
C. croceus	C108 *	23.06.2020	Liban	46.553983	25.520218	OP132850
C. croceus	C115	23.06.2020	Liban	46.553983	25.520218	OP099614
C. croceus	C116	23.06.2020	Liban	46.553983	25.520218	OP099615
C. croceus	C124	28.07.2020	Cheile Somesului Cald	46.6342004	22.7476044	OP099618
C. daulnoyae	C175	18.10.2020	Făgetul Clujului	46.738306	23.539194	OP099639
C. daulnoyae	C176	18.10.2020	Făgetul Clujului	46.738306	23.539194	OP099640
C. daulnoyae	C286	01.11.2020	Chinteni	46.895154	23.520302	OP099720
C. daulnoyae	C298 *	01.11.2020	Făgetul Clujului	46.732552	23.543583	OP142445
C. daulnoyae	C307	01.11.2020	Făgetul Clujului	46.729857	23.254839	OP099735
C. daulnoyae	C310	01.11.2020	Făgetul Clujului	46.734696	23.540594	OP099737
C. daulnoyae	C360	07.11.2020	Făgetul Clujului	46.6993848	23.5488839	OP099770
C. delibutus	C253	01.11.2020	Harghita	46.538506	25.612982	OP099699
C. delibutus	C254	01.11.2020	Harghita	46.538506	25.612982	OP099700
C. aff. *delibutus*	C110	23.06.2020	Liban	46.553983	25.520218	OP099612
C. elatior	C352	07.11.2020	Făgetul Clujului	46.6993848	23.5488839	OP099762
C. elegantior	C322	07.11.2020	Cheile Someșului Cald	46.67071058	22.81810123	OP099743
C. elegantissimus	C241	24.10.2020	Cheile Vârghişului	46.201743	25.560039	OP099690
C. eliae	C293	01.11.2020	Chinteni	46.895154	23.520302	OP099725
C. eliae	C294	01.11.2020	Chinteni	46.895154	23.520302	OP099726
C. eliae	C295	01.11.2020	Chinteni	46.895154	23.520302	OP099727
C. fraudulosus	C257	01.11.2020	Harghita	46.538506	25.612982	OP099702
C. fulminoides	C334	07.11.2020	Cheile Someșului Cald	46.64369044	22.7320642	OP099753
C. gallurae	C379	21.10.2020	Bătarci	48.014053	23.146755	OP099780
C. geniculatus	C357	07.11.2020	Făgetul Clujului	46.6993848	23.5488839	OP099767
C. gentilis	C051	22.09.2018	Valea Ierii	46.524888	23.274323	OP099585
C. glaucopus	C012	27.07.2018	Statiunea Muntele Baisoara	46.5382	23.3071099	OP099574
C. glaucopus	C248	02.11.2020	Mănăstireni	46.792118	23.153491	OP099697
C. glaucopus	C328	07.11.2020	Cheile Someșului Cald	46.64377853	22.73244046	OP099748

Table 1. *Cont.*

Species	DNA-Codes	Collection Date	Location	Latitude	Longitude	GenBank acc. no.
C. glaucopus	C329	07.11.2020	Cheile Someșului Cald	46.64377853	22.73244046	OP099749
C. glaucopus	C332	07.11.2020	Cheile Someșului Cald	46.64369044	22.7320642	OP099751
C. glaucopus	C335	07.11.2020	Cheile Someșului Cald	46.64420997	22.72738871	OP099754
C. aff. *glaucopus*	C326	07.11.2020	Cheile Someșului Cald	46.6310984	22.76807258	OP099746
C. aff. *glaucopus*	C331	07.11.2020	Cheile Someșului Cald	46.64369044	22.7320642	OP099750
C. aff. *glaucopus*	C341 *	07.11.2020	Cheile Someșului Cald	46.64060986	22.81455206	OP132857
C. hadrocroceus	C106	10.07.2020	Zetea	46.4733373	25.34546	OP099609
C. hadrocroceus	C140	08.08.2020	Statiunea Muntele Baisoara	46.51092527	23.27111959	OP099621
C. hillieri	C171	18.10.2020	Făgetul Clujului	46.738306	23.539194	OP099637
C. hinnuleus	C377	21.10.2020	Bătarci	48.014053	23.146755	OP099779
C. holoxanthus	C046	22.09.2018	Valea Ierii	46.524456	23.274586	OP099581
C. holoxanthus	C052	22.09.2018	Valea Ierii	46.524888	23.274323	OP099586
C. holoxanthus	C095	19.08.2017	Statiunea Muntele Baisoara	46.520139	23.271811	OP099601
C. huronensis	C016	15.08.2018	Statiunea Muntele Baisoara	46.5382	23.3071099	OP099575
C. huronensis	C021	15.08.2018	Statiunea Muntele Baisoara	46.537253	23.305079	OP099578
C. huronensis	C151	03.09.2020	Mărtiniș	46.2647	25.355	OP099629
C. hydrotelamonioides	C205	04.10.2020	Chiheru de Jos	46.679428	25.035528	OP099664
C. incognitus	C004	27.07.2018	Statiunea Muntele Baisoara	46.5382	23.3071099	OP099566
C. lacustris	C387	21.10.2020	Bătarci	48.014053	23.146755	OP099783
C. largus	C232	24.10.2020	Cheile Vârghişului	46.199804	25.583554	OP099684
C. largus	C242	24.10.2020	Cheile Vârghişului	46.630342	25.835442	OP099691
C. largus	C385 *	21.10.2020	Bătarci	48.014053	23.146755	OP132858
C. leproleptopus	C356	07.11.2020	Făgetul Clujului	46.6993848	23.5488839	OP099766
C. lilacinovelatus	C305	01.11.2020	Făgetul Clujului	46.729734	23.548487	OP099734
C. luridus	C390	21.10.2020	Bătarci	48.014053	23.146755	OP099785
C. luridus	C123 *	28.07.2020	Cheile Somesului Cald	46.643648	22.7281299	OP132851
C. luridus	C289 *	01.11.2020	Chinteni	46.895154	23.520302	OP132856
C. masseei	*C388*	21.10.2020	Bătarci	48.014053	23.146755	OP099784
C. aff. *magicus*	C285	01.11.2020	Chinteni	46.895154	23.520302	OP099719
C. aff. *magicus*	C303	01.11.2020	Făgetul Clujului	46.729599	23.548782	OP099732
C. multiformis	C001	27.07.2018	Statiunea Muntele Baisoara	46.5382	23.3071099	OP099564
C. multiformis	C008	27.07.2018	Statiunea Muntele Baisoara	46.5382	23.3071099	OP099570
C. multiformis	C138	08.08.2020	Statiunea Muntele Baisoara	46.50627098	23.26640838	OP099619

Table 1. *Cont.*

Species	DNA-Codes	Collection Date	Location	Latitude	Longitude	GenBank acc. no.
C. multiformis	C144	08.08.2020	Statiunea Muntele Baisoara	46.50358372	23.264886235	OP099624
C. multiformis	C181	04.10.2020	Sovata	46.699756	25.173806	OP099642
C. multiformis	C184	04.10.2020	Sovata	46.699756	25.173806	OP099645
C. multiformis	C096	19.08.2017	Statiunea Muntele Baisoara	46.520139	23.271811	OP099602
C. multiformis	C006	27.07.2018	Statiunea Muntele Baisoara	46.5382	23.3071099	OP099568
C. napus	C214	04.10.2020	Sovata	46.662003	25.215961	OP099672
C. neofurvolaesus	C047	22.09.2018	Valea Ierii	46.524456	23.274586	OP099582
C. neofurvolaesus	C049	22.09.2018	Valea Ierii	46.524456	23.274586	OP099584
C. ochraceopallescens	C055	22.09.2018	Valea Ierii	46.524076	23.274521	OP099587
C. ochraceopallescens	C056	18.11.2018	Făgetul Clujului	46.720612	23.5606	OP099588
C. ochraceopallescens	C283	01.11.2020	Făgetul Clujului	46.717317	23.536817	OP099717
C. odoratus	C284	01.11.2020	Borsa	46.951797	23.60519	OP099718
C. odoratus	C299	01.11.2020	Făgetul Clujului	46.734583	23.543577	OP099730
C. odoratus	C304	01.11.2020	Făgetul Clujului	46.729743	23.54856	OP099733
C. olearioides	C354	07.11.2020	Făgetul Clujului	46.6993848	23.5488839	OP099764
C. olearioides	C355	07.11.2020	Făgetul Clujului	46.6993848	23.5488839	OP099765
C. olidoamarus	C288	01.11.2020	Chinteni	46.895154	23.520302	OP099722
C. olidoamarus	C296	01.11.2020	Chinteni	46.895154	23.520302	OP099728
C. olidoamarus	C381	21.10.2020	Bătarci	48.014053	23.146755	OP099781
C. ominosus	C182	04.10.2020	Sovata	46.699756	25.173806	OP099643
C. pallidostriatus	C315	06.11.2020	Cheile Someșului Cald	46.67023386	22.81820682	OP099741
C. pelerinii	C219 *	24.10.2020	Brăduț	46.201631	25.598088	OP132854
C. persoonianus	C174	18.10.2020	Făgetul Clujului	46.738306	23.539194	OP099638
C. pilatii	C070	13.10.2019	Statiunea Muntele Baisoara	46.6776035	23.4564326	OP099591
C. pruinatus	C391	21.10.2020	Bătarci	48.014053	23.146755	OP099786
C. pseudodaulnoyae	C311	01.11.2020	Almașu	46.869822	23.146055	OP099738
C. pseudodaulnoyae	C386	21.10.2020	Bătarci	48.014053	23.146755	OP099782
C. pseudofervidus	C142	08.08.2020	Statiunea Muntele Baisoara	46.5113753	23.27106945	OP099623
C. pseudofervidus	C141	08.08.2020	Statiunea Muntele Baisoara	46.51101007	23.277106504	OP099622
C. pseudonaevosus	C007	27.07.2018	Statiunea Muntele Baisoara	46.5382	23.3071099	OP099569
C. purpurascens	C208	04.10.2020	Sovata	46.662003	25.215961	OP099667
C. purpurascens	C327	07.11.2020	Cheile Someșului Cald	46.63110059	22.76854548	OP099747
C. purpurascens	C367	18.11.2020	Romuli	47.560767	24.530413	OP099775
C. purpurascens	C370	19.11.2020	Parva	47.44473	24.64913	OP099777
C. purpurascens	C371	20.11.2020	Rebrișoara	47.44884	24.58896	OP099778

Table 1. *Cont.*

Species	DNA-Codes	Collection Date	Location	Latitude	Longitude	GenBank acc. no.
C. radicosissimus	C313A	06.11.2020	Cheile Someșului Cald	46.67023386	22.81820682	OP099740
C. renidens	C019	15.08.2018	Statiunea Muntele Baisoara	46.538338	23.305955	OP099577
C. renidens	C109	23.06.2020	Liban	46.553983	25.520218	OP099611
C. renidens	C119	28.07.2020	Cheile Somesului Cald	46.626862	22.788439	OP099616
C. rubellus	C098	05.08.2017	Ponok	46.638228	22.815111	OP099604
C. rubricosus	C281	01.11.2020	Făgetul Clujului	46.717317	23.536817	OP099715
C. rubrophyllus	C102	10.07.2020	Zetea	46.56182	25.3733821	OP099606
C. rufoallutus	C005	27.07.2018	Statiunea Muntele Baisoara	46.5382	23.3071099	OP099567
C. saginus	C150	03.09.2020	Mărtiniș	46.2647	25.355	OP099628
C. salor	C297	01.11.2020	Rediu	46.733722	26.537746	OP099729
C. saporatus	C369	19.11.2020	Parva	47.440112	24.650527	OP099776
C. scaurocaninus	C222	24.10.2020	Cheile Vârghişului	46.203991	25.558544	OP099679
C. scaurocaninus	C230	24.10.2020	Cheile Vârghişului	46.20322	25.559007	OP099683
C. scaurocaninus	C239	24.10.2020	Cheile Vârghişului	46.205054	25.558222	OP099688
C. scaurocaninus	C240	24.10.2020	Cheile Vârghişului	46.215552	25.5473	OP099689
C. semisanguineus	C010	27.07.2018	Statiunea Muntele Baisoara	46.5382	23.3071099	OP099572
C. semivelatus	C350	07.11.2020	Făgetul Clujului	46.6993848	23.5488839	OP099761
C. sodagnitus	C280	01.11.2020	Făgetul Clujului	46.717317	23.536817	OP099714
C. sodagnitus	C282	01.11.2020	Făgetul Clujului	46.717317	23.536817	OP099716
C. spadicellus	C154	10.10.2020	Statiunea Stana de Vale	46.697647	22.626014	OP099631
C. spadicellus	C155	10.10.2020	Statiunea Stana de Vale	46.697647	22.626014	OP099632
C. spadicellus	C157	10.10.2020	Statiunea Stana de Vale	46.697647	22.626014	OP099633
C. spadicellus	C158	10.10.2020	Statiunea Stana de Vale	46.697647	22.626014	OP099634
C. spadicellus	C211	04.10.2020	Sovata	46.662003	25.215961	OP099670
C. spilomeus	C196	04.10.2020	Chiheru de Jos	46.672231	25.051561	OP099655
C. subargyronotus	C358	07.11.2020	Făgetul Clujului	46.6993848	23.5488839	OP099768
C. subdecolorans	C229	24.10.2020	Cheile Vârghişului	46.205947	25.557701	OP099682
C. subfoetens	C333	07.11.2020	Cheile Someșului Cald	46.64369044	22.7320642	OP099752
C. aff. *sublilacinopes*	C309	01.11.2020	Făgetul Clujului	46.738275	23.537126	OP099736
C. subparvannulatus	C071	13.10.2019	Statiunea Muntele Baisoara	46.6776035	23.4564326	OP099592
C. subporphyropus	C178	10.10.2020	Groșii Țibleșului	47.52174	24.15647	OP099641
C. subpurpurascens	C091	30.11.2019	Făgetul Clujului	46.698086	23.587791	OP099597
C. subpurpurascens	C195	04.10.2020	Chiheru de Jos	46.672231	25.051561	OP099654
C. subpurpurascens	C216	19.10.2020	Feleacu	46.695459	23.58806	OP099674
C. subpurpurascens	C217	19.10.2020	Feleacu	46.695459	23.58806	OP099675
C. subpurpurascens	C237	24.10.2020	Cheile Vârghişului	46.199842	25.583542	OP099686

Table 1. *Cont.*

Species	DNA-Codes	Collection Date	Location	Latitude	Longitude	GenBank acc. no.
C. subtortus	C017	15.08.2018	Statiunea Muntele Baisoara	46.5382	23.3071099	OP099610
C. subtortus	C075	13.10.2019	Statiunea Muntele Baisoara	46.522375	23.273463	OP099595
C. subtortus	C099	19.08.2017	Statiunea Muntele Baisoara	46.520139	23.271811	OP099605
C. subtortus	C139	08.08.2020	Statiunea Muntele Baisoara	46.50358372	23.262648623	OP099620
C. subtortus	C243	24.10.2020	Statiunea Muntele Baisoara	46.50627098	23.26640838	OP099692
C. subtortus	C245	24.10.2020	Statiunea Muntele Baisoara	46.50627098	23.26640838	OP099694
C. subtortus	C247	24.10.2020	Statiunea Muntele Baisoara	46.50627098	23.26640838	OP099696
C. subtortus	C073	13.10.2019	Statiunea Muntele Baisoara	46.522375	23.273463	OP099593
C. sulphurinus	C318	06.11.2020	Cheile Someșului Cald	46.67071058	22.81810123	OP099742
C. talimultiformis	C185	04.10.2020	Sovata	46.699756	25.173806	OP099646
C. talimultiformis	C201	04.10.2020	Chiheru de Jos	46.679428	25.035528	OP099660
C. talimultiformis	C312	06.11.2020	Cheile Someșului Cald	46.67023386	22.81820682	OP099739
C. talimultiformis	C323	07.11.2020	Cheile Someșului Cald	46.67071058	22.81810123	OP099744
C. testaceomicaceus	C342	07.11.2020	Făgetul Clujului	46.6993848	23.5488839	OP099758
C. tirolianus	C218	24.10.2020	Cheile Vârghişului	46.204486	25.558374	OP099676
C. tirolianus	C259 *	01.11.2020	Harghita	46.538506	25.612982	OP132855
C. tirolianus	C268	01.11.2020	Ghimes	46.528277	26.033167	OP099707
C. tirolianus	C269	01.11.2020	Ghimes	46.528277	26.033167	OP099708
C. tirolianus	C276	01.11.2020	Ghimes	46.528277	26.033167	OP099712
C. tirolianus	C277	01.11.2020	Ghimes	46.528277	26.033167	OP099713
C. torvus	C353	07.11.2020	Făgetul Clujului	46.6993848	23.5488839	OP099763
C. traganus	C183	04.10.2020	Sovata	46.699756	25.173806	OP099644
C. traganus	C210A	04.10.2020	Sovata	46.662003	25.215961	OP099669
C. traganus	C244	24.10.2020	Statiunea Muntele Baisoara	46.50627098	23.26640838	OP099693
C. trivialis	C220	24.10.2020	Brăduț	46.201631	25.598088	OP099677
C. trivialis	C366	07.11.2020	Făgetul Clujului	46.6993848	23.5488839	OP099774
C. aff. *trivialis*	C225	24.10.2020	Cheile Vârghişului	46.203991	25.5585445	OP099681
C. aff. *trivialis*	C362	07.11.2020	Făgetul Clujului	46.6993848	23.5488839	OP099772
C. turgidus	C064	04.10.2019	Valea Bratcutei	46.886256	22.58647	OP099590
C. turgidus	C163	19.10.2020	Făgetul Clujului	46.738306	23.539194	OP099636
C. turmalis	C203	04.10.2020	Chiheru de Jos	46.679428	25.035528	OP099662
C. turmalis	C209	04.10.2020	Sovata	46.662003	25.215961	OP099668
C. turmalis	C215	19.10.2020	Feleacu	46.695459	23.58806	OP099673
C. ultrodistortus	C107	10.07.2020	Zetea	46.4733373	25.34546	OP099610

Table 1. *Cont.*

Species	DNA-Codes	Collection Date	Location	Latitude	Longitude	GenBank acc. no.
C. uraceonemoralis	C238	24.10.2020	Cheile Vârghişului	46.199056	25.573537	OP099687
C. variicolor	C057	06.09.2019	Demsus	45.5623456	22.700035	OP099589
C. variicolor	C186	04.10.2020	Chiheru de Jos	46.686903	25.040178	OP099647
C. variicolor	C197	04.10.2020	Chiheru de Jos	46.672231	25.051561	OP099656
C. variicolor	C198	04.10.2020	Chiheru de Jos	46.672231	25.051561	OP099657
C. variicolor	C204	04.10.2020	Chiheru de Jos	46.679428	25.035528	OP099663
C. varius	C275	01.11.2020	Ghimes	46.528277	26.033167	OP099711
C. varius	C337	07.11.2020	Cheile Someșului Cald	46.62928894	22.7806518	OP099755
C. varius	C340	07.11.2020	Cheile Someșului Cald	46.62928894	22.7806518	OP099757
C. venetus	C251	01.11.2020	Harghita	46.538506	25.612982	OP099698
C. venetus	C260	01.11.2020	Harghita	46.538506	25.612982	OP099703
C. venetus	C265	01.11.2020	Ghimes	46.528277	26.033167	OP099705
C. venetus	C266	01.11.2020	Ghimes	46.528277	26.033167	OP099706
C. venetus	C271	01.11.2020	Ghimes	46.528277	26.033167	OP099709
C. aff. *vibratilis*	C188	04.10.2020	Chiheru de Jos	46.686903	25.040178	ON832643
C. violaceus	C190	04.10.2020	Chiheru de Jos	46.686903	25.040178	OP099649
C. violaceus	C191A	04.10.2020	Pădurea Bunetilor	46.081617	25.034936	OP099650
C. xanthochlorus	C287	01.11.2020	Chinteni	46.895154	23.520302	OP099721
C. xanthochlorus	C301	01.11.2020	Făgetul Clujului	46.729576	23.548556	OP099731

* Short sequences excluded from the phylogenetic analyses.

Figure 1. Basidiomata of some *Cortinarius* s.l. species new to Romania: (**A**,**B**) *C. catharinae* C089 (sect. *Calochroi*), (**C**) *C. daulnoyae* C298 (sect. *Phlegmacioides*), (**D**) *C. fulminoides* C334 (sect. *Aureocistophili*), (**E**) *C. hadrocroceus* C140 (sect. *Dermocybe*), (**F**) *C. lacustris* C387 (sect. *Hinnulei*), (**G**,**H**) *C. masseei* C388 (sect. *Obtusi*), (**I**) *C. subdecolorans* C229 (singleton), (**J**) *C. subfoetens* C333 (sect. *Glaucopodes*), Photos. (**A**,**C**,**E**,**F**,**I**,**J**) E. Szabó; (**B**,**D**) A. Dénes, (**G**,**H**) K. Babos and K. Fehér.

2.2. Micromorphological Study

We studied the basidiospores of our collections (Figure 2). In general, the size of the spores are 5–15 × 3–8 µm. Basidiospores were studied in 3% KOH or Melzer's reagent, the latter to observe the dextrinoid reaction of the spore wall in some groups [98]. The analyses were performed with an Olympus CX23 microscope with 1000× magnification using immersion oil and an oil immersion lens. The photographs were taken with a Canon 700D camera attached to the microscope. Photos were stacked with Zerene Stacker (zerenesystems.com/cms/home), and the size of the spores was measured with Piximètre (http://ach.log.free.fr/Piximetre/) software. In addition to the basidiospore sizes, coloration and the ornamentation were also noted. From the measured parameters, the Q value (length/width ratio) was calculated, which indicates the shape of the spores (Q = 1.01–1.05: globose, Q = 1.05–1.15: subglobose, Q = 1.15–1.30: broadly ellipsoid, Q = 1.30–1.60: ellipsoid, Q = 1.60–2.0: oblong, Q = 2.0–3.0: cylindrical). Basidiospore ranges for the species new to Romania are given in Table 2. To exclude aberrant spores, the values are based on spores within the 0.75 confidence interval.

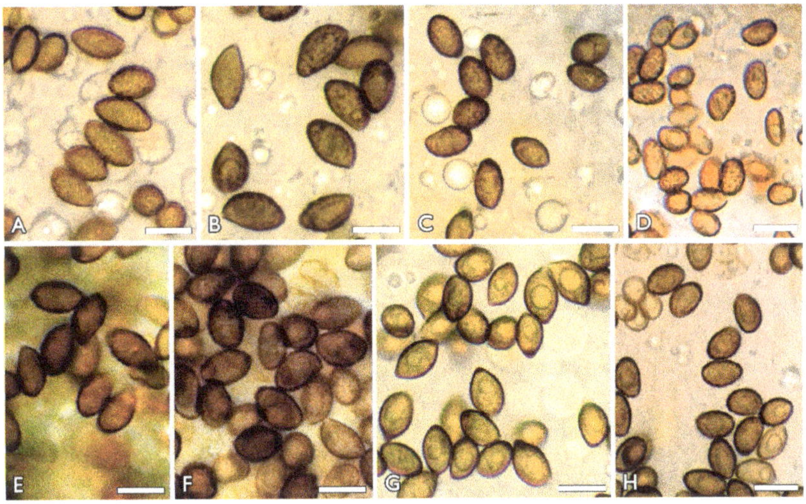

Figure 2. Basidiospores of some *Cortinarius* s.l. species new to Romania: (**A**) *C. catharinae* C089 (sect. *Calochroi*), (**B**) *C. daulnoyae* C298 (sect. *Phlegmacioides*), (**C**) *C. fulminoides* C334 (sect. *Aureocistophili*), (**D**) *C. subdecolorans* C229 (singleton), (**E**) *C. subfoetens* C333 (sect. *Glaucopodes*), (**F**) *C. hadrocroceus* C140 (sect. *Dermocybe*), (**G**) *C. masseei* C388 (sect. *Obtusi*), (**H**) *C. lacustris* C387 (sect. *Hinnulei*). Scale bar: 10 µm. Photos: E. Szabó.

Table 2. *Cortinarius* species discovered new to Romania. Infrageneric classification, habitat types as well as basidiospore measurements and Q values are given for each species. BLf–broad-leaved forest; Cf–coniferous forest; Mf–mixed forest; P–pastures; TWS–transitional woodland–shrub; NG–natural grasslands.

Species	Section/Clade	Habitat	No. of Collected Samples	Basidiospores
Cortinarius anfractoides Rob. Henry and Trescol 1987	*Infracti*	BLf	2	8.8–9.5 × 7.2–7.4 µm Q = 1.2–1.3
Cortinarius aureopulverulentus M.M. Moser 1952	*Calochroi*	Cf	1	12.4–13.1 × 8.0–8.7 µm Q = 1.5–1.6

Table 2. *Cont.*

Species	Section/Clade	Habitat	No. of Collected Samples	Basidiospores
Cortinarius balteatocumatilis Rob. Henry 1939	*Phlegmacioides*	P	2	8.4–9.2 × 5.2–5.9 μm Q = 1.5–1.7
Cortinarius brunneus (Pers.) Fr. 1838	*Brunnei*	Cf	1	8.2–9.5 × 6.1–6.8 μm Q = 1.2–1.5
Cortinarius caesiophylloides Kytöv., Liimat., Niskanen, Brandrud and Frøslev 2014	*Multiformes*	Cf	1	10.4–11.0 × 6.2–6.6 μm Q = 1.57–1.73
Cortinarius catharinae Consiglio 1997	*Calochroi*	BLf	1	9.7–10.5 × 5.8–6.3 μm Q = 1.6–1.8
Cortinarius cinereobrunneolus Chevassut and Rob. Henry 1982	*Urbici*	BLf	1	7.7–8.4 × 5.0–5.2 μm Q = 1.5–1.7
Cortinarius claricolor (Fr.) Fr. 1838	*Claricolores*	Cf	1	6.4–7.2 × 3.5–3.7 μm Q = 1.8–1.9
Cortinarius colymbadinus Fr. 1838	*Uracei*	Mf	1	8.5–9.4 × 5.4–5.8 μm Q = 1.5–1.7
Cortinarius corrosus Fr. 1838	*Calochroi*	Cf	1	12.5–14.1 × 7.5–7.9 μm Q = 1.6–1.9
Cortinarius daulnoyae (Quél.) Sacc. 1910	*Phlegmacioides*	Cf	7	12.4–13.2 × 7.3–7.7 μm Q = 1.7–1.8
Cortinarius aff. *delibutus*	*Delibuti*	Mf	1	8.5–8.9 × 7.0–7.4 μm Q = 1.2–1.3
Cortinarius eliae Bidaud, Moënne-Locc. and Reumaux 1996	*Phlegmacioides*	P	3	11.5–12.5 × 6.8–7.1 μm Q = 1.7–1.8
Cortinarius fraudulosus Britzelm. 1885	*Arguti*	Cf	1	15.9–16.6 × 8.6–9.0 μm Q = 1.8–1.9
Cortinarius fulminoides (M.M. Moser) M.M. Moser 1967	*Aureocistophili*	Mf	1	9.4–10.2 × 5.9–6. 5 μm Q = 1.5–1.6
Cortinarius gallurae D. Antonini, M. Antonini and Consiglio 2005	/Gallurae	BLf	1	8.6–9.4 × 5.7–6.5 μm Q = 1.4–1.6
Cortinarius geniculatus Bidaud 2014	*Bovini*	BLf	1	10.7–12.0 × 6.4–6.7 μm Q = 1.6–1.8
Cortinarius glaucopus aff.	*Glaucopodes*	Cf	2	8.5–9.3 × 5.4–5.7 μm Q = 1.6–1.7
Cortinarius hadrocroceus Ammirati, Niskanen, Liimat. and Bojantchev 2014	*Dermocybe*	Cf	2	7.3–7.8 × 4.3–4.7 μm Q = 1.6–1.8
Cortinarius hillieri Rob. Henry 1938	*Bovini*	BLf	1	10.7–11.8 × 6.1–6.7 μm Q = 1.7–1.8
Cortinarius holoxanthus (M.M. Moser and I. Gruber) Nezdojm. 1980	*Dermocybe*	NG	3	9.3–9.9 × 4.8–5.2 μm Q = 1.8–2.1
Cortinarius huronensis Ammirati and A.H. Sm. 1972	*Dermocybe*	Cf	3	8.0–8.5 × 4.9–5.4 μm Q = 1.47–1.65
Cortinarius hydrotelamonioides Rob. Henry 1970	*Firmiores*	BLf	1	9.3–10.3 × 5.1–5.8 μm Q = 1.7–1.9
Cortinarius incognitus Ammirati and A.H. Sm. 1972	*Dermocybe*	Cf	1	7.4–8.0 × 5.2–5.4 μm Q = 1.3–1.6
Cortinarius lacustris Moënne-Locc. and Reumaux 1997	*Hinnulei*	BLf	1	9.6–11.1 × 5.9–6.6 μm Q = 1.5–1.8

Table 2. *Cont.*

Species	Section/Clade	Habitat	No. of Collected Samples	Basidiospores
Cortinarius leproleptopus Chevassut and Rob. Henry 1988	*Leprocybe*	BLf	1	7.9–8.5 × 6.8–7.3 µm Q = 1.1–1.2
Cortinarius lilacinovelatus Reumaux and Ramm 2001	*Calochroi*	BLf	1	11.1–11.7 × 6.1–6.5 µm Q = 1.7–1.8
Cortinarius luridus Rob. Henry 1969	*Hinnulei*	BLf	3	8.9–9.7 × 6.3–6.7 µm Q = 1.4–1.5
Cortinarius masseei Bidaud, Moënne-Locc. and Reumaux 1993	*Obtusi*	BLf	1	7.0–7.8 × 4.9–5.3 µm Q = 1.4–1. 6
Cortinarius aff. *magicus*	*Glaucopodes*	Cf	2	7.6–8.0 × 5.0–5.3 µm Q = 1.5–1.6
Cortinarius neofurvolaesus Kytöv., Niskanen, Liimat. and H. Lindstr. 2005	*Bovini*	NG	2	8.5–9.5 × 5.1–5.9 µm Q = 1.5–1.8
Cortinarius ochraceopallescens Moënne-Locc. and Reumaux 2001	*Calochroi*	BLf	3	12.4–13.2 × 6.4–6.9 µm Q = 1.8–2.0
Cortinarius odoratus (Joguet ex M.M. Moser) M.M. Moser 1967	*Calochroi*	BLf	3	11.6–12.5 × 6.8–7.2 µm Q = 1.7–1.8
Cortinarius olidoamarus A. Favre 1986	*Glaucopodes*	P	3	8.4–9.2 × 5.2–5.9 µm Q = 1.5–1.7
Cortinarius ominosus Bidaud 1994	*Dermocybe*	TWS	1	6.8–7.7 × 4.4–4.7 µm Q = 1.5–1.6
Cortinarius pallidostriatus Rob. Henry 1968	*Hydrocybe*	Cf	1	8.6–9.6 × 5.5–6.0 µm Q = 1.5–1.7
Cortinarius pelerinii Bellanger, Carteret and Reumaux 2013	*Anomali*	BLf	1	8.8–9.4 × 6.3–7.0 µm Q = 1.3–1.4
Cortinarius persoonianus Bidaud 2009	*Infracti*	BLf	1	8.7–9.5 × 7.2–7.6 µm Q = 1.2–1.3
Cortinarius pilatii Svrček 1968	*Flexipedes*	Cf	1	9.0–9.7 × 6.0–6.5 µm Q = 1.4–1.6
Cortinarius pruinatus Bidaud, Moënne-Locc. and Reumaux 1993	*Obtusi*	BLf	1	10.5–11.7 × 6.2–7.3 µm Q = 1.5–1.7
Cortinarius pseudodaulnoyae Rob. Henry and Ramm 1991	*Phlegmacioides*	BLf	2	12.7–13.2 × 7.2–7.7 µm Q = 1.7–1.8
Cortinarius pseudofervidus Niskanen, Liimat., Ammirati and Kytöv. 2014	*Dermocybe*	NG	2	6.9–7.6 × 4.3–4.9 µm Q = 1.4–1.7
Cortinarius pseudonaevosus Rob. Henry 1957	*Phlegmacioides*	Cf	1	12.9–14.4 × 7.5–8.3 µm Q = 1.7–1.8
Cortinarius radicosissimus Moënne-Locc. 1997	*Hinnulei*	Cf	1	8.6–9.4 × 6.8–8.0 µm Q = 1.2–1.3
Cortinarius renidens Fr. 1838	*Renidentes*	Cf	3	7.8–8.1 × 5.6–6.0 µm Q = 1.3–1.4
Cortinarius rubricosus (Fr.) Fr. 1838	*Rubricosi*	BLf	1	10.0–10.9 × 7.0–7.5 µm Q = 1.4–1.6
Cortinarius rubrophyllus (Moënne-Locc.) Liimat., Niskanen, Ammirati and Dima 2014	*Dermocybe*	Cf	1	6.0–6.3 × 3.8–4.3 µm Q = 1.5–1.6
Cortinarius rufoallutus Rob. Henry ex Bidaud and Reumaux 2006	*Multiformes*	Cf	1	9.7–10.3 × 5.7–5.93 µm Q = 1.7–1.8

Table 2. *Cont.*

Species	Section/Clade	Habitat	No. of Collected Samples	Basidiospores
Cortinarius saginus (Fr.) Fr. 1838	*Phlegmacium*	TWS	1	10.5–10.8 × 6.0–6.5 µm Q = 1.7–1.8
Cortinarius saporatus Britzelm. 1897	*Calochroi*	Mf	1	10.5–11.7 × 6.4–7.6 µm Q = 1.5–1.8
Cortinarius scaurocaninus Chevassut and Rob. Henry 1982	*Glaucopodes*	Cf	4	8.7–9.3 × 5.2–5.4 µm Q = 1.6–1.8
Cortinarius semivelatus Rob. Henry 1970	*Squalidi*	BLf	1	8.0–8.7 × 4.9–5.7 µm Q = 1.4–1.7
Cortinarius sodagnitus Rob. Henry 1935	*Calochroi*	BLf	2	11.5–12.5 × 6.3–7.1 µm Q = 1.6–1.9
Cortinarius spadicellus Brandrud 1997	*Phlegmacioides*	Mf	5	10.9–11.5 × 6.9–7.1 µm Q = 1.5–1.7
Cortinarius subargyronotus Niskanen, Liimat. and Kytöv. 2014	*Uracei*	BLf	1	9.5–10.3 × 5.9–6.4 µm Q = 1.5–1.8
Cortinarius subdecolorans M. Langl. and Reumaux 2000	/Dionysae	BLf	1	9.8–10.6 × 6.2–6.6 µm Q = 1.5–1.6
Cortinarius subfoetens M.M. Moser and McKnight 1995	*Glaucopodes*	Mf	1	8.3–8.8 × 5.4–5.7 µm Q = 1.5–1.6
Cortinarius aff. *sublilacinopes*	*Calochroi*	BLf	1	11.7–12.2 × 7.0–7.4 µm Q = 1.6–1.7
Cortinarius subporphyropus Pilát 1954	*Purpurascentes*	Mf	1	11.1–12.1 × 6.6–7.3 µm Q = 1.6–1.7
Cortinarius subpurpurascens (Batsch) J. Kickx f. 1867	*Purpurascentes*	BLf	5	10.6–11.2 × 6.0–6.5 µm Q = 1.7–1.8
Cortinarius testaceomicaceus Bidaud 2014	*Exsulares*	BLf	1	9.5–11.2 × 6.8–7.5 µm Q = 1.3–1.6
Cortinarius tirolianus Bidaud, Moënne-Locc. and Reumaux 2005	*Glaucopodes*	Cf	6	7.5–7.9 × 5.0–5.2 µm Q = 1.5–1.6
Cortinarius aff. *trivialis*	*Myxacium*	BLf	2	13.4–14.7 × 8.0–8.8 µm Q = 1.6–1.7
Cortinarius ultrodistortus Rob. Henry and Vagnet 1992	/Ultrodistortus	Cf	1	7.4–8.8 × 4.5–5.0 µm Q = 1.5–1.9
Cortinarius uraceonemoralis Niskanen, Liimat., Dima, Kytöv., Bojantchev and H. Lindstr. 2014	*Uracei*	BLf	1	9.1–10.3 × 5.5–6.0 µm Q = 1.58–1.87
Cortinarius aff. *vibratilis*	*Vibratiles*	BLf	1	8.4–8.9 × 5.2–5.6 µm Q = 1.6–1.7

2.3. Molecular Genetic Analysis

DNA extractions, PCR amplifications, gel electrophoreses, and the purification of the PCR products were performed in the molecular laboratory at the Interdisciplinary Research Institute on Bio-Nano-Sciences of Babeș-Bolyai University, Cluj-Napoca, Romania.

Genomic DNA was extracted from a small piece of the lamella (10–15 mg) under sterile conditions using the ISOLATE II Genomic DNA Kit (Bioline Meridian Bioscience, Inc. Cincinnati, OH, USA) following a modified protocol: (1) After the crushing the dried samples, 180 µL of Lysis Buffer GL and 25 µL of Proteinase K solution were added to the sample and were mixed using Thermoblock for 3 h at 56 °C for better amalgamation; (2) the elution of the DNA was performed in two steps with 30–30 µL of Elution Buffer G solution (before spinning, the elution buffer stayed in the column for 3 min). The concentration of the DNA solutions was measured with a NanoDrop ND-1000 Spectrophotometer (NanoDrop Technologies, Wilmington, DE, USA). All DNA concentrations fell between 30 and 100 ng/µL.

The target region was amplified using the following primers: ITS1F (5′-CTTGGTCATTTAGAGGAAGTAA-3′) and ITS4 (5′-TCCTCCGCTTATTGATATGC-3′) [94]. The PCR was performed in a volume of 50 µL of the reaction mixture, which contained 0.5 µL of MyTaq™ DNA Polymerase (Bioline Reagents Ltd., London, UK), 10 µL of 5× MyTaq™ Reaction Buffer, 1 µL of the primer pairs (20 µM each, from the following primer mix: 10 µL ITS1F + 10 µL ITS4 + 80 µL PCR-grade water), 1 µL of DNA solution and 37.5 µL of PCR-grade water. The PCR conditions included an initial 5 min denaturation step at 95 °C, followed by 35 cycles of denaturation of 1 min at 95 °C, 1 min of annealing at 52 °C and 1 min of elongation at 72 °C. Finally, there was a 10 min final extension at 72 °C.

The success of the PCR was confirmed with a 1% agarose gel electrophoresis of 2 µL from each sample, performed at 120 V for 20 min. The gel was dyed with ethidium-bromide and was examined under UV light. After the run, the positive samples were purified using Wizard SV Gel and a PCR Clean-Up System (Promega, Madison, Wisconsin, USA USA) following the provided protocol, with one modification: for a more concentrated solution of the PCR product, the elution of the PCR product was performed in two steps with 20–20 µL of nuclease free water (before spinning, the buffer stayed in the column for 3 min). The purity of the PCR product and the concentration of the solution was measured with a spectrophotometer.

The PCR products were sent to Macrogen Europe (Amsterdam, The Netherlands) for Sanger sequencing with ITS1F and ITS4 primers.

2.4. Phylogenetic Analysis

The results of the sequencing were downloaded from the Macrogen Europe website. The quality of the sequences was analyzed using Trev from the Staden Program Package [99], and the primers' connection region was cut manually. Homologous sequences were searched by using the BLASTn algorithm [100] both in public, non-curated database (GenBank; http://www.ncbi.nlm.nih.gov/, accessed on 12 November 2021), and in curated public databases (UNITE; http://unite.ut.ee/, BOLD; http://www.boldsystems.org/). Reference and type sequences (Table S1) were downloaded and added to our own dataset.

The sequences were aligned in BioEdit [101] using MAFFT with the E-INS-I strategy [102]. To refine the nucleotide alignment, the phylogenetically informative indel positions were coded in FastGap 1.2 [103], and the binary matrix was added to the nucleotide alignment (ITS1, 5.8S, ITS2) in SeaView 4 [104]. Our dataset was analyzed based on Maximum Likelihood (ML) and Bayesian Inference (BI) methods. The ML phylogenetic reconstruction was performed in raxmlGUI [105] using rapid bootstrap analysis with 1000 replicates. Three nucleotide partitions (ITS1, 5.8S, ITS2) were set to the GTRGAMMA substitution model in addition to one binary partition (indel characters) that was set to default. The BI analysis was inferred in MrBayes 3.2.6 [106], with a MCMC (Markov chain Monte Carlo) algorithm, and the GTR + Γ substitution model with gamma distribution. The nucleotide partitions and the binary matrix from gap coding were treated as a mixed data type. The analysis ran for 10,000,000 generations with 25% burn-in. The resulting phylogenetic tree was visualized in FigTree v1.4.2 [107], MEGA 7 [108] and Adobe Illustrator CS4.

3. Results

In total, 234 sequences were analyzed (Table 1). BI and ML analyses of the nrDNA ITS region revealed topologically similar phylograms. A preliminary survey was conducted to see whether the 10 different genera of *Cortinariaceae* established by Liimatainen et al. [76] could be reconstructed based only on single gene analysis, but ITS was not found to be suitable for recognizing these genera. Therefore, we decided to treat *Cortinarius* in a classical sense. In order to achieve a better overview of the morphological similarities of the studied species, the dataset was divided into two parts. The first alignment was composed of 760 characters. After gap coding, a binary set of 267 characters was added to the nucleotide alignment, leading to a mixed data matrix containing 1027 characters. The alignment contained ITS sequences of leprocyboid, dermocyboid, rozitoid, telamonioid, and myxacioid species, i.e., from the classical morphological groups such as subgen. *Cortinarius*, subgen. *Telamonia*, and subgen. *Myxacium*. Our results showed (Figure 3) that the studied species belonged to sections *Anomali, Bovini, Brunnei, Camphorati, Cortinarius, Defibulati, Delibuti, Dermocybe, Exulares, Firmiores, Flexipedes, Hinnulei, Hydrocybe, Leprocybe, Myxacium, Obtusi, Orellani, Renidentes, Rozites, Rubricosi, Spilomei, Squalidi, Telamonia, Tragani, Uracei, Urbici* and *Vibratiles*, as well as to the clades /Gallurae and /Ultrodistortus. The second alignment included ITS sequences of species from the morphological subgen. *Phlegmacium* and was composed of 712 characters. In this case, 185 binary characters from the gap coding were added to the nucleotide characters, resulting in a final alignment of 897 characters. Our results showed (Figure 4) that the studied species belonged to sections *Arguti, Aureocistophili, Calochroi, Claricolores, Glaucopodes, Infracti, Multiformes, Phlegmacioides, Phlegmacium, Purpurascentes, Subtorti, Turmales* and *Varii*, and the clade /Dionysae (= *Phlegmacium* sect. *Dionysae*). Most sections and clades are well-supported with ML bootstrap support values between 80–100% and BI posterior probabilities between 0.92–1. For the sectional names, the taxonomic work of Liimatainen et al. [3,56,76] were followed.

Based on our study, we were able to recognize 109 *Cortinarius* s.l. species from Romania, belonging to 40 sections and 3 clades. Out of these, 43 species have been previously documented, but their identifications were based only on morphology (Table S2). The remaining 66 species are reported here as new to Romania (Table 2). Among these, we were not able to link any of the described *Cortinarius* names to six phylogenetically well-separated species; therefore, we used the 'aff.' prefix before the epithets which link them to their closest phylogenetic or morphological species (Figures 3 and 4). These are C. aff. *delibutus* in sect. *Delibuti*, C. aff. *glaucopus* and C. aff. *magicus* in sect. *Glaucopodes*, C. aff. *sublilacinopes* in sect. *Calochroi*, C. aff. *trivialis* in sect. *Mycaxium* and C. aff. *vibratilis* in sect. *Vibratiles*. Concerning the ongoing *Cortinarius* studies in Europe, our unpublished results (data not shown) indicate that these species are likely taxonomic novelties, but unveiling their taxonomy and nomenclature needs further analyses and the results will be included in different publications.

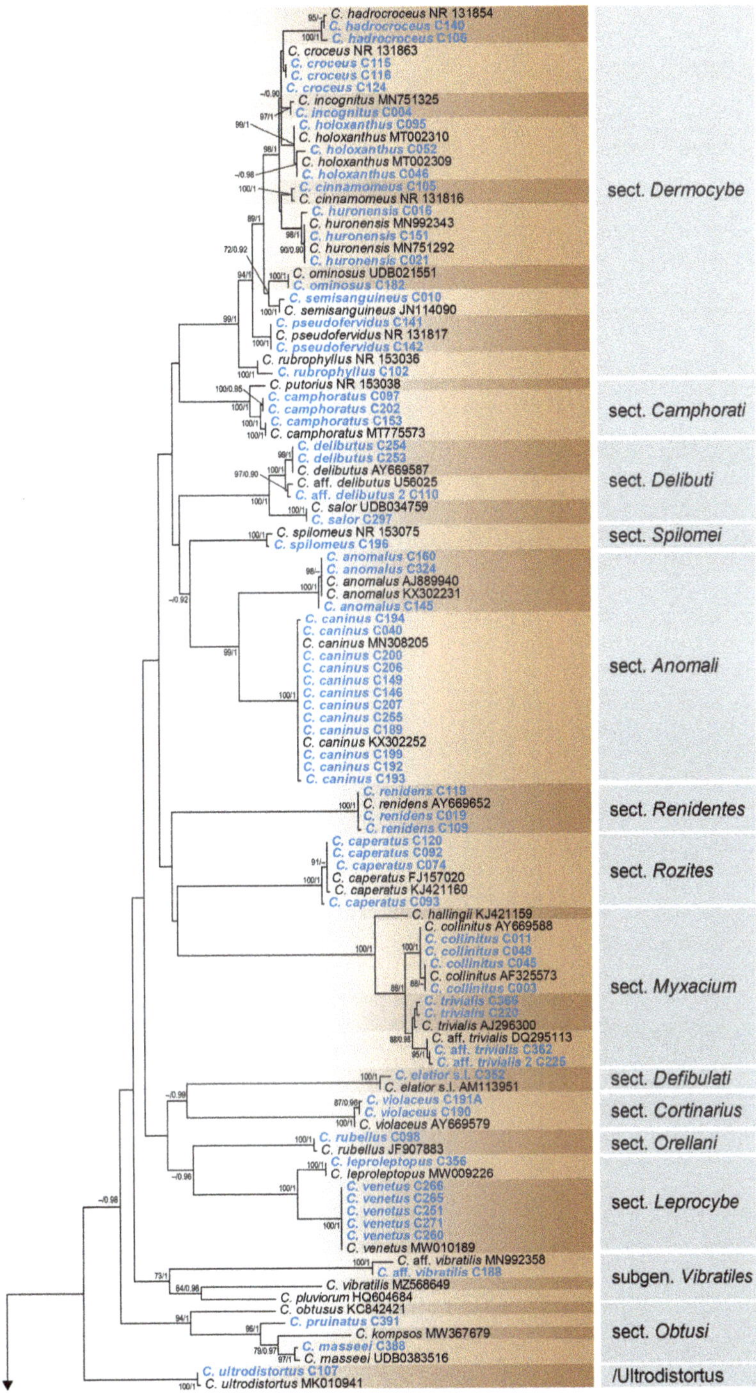

Figure 3. *Cont.*

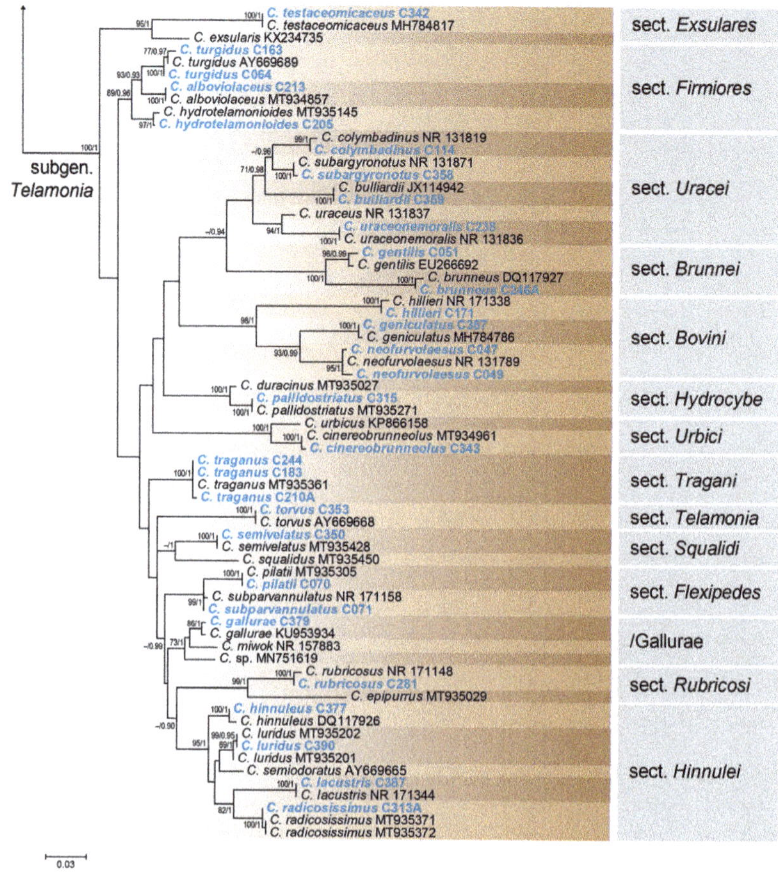

Figure 3. Maximum likelihood phylogenetic tree of the dermocyboid, leprocyboid, rozitoid, telamonioid, and myxacioid sections/clades of the genus *Cortinarius* s.l. based on nrDNA ITS sequence analyses with gap coding. Sequences produced in this study are in blue and boldface and labelled with their voucher numbers. Sequences from public repositories are marked with their GenBank/UNITE accession numbers. ML bootstrap values of >70% as well as Bayesian posterior probabilities of >0.9 are placed above or below branches. The scale bar indicates 0.03 expected change per site per branch.

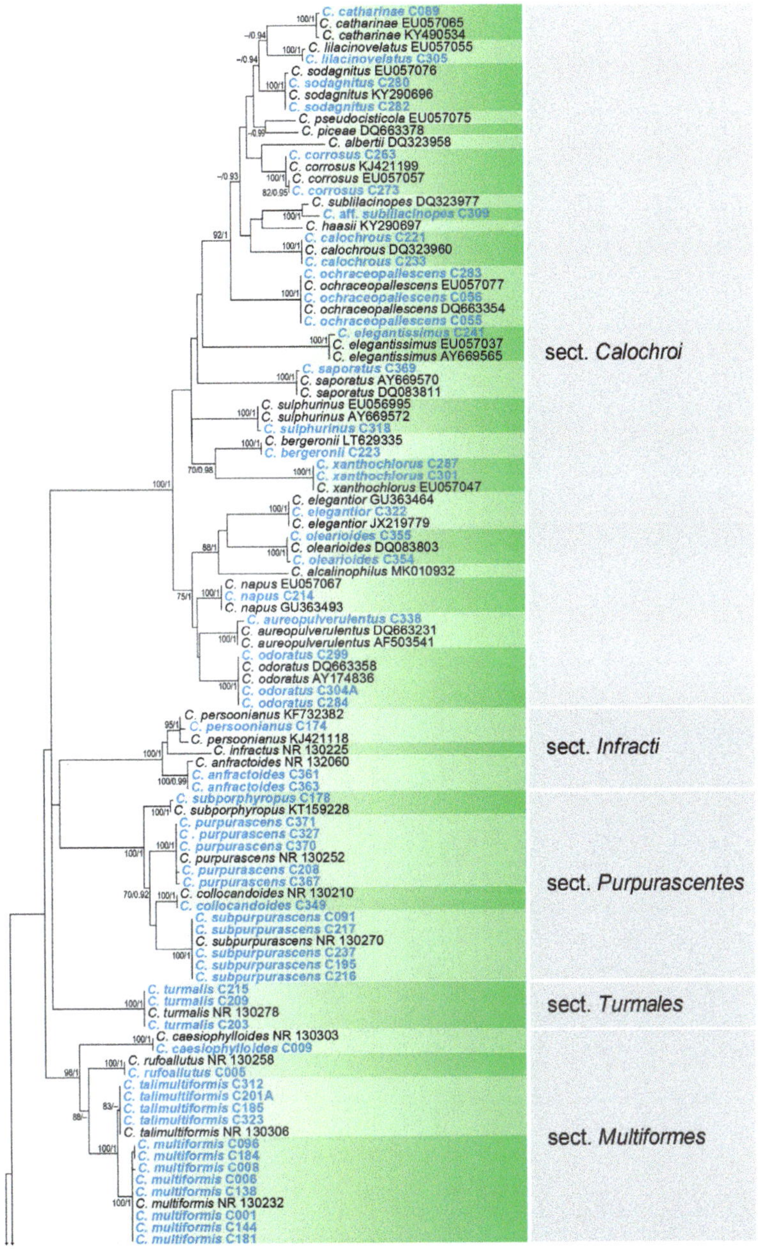

Figure 4. *Cont.*

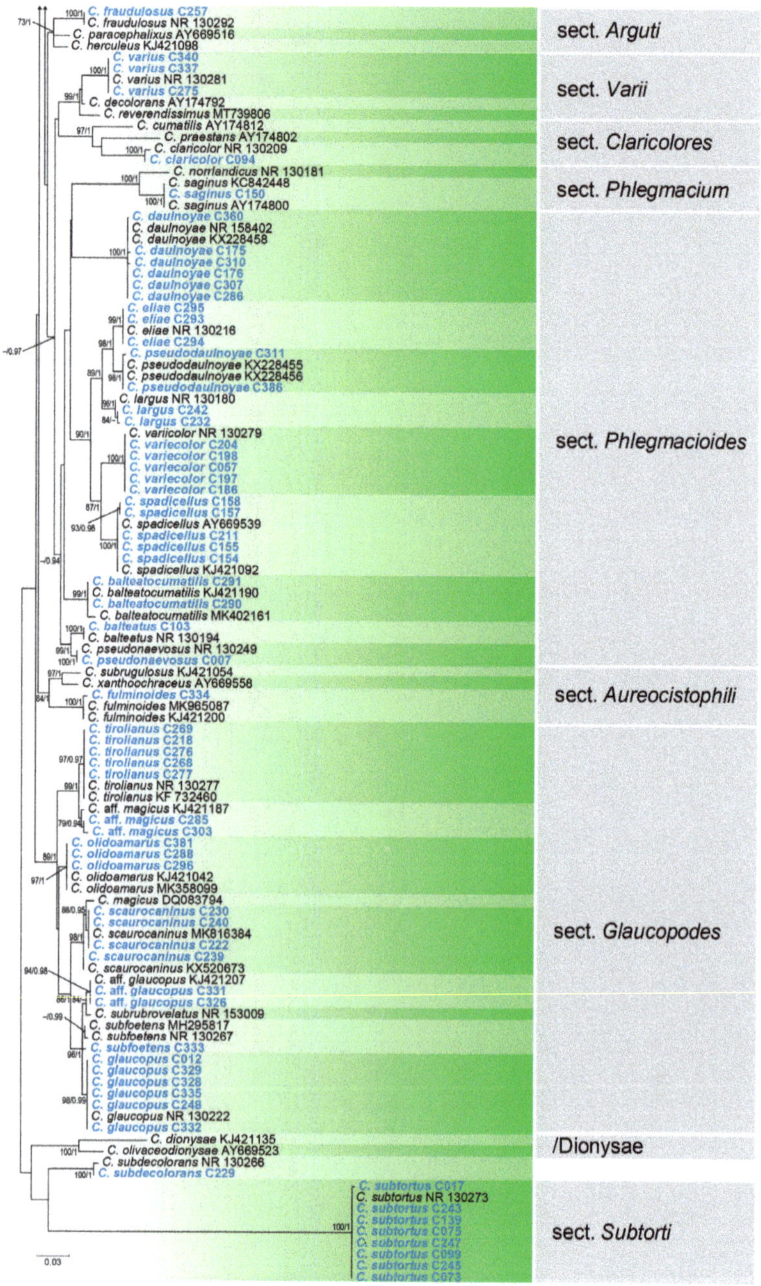

Figure 4. Maximum likelihood phylogenetic tree of the phlegmacioid sections/clades of the genus *Cortinarius* s.l. based on nrDNA ITS sequence analyses with gap coding. Sequences produced in this study are in blue and boldface and labelled with their voucher numbers. Sequences from public repositories are marked with their GenBank/UNITE accession numbers. ML bootstrap values of >70% as well as Bayesian posterior probabilities of >0.9 are placed above or below branches. The scale bar indicates 0.03 expected change per site per branch.

4. Discussion

In this study, we updated the checklist of the largest agaric genus, *Cortinarius* s.l., in Romania, employing nrDNA ITS barcoding. Macrofungi, including *Cortinarius*, were only studied using classical morphological methods in Romania to date; thus, our work presents the first biodiversity study using the combination of molecular phylogenetic and morphological methods in the country. Similar to the results of other national barcoding projects in Europe, e.g., in Norway [109–111], Finland [112], the Netherlands [113] or in Austria (https://www.abol.ac.at/en/project/higher_fungi/, accessed on 1 July 2022), our results showed a high number of previously not reported taxa at the country level. From the 109 *Cortinarius* s.l. species identified in the course of this study, 66 species are reported here as new to Romania, raising the known number of species up to ca. 190. This number is, however, lagging far behind that from Western and Northern European countries, where species observation activities and DNA barcoding campaigns and databases are in a more advanced stage compared to Romania or its neighboring countries, e.g., in Bulgaria or Hungary. For example, in Norway, the listed number of *Cortinarius* species is over 550 [114], while based on data from the older literature, the number of species in Bulgaria is 105 [115], and in Hungary, there are 169 species [116]; however, these numbers are hardly comparable with each other due to the different methods used in species identification (i.e., integrative taxonomy vs. morphological species recognition).

The majority of the species discovered as new to Romania belongs to phlegmacioid lineages/sections, especially to sect. *Calochroi* (nine species) as well as sect. *Glaucopodes* and sect. *Phlegmacioides* (six species in either section). Sect. *Dermocybe* is represented with 7 new species to the country, whereas 29 other sections/clades are represented by only 1–3 species in our dataset. Altogether, the 109 identified species in this study belong to 43 sections/clades. After actualizing the species list of genus *Cortinarius* s.l. with our new data, we assessed ca. 190 species now known in Romania. However, we are aware of the fact that among the previously published data, there could be wrongly identified and named collections; thus, without taxonomic revision of these materials, the correct number of *Cortinarius* species cannot be accurately established. Based on the various valuable habitats and nature types in the Carpathian Mountains and adjacent regions in Romania (out of which we only conducted samplings from the Apuseni Mts, and few sites in the Eastern Carpathians), we anticipate discovering an even larger diversity of *Cortinarius* s.l. species than that established in this work, when the sampling is extended to all areas with suitable habitats for *Cortinarius* in Romania. We also believe that other groups of macrofungi will be shown to be more species-rich when accurate investigations combining molecular techniques (e.g., DNA barcoding) together with morphological identification methods will be applied in the future.

Supplementary Materials: The following supporting information can be downloaded at: https://www.mdpi.com/article/10.3390/d15040553/s1, Table S1: *Cortinarius* sequences from public databases used in this study, Table S2: *Cortinarius* taxa published to date from Romania and identified with morphological methods. References [53,79,81,85,87,117–129] are cited in Supplementary Materials.

Author Contributions: Conceptualization, E.S., B.D. and L.K.; methodology, E.S. and B.D.; molecular analysis, E.S., B.D. and A.L.D., resources, L.K. and V.P., data curation, E.S. and B.D., writing—original draft preparation, E.S. and B.D., writing—review and editing, E.S., B.D., A.L.D., L.K. and V.P., supervision B.D. and L.K., funding acquisition L.K. and V.P. All authors have read and agreed to the published version of the manuscript.

Funding: This research was partially funded by the Collegium Talentum Program of Hungary. The work of B.D. was funded by the János Bolyai Research Scholarship of the Hungarian Academy of Sciences; the National Research, Development and Innovation Office of Hungary (OTKA FK-143061); and the ELTE Thematic Excellence Programme 2020 (TKP2020-IKA-05), financed by the National Research, Development and Innovation Office of Hungary. The support of the János Bolyai Research Scholarship of the Hungarian Academy of Sciences to V.P. is highly appreciated.

Institutional Review Board Statement: Not applicable.

Data Availability Statement: Data are contained within the article and Supplementary Materials. Some data can also be found in publicly available datasets: https://www.ncbi.nlm.nih.gov/; http://www.mycobank.org/; http://www.indexfungorum.org/, accessed on 2 March 2023.

Acknowledgments: We would like to thank Krisztina Babos, László Bartha, Gábor Bélfenyéri, Anna Dénes, Réka Erös, Kinga Fehér, László Gál, Boróka Jancsó, Kunigunda Macalik and Csilla Szabó for their help with the sample collection, Boróka Jancsó for her help with the microscopical spore photos, and Ágota Szabó for her help in recording habitat composition. We are grateful to the three anonymous reviewers (especially to one of them) for their valuable comments which greatly improved our manuscript.

Conflicts of Interest: The authors declare no conflict of interest.

References

1. Kirk, P.M.; Cannon, P.F.; Minter, D.W.; Stalpers, J.A. *Dictionary of the Fungi*, 10th ed.; CAB International: Wallingford, UK, 2008.
2. Roskov, Y.; Ower, G.; Orrell, T.; Nicolson, D.; Bailly, N.; Kirk, P.M.; Bourgoin, T.; DeWalt, R.E.; Decock, W.; Nieukerken, E.; et al. Species 2000 & ITIS Catalogue of Life, 2019 Annual Checklist, 2019, (2405-884X). Available online: http://www.catalogueoflife.org/annual-checklist/2019/ (accessed on 1 July 2022).
3. Liimatainen, K.; Niskanen, T.; Dima, B.; Kytövuori, I.; Ammirati, J.F.; Frøslev, T.G. The largest type study of *Agaricales* species to date: Bringing identification and nomenclature of *Phlegmacium* (*Cortinarius*) into the DNA era. *Pers. Mol. Phylogeny Evol. Fungi* **2014**, *33*, 98–140. [CrossRef] [PubMed]
4. Peintner, U.; Moncalvo, J.M.; Vilgalys, R. Toward a better understanding of the infrageneric relationships in *Cortinarius*. *Mycologia* **2004**, *96*, 4. [CrossRef]
5. Singer, R. *The Agaricales in Modern Taxonomy*; Koeltz Scientific Books: Koenigstein, Germany, 1986; pp. 859–889.
6. Peintner, U.; Bougher, N.L.; Catellano, M.A.; Moncalvo, J.M.; Moser, M.M.; Trappe, J.M.; Vilgalys, R. Multiple origins of sequestrate fungi related to *Cortinarius* (*Cortinariaceae*). *Am. J. Bot.* **2001**, *88*, 2168–2179. [CrossRef]
7. Fries, E. *Systema Mycologicum*; Lundae: Ex Officina Berlingiana: Uppsala, Sweden, 1821; Volume 1.
8. Fries, E. *Epicrisis Systematis Mycologici seu Synopsis Hymenomycetum*; Typographia Academica: Uppsala, Sweden, 1838.
9. Fries, E. *Monographia Cortinariorum Sueciae*; Excudebat Reg. Acad. Typograph: Uppsala, Sweden, 1851.
10. Henry, R. Étude de quelques cortinaires du groupe des Scauri. Deux espèces nouvelles. *Bull. De La Société Mycol. De Fr.* **1951**, *51*, 34–100.
11. Henry, R. Suite à l'étude des Cortinaires. *Bull. De La Société Mycol. De Fr.* **1958**, *74*, 365–422.
12. Henry, R. Les Cortinaires. *Bull. De La Société Mycol. De Fr.* **1981**, *97*, 157–279.
13. Henry, R. Suite à l'étude des Cortinaires. *Bull. De La Société Mycol. De Fr.* **1986**, *102*, 19–96.
14. Moser, M. *Die Röhrlinge und Blätterpilze*; Kleine Kryptogamenflora: Stuttgart, Germany, 1983; Volume IIb.
15. Moser, M. *Die Gattung Phlegmacium (Schleimköpfe). Die Pilze Mitteleuropas*; Julius Klinikhalt: Bad Heilbrunn, Germany, 1960; Volume IV.
16. Moser, M.M. *Keys to Agarics and Boleti: Polyporales, Boletales, Agaricales, Russulales*, 4th ed.; Mad River Pr Inc.: London, UK, 1985.
17. Orton, P.D. *The Genus Cortinarius I. Myxacium and Phlegmacium*; The Naturalist: Leeds, UK, 1955.
18. Velenovský, J. *České Houby*; Česká Botanická Společnost: Praha, Czech Republic, 1921; Volume 3, pp. 425–632.
19. Moënne-Loccoz, P.; Reumaux, P.; Bidaud, A. *Atlas des Cortinaires*; Éditions Fédération Mycologique Dauphiné-Savoie: Annecy, France, 1990; Volume I-1.
20. Moënne-Loccoz, P.; Reumaux, P.; Bidaud, A. *Atlas des Cortinaires*; Éditions Fédération Mycologique Dauphiné-Savoie: Annecy, France, 1990; Volume I-2.
21. Moënne-Loccoz, P.; Reumaux, P.; Bidaud, A. *Atlas des Cortinaires*; Éditions Fédération Mycologique Dauphiné-Savoie: Annecy, France, 1991; Volume II.
22. Bidaud, A.; Moënne-Loccoz, P.; Reumaux, P. *Atlas des Cortinaires*; Éditions Fédération Mycologique Dauphiné-Savoie: Annecy, France, 1991; Volume III.
23. Bidaud, A.; Moënne-Loccoz, P.; Reumaux, P.; Henry, R. *Atlas des Cortinaires*; Éditions Fédération Mycologique Dauphiné-Savoie: Annecy, France, 1992; Volume IV-1.
24. Bidaud, A.; Moënne-Loccoz, P.; Reumaux, P.; Henry, R. *Atlas des Cortinaires*; Éditions Fédération Mycologique Dauphiné-Savoie: Annecy, France, 1992; Volume IV-2.
25. Bidaud, A.; Moënne-Loccoz, P.; Reumaux, P.; Henry, R. *Atlas des Cortinaires*; Éditions Fédération Mycologique Dauphiné-Savoie: Annecy, France, 1994; Volume V-1.
26. Bidaud, A.; Moënne-Loccoz, P.; Reumaux, P.; Henry, R. *Atlas des Cortinaires*; Éditions Fédération Mycologique Dauphiné-Savoie: Annecy, France, 1994; Volume V-2.
27. Bidaud, A.; Moënne-Loccoz, P.; Reumaux, P. *Atlas des Cortinaires*; Éditions Fédération Mycologique Dauphiné-Savoie: La Roche-sur-Foron, France, 1995; Volume VI.
28. Bidaud, A.; Moënne-Loccoz, P.; Reumaux, P. *Atlas des Cortinaires*; Éditions Fédération Mycologique Dauphiné-Savoie: La Roche-sur-Foron, France, 1996; Volume VII.

29. Bidaud, A.; Moënne-Loccoz, P.; Reumaux, P. *Atlas des Cortinaires*; Éditions Fédération Mycologique Dauphiné-Savoie: La Roche-sur-Foron, France, 1997; Volume VIII.
30. Bidaud, A.; Moënne-Loccoz, P.; Reumaux, P. *Atlas des Cortinaires*; Éditions Fédération Mycologique Dauphiné-Savoie: Seynod, France, 1999; Volume IX.
31. Bidaud, A.; Moënne-Loccoz, P.; Reumaux, P. *Atlas des Cortinaires*; Éditions Fédération Mycologique Dauphiné-Savoie: Lyon, France, 2000; Volume X.
32. Bidaud, A.; Carteret, X.; Eyssartier, G.; Moënne-Loccoz, P.; Reumaux, P.; Carteret, X.; Eyssartier, G. *Atlas des Cortinaires*; Éditions Fédération Mycologique Dauphiné-Savoie: Marlioz, France, 2001; Volume XI.
33. Bidaud, A.; Carteret, X.; Eyssartier, G.; Moënne-Loccoz, P.; Reumaux, P. *Atlas des Cortinaires*; Éditions Fédération Mycologique Dauphiné-Savoie: Marlioz, France, 2002; Volume XII.
34. Bidaud, A.; Moënne-Loccoz, P.; Reumaux, P.; Carteret, X.; Eyssartier, G. *Atlas des Cortinaires*; Éditions Fédération Mycologique Dauphiné-Savoie: Marlioz, France, 2003; Volume XIII.
35. Bidaud, A.; Carteret, X.; Eyssartier, G.; Moënne-Loccoz, P.; Reumaux, P. *Atlas des Cortinaires*; Éditions Fédération Mycologique Dauphiné-Savoie: Marlioz, France, 2004; Volume XIV.
36. Bidaud, A.; Carteret, X.; Eyssartier, G.; Moënne-Loccoz, P.; Reumaux, P. *Atlas des Cortinaires*; Éditions Fédération Mycologique Dauphiné-Savoie: Marlioz, France, 2005; Volume XV.
37. Bidaud, A.; Moënne-Loccoz, P.; Reumaux, P.; Carteret, X.; Eyssartier, G. *Atlas des Cortinaires*; Éditions Fédération Mycologique Dauphiné-Savoie: Marlioz, France, 2006; Volume XVI.
38. Bidaud, A.; Moënne-Loccoz, P.; Reumaux, P.; Carteret, X.; Eyssartier, G. *Atlas des Cortinaires*; Éditions Fédération Mycologique Dauphiné-Savoie: Marlioz, France, 2008; Volume XVII.
39. Bidaud, A.; Moënne-Loccoz, P.; Reumaux, P.; Carteret, X. *Atlas des Cortinaires*; Éditions Fédération Mycologique Dauphiné-Savoie: Marlioz, France, 2009; Volume XVIII.
40. Bidaud, A.; Moënne-Loccoz, P.; Reumaux, P.; Carteret, X. *Atlas des Cortinaires*; Éditions Fédération Mycologique Dauphiné-Savoie: Marlioz, France, 2010; Volume XIX.
41. Bidaud, A.; Carteret, X.; Reumaux, P.; Moënne-Loccoz, P. *Atlas des Cortinaires*; Éditions Fédération Mycologique Dauphiné-Savoie: Marlioz, France, 2012; Volume XX.
42. Bidaud, A.; Carteret, X.; Reumaux, P.; Moënne-Loccoz, P.; Bellanger, J.M. *Atlas des Cortinaires*; Éditions Fédération Mycologique Dauphiné-Savoie: Marlioz, France, 2013; Volume XXI.
43. Bidaud, A.; Bellanger, J.M.; Carteret, X.; Reumaux, P.; Moënne-Loccoz, P. *Atlas des Cortinaires*; Éditions Fédération Mycologique Dauphiné-Savoie: Marlioz, France, 2014; Volume XXII.
44. Bidaud, A.; Bellanger, J.M.; Carteret, X.; Reumaux, P.; Moënne-Loccoz, P. *Atlas des Cortinaires*; Éditions Fédération Mycologique Dauphiné-Savoie: Marlioz, France, 2015; Volume XXIII.
45. Bidaud, A.; Bellanger, J.M.; Carteret, X.; Reumaux, P.; Moënne-Loccoz, P. *Atlas des Cortinaires*; Éditions Fédération Mycologique Dauphiné-Savoie: Marlioz, France, 2017; Volume XXIV.
46. Brandrud, T.E.; Lindström, H.; Marklund, H.; Melot, J.; Muskos, S. *Cortinarius, Flora Photographica*; Cortinarius HB: Matfors, Sweden, 1990; Volume 1.
47. Brandrud, T.E.; Lindström, H.; Marklund, H.; Melot, J.; Muskos, S. *Cortinarius, Flora Photographica*; Cortinarius HB: Matfors, Sweden, 1992; Volume 2.
48. Brandrud, T.E.; Lindström, H.; Marklund, H.; Melot, J.; Muskos, S. *Cortinarius, Flora Photographica*; Cortinarius HB: Matfors, Sweden, 1995; Volume 3.
49. Brandrud, T.E.; Lindström, H.; Marklund, H.; Melot, J.; Muskos, S. *Cortinarius, Flora Photographica*; Cortinarius HB: Matfors, Sweden, 1998; Volume 4.
50. Brandrud, T.E. *Cortinarius* subgenus *Phlegmacium* section *Phlegmacium* in Europe: Descriptive part. *Edinb. J. Bot.* **1996**, *53*, 331–400. [CrossRef]
51. Brandrud, T.E. *Cortinarius* subgenus *Phlegmacium* section *Phlegmacioides* (=*Variecolores*) in Europe. *Edinb. J. Bot.* **1998**, *55*, 65–156. [CrossRef]
52. Brandrud, T.E.; Dima, B. Cortinarius subgenus Phlegmacium section Multiformes in Europe. *J. Des. J.E.C.* **2014**, *16*, 162–199.
53. Brandrud, T.E.; Scmidt-Stohn, G.; Liimatainen, K.; Niskanen, T.; Frøslev, T.G.; Soop, K.; Bojantchev, D.; Kytövuori, I.; Jeppesen, T.S.; Bellù, F.; et al. *Cortinarius* sect. *Riederi: Taxonomy and phylogeny of the new section with European and North American distribution. Mycol. Prog.* **2018**, *17*, 1323–1354. [CrossRef]
54. Brandrud, T.E.; Frøslev, T.G.; Dima, B. Rare, whitish–pale ochre *Cortinarius* species of sect. *Calochroi from calcareous Tilia forests in South East Norway. Agarica* **2018**, *38*, 3–20.
55. Liimatainen, K. Towards a Better Understanding of the Systematics and Diversity of *Cortinarius*, with an Emphasis on Species Growing in Boreal and Temperate Zones of Europe and North America. Ph.D. Thesis, Faculty of Biological and Environmental Sciences of the University of Helsinki, Helsinki, Finland, 2013.
56. Liimatainen, K.; Niskanen, T.; Dima, B.; Ammirati, J.F.; Kirk, P.M.; Kytövuori, I. Mission impossible completed: Unlocking the nomenclature of the largest and most complicated subgenus of *Cortinarius*, *Telamonia*. *Fungal Divers.* **2020**, *104*, 291–331. [CrossRef]

57. Niskanen, T.; Liimatainen, K.; Mahiques, R.; Ballarà, J.; Kytövuori, I. *Cortinarius badiolaevis*, a new conifer-associated, darkening species in the subgenus *Telamonia* (Basidiomycota, Agaricales). *Mycol. Prog.* **2011**, *10*, 101–105. [CrossRef]
58. Niskanen, T.; Kytövuori, I.; Liimatainen, K. *Cortinarius* sect. *Armillati in northern Europe. Mycologia* **2011**, *103*, 1080–1101. [CrossRef]
59. Niskanen, T.; Kytövuori, I.; Liimatainen, K.; Lindström, H. The species of *Cortinarius*, section *Bovini*, associated with conifers in northern Europe. *Mycologia* **2013**, *105*, 977–993. [CrossRef]
60. Niskanen, T.; Laine, S.; Liimatainen, K.; Kytövuori, I. *Cortinarius sanguineus* and equally red species in Europe with an emphasis on Northern European material. *Mycologia* **2012**, *104*, 242–253. [CrossRef]
61. Niskanen, T.; Liimatainen, K.; Kytövuori, I. Two new species in *Cortinarius* subgenus *Telamonia*, *Cortinarius brunneifolius* and *C. leiocastaneus*, from Fennoscandia (*Basidiomycota*, *Agaricales*). *Mycol. Prog.* **2008**, *7*, 239–247. [CrossRef]
62. Niskanen, T.; Liimatainen, K.; Kytövuori, I.; Ammirati, J.F. New *Cortinarius* species from conifer-dominated forests of North America and Europe. *Botany* **2012**, *90*, 743–754. [CrossRef]
63. Niskanen, T.; Liimatainen, K.; Kytövuori, I.; Lindström, H.; Dentinger, B.T.M.; Ammirati, J.F. *Cortinarius* subgenus *Callistei* in North America and Europe-type studies, diversity, and distribution of species. *Mycologia* **2016**, *108*, 1018–1027. [CrossRef]
64. Dima, B.; Liimatainen, K.; Niskanen, T.; Kytövuori, I.; Bojantchev, D. Two new species of *Cortinarius*, subgenus *Telamonia*, sections *Colymbadini* and *Uracei*, from Europe. *Mycol. Prog.* **2014**, *13*, 867–879. [CrossRef]
65. Dima, B.; Lindström, H.; Liimatainen, K.; Olson, Å.; Soop, K.; Kytövuori, I.; Dahlberg, A.; Niskanen, T. Typification of Friesian names in *Cortinarius* sections *Anomali*, *Spilomei*, and *Bolares*, and description of two new species from northern Europe. *Mycol. Prog.* **2016**, *15*, 903–919. [CrossRef]
66. Frøslev, T.G.; Jeppesen, T.S.; Dima, B. *Cortinarius koldingensis*—A new species of *Cortinarius*, subgenus *Phlegmacium* related to *Cortinarius sulfurinus*. *Mycol. Prog.* **2015**, *14*, 73. [CrossRef]
67. Frøslev, T.G.; Jeppesen, T.S.; Læssøe, T. Seven new calochroid and fulvoid species of *Cortinarius*. *Mycol. Res.* **2006**, *110*, 1046–1058. [CrossRef]
68. Frøslev, T.G.; Jeppesen, T.S.; Læssøe, T.; Kjøller, R. Molecular phylogenetics and delimitation of species in *Cortinarius* section *Calochroi* (*Basidiomycota*, *Agaricales*) in Europe. *Mol. Phylogenetics Evol.* **2007**, *44*, 217–227. [CrossRef] [PubMed]
69. Garnica, S.; Weiß, M.; Oertel, B.; Oberwinkler, F. Phylogenetic relationships of European *Phlegmacium* species (*Cortinarius*, *Agaricales*). *Mycologia* **2003**, *95*, 1155–1170. [CrossRef] [PubMed]
70. Garnica, S.; Weiß, M.; Oertel, B.; Oberwinkler, F. A framework for a phylogenetic classification in the genus *Cortinarius* (*Basidiomycota*, *Agaricales*) derived from morphological and molecular data. *Can. J. Bot.* **2005**, *83*, 1457–1477. [CrossRef]
71. Bidaud, A.; Loizides, M.; Armada, F.; Reyes, J.D.D.; Carteret, X.; Corriol, G.; Consiglio, G.; Reumaux, P.; Bellanger, J.-M. *Cortinarius* subgenus *Leprocybe* in Europe: Expanded Sanger and Next Generation Sequencing unveil unexpected diversity in the Mediterranean. *Persoonia* **2021**, *46*, 188–215. [CrossRef]
72. Garrido-Benavent, I.; Ballara, J.; Mahiques, R. New insights into subg. *Phlegmacium* sect. *Calochroi*: Adding morphological and molecular data from Mediterranean representatives, with special regard to *Cortinarius prasinus*, *C. natalis* and *C. murellensis* species complexes. *J. Des. J.E.C.* **2015**, *17*, 38–78.
73. Suárez-Santiago, V.N.; Ortega, A.; Peintner, U.; López-Flores, I. Study on *Cortinarius* subgenus *Telamonia* section *Hydrocybe* in Europe, with especial emphasis on Mediterranean taxa. *Mycol. Res.* **2009**, *113*, 1070–1090. [CrossRef]
74. Garnica, S.; Schön, M.E.; Abarenkov, K.; Riess, K.; Liimatainen, K.; Niskanen, T.; Dima, B.; Soop, K.; Frøslev, T.G.; Jeppesen, T.S.; et al. Determining threshold values for barcoding fungi: Lessons from *Cortinarius* (*Basidiomycota*), a highly diverse and widespread ectomycorrhizal genus. *FEMS Microbiol. Ecol.* **2016**, *92*, fiw045. [CrossRef]
75. Soop, K.; Dima, B.; Cooper, J.A.; Park, D.; Oertel, B. A phylogenetic approach to a global supraspecific taxonomy of *Cortinarius* (*Agaricales*) with an emphasis on the southern mycota. *Persoonia* **2019**, *42*, 261–290. [CrossRef]
76. Liimatainen, K.; Kim, J.T.; Pokorny, L.; Kirk, P.M.; Dentinger, B.; Niskanen, T. Taming the beast: A revised classification of *Cortinariaceae* based on genomic data. *Fungal Divers.* **2022**, *112*, 89–170. [CrossRef]
77. Bontea, V. *Ciuperci Parazite și Saprofite din România*; Academia Republicii Socialiste Romania: Bucharest, Romania, 1985; Volume 1.
78. Bontea, V. *Ciuperci Parazite și Saprofite din România*; Academia Republicii Socialiste Romania: Bucharest, Romania, 1986.
79. Bellù, F.; Zsigmond, G.Y.; Szász, B. Liste der Cortinarienfunde während den Journées européennes du Cortinaire in Bálványos. *J. Des. J. E. C.* **2018**, *21*, 80–81.
80. Bîrsan, C.; Tănase, C.; Mardari, C. Variation of Macromycetes Species Composition in Two Forest Habitats From Giumalău Massif (Eastern Carpathians, Romania). *J. Plant Dev.* **2013**, *20*, 79–103.
81. Bucșa, L. Macromycetes of the Breite Nature Reserve of ancient oaks (Transylvania, Romania). *Transylv. Rev. Sys-Temat. Ecol. Res.* **2007**, *4*, 33–42.
82. Chinan, V.C.; Mânzu, C. Macrofungal diversity of a peat bog from Dorna Depression (eastern Carpathians, Romania). *An. Științifice Ale Univ. Al. I. Cuza Iași S. II. A Biol. Veg.* **2014**, *60*, 43–52.
83. Chinan, V.C.; Tanase, C. Coenological observations referring to macromycetes from peat bogs situated in eastern carpathians (Romania). In Proceedings of the 8th International Scientific Conference on Modern Management of Mine Producing, Geology and Environmental Protection, Albena, Bulgaria, 16–20 June 2008.
84. Ciortan, I. Contributions to the mycobiota knowledge of spruce forests from Obârşia Lotrului health resort (Romania). *For. Biotechnol.* **2013**, *17*, 16–21.
85. Ciortan, I. The taxonomic diversity of the macromycetes from Căpăţânii mountains (Romania). *For. Biotechnol.* **2013**, *17*, 41–50.

86. Elekes, C.C.; Busuioc, G. The mycoremediation of metals polluted soils using wild growing species of mushrooms. In Proceedings of the International Conference on Engineering Education and International Conference on Education and Educational Technologies-Proceedings, Changzhou, China, 18–19 December 2010.
87. Elekes, C.C.; Busuioc, G.; Dumitriu, I. Bioaccumulation and Translocation Factors of Some Wild Growing Mushrooms Species of Cortinarius Genus. *Adv. Biomed. Res.* **2010**, *61*, 464–468.
88. Oross, K.; Benedek, L.; Pál-Fám, F. Macrofungi from Apuseni Mountains, Transylvania, Romania. *Acta Biol. Plant. Agriensis* **2017**, *5*, 38. [CrossRef]
89. Pál-Fám, F. Adatok a Baróti-hegység nagygombáinak ismeretéhez. In *Acta: A Csíki Székely Múzeum és a Székely Nemzeti Múzeum évkönyve*; Székely Nemzeti Múzeum: Sfântu Gheorghe, Romania, 1997; pp. 61–68.
90. Garnica, S.; Spahn, P.; Oertel, B.; Ammirati, J.; Oberwinkler, F. Tracking the evolutionary history of *Cortinarius* species in section *Calochroi*, with transoceanic disjunct distributions. *BMC Evol. Biol.* **2011**, *11*, 213. [CrossRef]
91. Schoch, C.L.; Seifert, K.A.; Huhndorf, S.; Robert, V.; Spouge, J.L.; Levesque, C.A.; Chen, W.; Fungal Barcoding Consortium. Nuclear ribosomal internal transcribed spacer (ITS) region as a universal DNA barcode marker for Fungi. *Proc. Natl. Acad. Sci. USA* **2012**, *109*, 6241–6246. [CrossRef] [PubMed]
92. Nilsson, R.H.; Ryberg, M.; Abarenkov, K.; Sjökvist, E.; Kristiansson, E. The ITS region as a target for characterization of fungal communities using emerging sequencing technologies. *FEMS Microbiol. Lett.* **2009**, *296*, 97–101. [CrossRef] [PubMed]
93. Bellemain, E.; Carlsen, T.; Brochmann, C.; Coissac, E.; Taberlet, P.; Kauserud, H. ITS as an environmental DNA barcode for fungi: An in silico approach reveals potential PCR biases. *BMC Microbiol.* **2010**, *10*, 189. [CrossRef]
94. White, T.J.; Bruns, T.; Lee, S.; Taylor, J. Amplification and Direct Sequencing of Fungal Ribosomal RNA Genes for Phylogenetics. In *PCR Protocols*; Innis, M.A., Gelfand, D.H., Sninsky, J.J., White, T.J., Eds.; Academic Press, Inc: Cambridge, UK, 1990; pp. 315–322. [CrossRef]
95. Kern, Z.; Popa, I. Climate–growth relationship of tree species from a mixed stand of Apuseni Mts., Romania. *Dendrochronologia* **2007**, *24*, 109–115. [CrossRef]
96. Buza, M.; Dimen, L.; Pop, G.; Turnock, D. Environmental protection in the Apuseni Mountains: The role of Environmental Non-Governmental Organisation (ENGOs). *GeoJournal* **2001**, *54*, 631–653. [CrossRef]
97. Pop, A.A.; Pop, V.V.; Csuzdi, C.S. Significance of the Apuseni Mountains (the Carpathians) in the origin and distribution of Central European earthworm fauna (Oligochaeta: Lumbricidae). *Zool. Middle East Suppl.* **2010**, *2*, 89–110. [CrossRef]
98. Leonard, L.M. Melzer's, Lugol's or Iodine for Identification of white-spored Agaricales? *McIlvainea* **2006**, *16*, 43–51.
99. Bonfield, J.; Beal, K.; Jordan, M.; Cheng, Y.; Staden, R. *The Staden Package Manual*; Medical Research Council, Laboratory of Molecular Biology: Cambridge, UK, 2002; pp. 1–514.
100. Altschul, S.F.; Gish, W.; Miller, W.; Myers, E.W.; Lipman, D.J. Basic Local Alignment Search Tool. *J. Mol. Biol.* **1990**, *215*, 403–410. [CrossRef]
101. Hall, A. BioEdit: A User-Friendly Biological Sequence Alignment Editor and Analysis Program for Windows 95/98/NT. *Nucleic Acids Symp. Ser.* **1999**, *41*, 95–98.
102. Katoh, K.; Misawa, K.; Kuma, K.; Miyata, T. MAFFT: A novel method for rapid multiple sequence alignment based on fast Fourier transform. *Nucleic Acids Res.* **2002**, *30*, 3059–3066. [CrossRef] [PubMed]
103. Borchsenius, F. FastGap. Department of Biosciences, Aarhus University, Denmark. 2009. Available online: http://www.aubot.dk/FastGap_home.htm (accessed on 13 March 2023).
104. Gouy, M.; Guindon, S.; Gascuel, O. SeaView version 4: A multiplatform graphical user interface for sequence alignment and phylogenetic tree building. *Mol. Biol. Evol.* **2010**, *27*, 221–224. [CrossRef] [PubMed]
105. Silvestro, D.; Michalak, I. RaxmlGUI: A graphical front-end for RAxML. *Org. Divers. Evol.* **2012**, *12*, 335–337. [CrossRef]
106. Ronquist, F.; Teslenko, M.; Van der Mark, P.; Ayres, D.L.; Darling, A.; Höhna, S.; Larget, B.; Liu, L.; Suchard, M.A.; Huelsenbeck, J.P. MrBayes 3.2: Efficient bayesian phylogenetic inference and model choice across a large model space. *Syst. Biol.* **2012**, *61*, 539–542. [CrossRef] [PubMed]
107. Rambault, A. FigTree v1.4.2, A Graphical Viewer of Phylogenetic Trees. 2014. Available online: http://tree.bio.ed.ac.uk/software/figtree/ (accessed on 13 March 2023).
108. Kumar, S.; Stecher, G.; Tamura. MEGA7: Molecular Evolutionary Genetics Analysis Version 7.0 for Bigger Datasets. *Mol. Biol. Evol.* **2016**, *33*, 1870–1874. [CrossRef]
109. Brandrud, T.E.; Bendiksen, E.; Dima, B. Some new and little known telamonioid *Cortinarius* species from Norway. *Agarica* **2015**, *36*, 11–42.
110. Brandrud, T.E.; Bendiksen, E.; Jordal, J.B.; Weholt, Ø.; Eidissen, S.E.; Lorås, J.; Dima, B.; Noordeloos, M.E. *Entoloma* species of the rhodopolioid clade (subgenus *Entoloma*; *Tricholomatinae, Basidiomycota*) in Norway. *Agarica* **2018**, *38*, 21–46.
111. Brandrud, T.E.; Bendiksen, E.; Jordal, J.B.; Weholt, Ø.; Dima, B.; Morozova, O.; Noordeloos, M.E. On some *Entoloma* species (*Tricholomatinae, Basidiomycota*) little known or new to Norway. *Agarica* **2019**, *39*, 31–52.

112. von Bonsdorff, T.; Niskanen, T.; Liimatainen, K.; Kytövuori, I.; Huhtinen, S.; Vauras, J.; Höijer, P.; Kekki, T.; Lahti, M.; Puolasmaa, A.; et al. New national and regional biological records for Finland 8. Contributions to agaricoid, gastroid and ascomycetoid taxa of fungi 5. *Memo. Soc. Pro Fauna Et Flora Fenn.* **2016**, *92*, 120–128. Available online: http://ojs.tsv.fi/index.php/msff/article/view/60364/21223 (accessed on 13 March 2023).
113. Noordeloos, M.E.; Nuytinck, J.; Dima, B. Some new results of the ITS barcoding project—About *Boletales*, *Paxillus* and *Entoloma*. *Coolia* **2019**, *62*, 127–140.
114. Available online: https://artsdatabanken.no/ (accessed on 2 April 2023).
115. Dentchev, C.M.; Assyov, B. Checklist of the larger basidiomycetes in Bulgaria. *Mycotaxon* **2010**, *111*, 279–282. [CrossRef]
116. Rimóczi, I. Gefährdete Cortinariaceae-Taxa in Ungarn. *J. Des. J. E. C.* **1998**, 16–27.
117. Eliade, E. Conspectul Macromicetelor din România. *Acta Bot. Horti București* **1965**, *5*, 185–324.
118. Chifu, T.; Vițălariu, G. Contribuții la cunoașterea macromicetelor din bazinul Crasnei. *Stud. Și Com. Bacău* **1971**, 637–640.
119. Pop, A.; Tănase, C.; Negrean, G.; Buscă, L.; Toma, D.; Robu, T. Contribuţii la studiul macromicetelor din Masivul Rarău. *Bul. Grădinii Bot. Iași* **1999**, *8*, 37–44.
120. Todorescu, V. Macromicetele din Munții Ciucaș. *An. Univ. București Biol. Veg.* **1972**, *21*, 159–166.
121. Popovici, A. Contribution à la flore micologique de la Romanie. *Ann. Sci. Univ. De Jassy* **1903**, *2*, 31–44.
122. Czihac, J.; Szabó, J. Heil- und Nahrungsmittel, Farbstoffe, Nutz-und Hausgeräthe welche Ost-Romänien, Moldauer und Valachen aus dem Pflanzenreiche gewinnen. *Flora Oder Allg. Bot. Ztg.* **1975**, *21*, 308–315.
123. Tănase, C.; Pop, A. Red List of Romanian Macrofungi Species. In *Bioplatform-Romanian National Platform for Biodiversity*; Editura Academiei Române: Bucharest, Romania, 2001; pp. 101–107.
124. László, K. Noi contribuții la cunoașterea macromicetelor din R.S. România. Muz. Sf. Gheorghe. *Aluta* **1972**, 41–60.
125. Tănase, C.; Chifu, T. Biodiversitatea speciilor de macromicete în păduri din Bazinul Superior al Râului Jijia. *Bul. Grădinii Bot. Iaşi* **2000**, *9*, 79–82.
126. Sălăgeanu, A. Macromicete din Grădina Botanică a Universității din Cluj. In *Contribuții Botanice*; Universitatea "Babeș-Bolyai": Cluj, Romania, 1966; pp. 95–108.
127. Lupoi, A. Materiale pentru flora micologică a Munților Lăpuş. In *Contribuții Botanice*; Universitatea "Babeș-Bolyai": Cluj, Romania, 1965; pp. 71–74.
128. Bohlin, A.; Senn-Irlet, B. (Eds.) *Secretary Newsletter*; European Council for the Conservation of Fungi: Brussels, Belgium, 13 December 2003. Available online: http://www.eccf.eu/newsletter13.pdf (accessed on 13 March 2023).
129. Chinan, V. Macrofungi from "Tinovul de la Româneşti" peat bog (Dornelor Depression, Romania). *AAB Bioflux* **2010**, *2*, 65–70.

diversity

Article

Molecular Identification of an Invasive *Sarotherodon* Species from the Atchakpa Freshwater Reservoir (Ouémé River Basin, Benin) and Comparison within *S. melanotheron* Using COI Markers

Rodrigue Orobiyi Edéya Pèlèbè [1,2,*], Ibrahim Imorou Toko [1], Erik Verheyen [3,4] and Maarten Van Steenberge [3]

1 Research Laboratory in Aquaculture and Aquatic Ecotoxicology (LaRAEAq), Faculty of Agronomy (FA), University of Parakou (UP), P.O. Box 61 Parakou-University, Parakou, Benin; iimorou_toko@hotmail.com
2 World Bank Africa Centre of Excellence in Coastal Resilience (ACECoR), University of Cape Coast (UCC), Ground Floor C. A. Ackah Lecture Theatre Complex, Cape Coast, Ghana
3 Operational Directorate (OD) Taxonomy and Phylogeny, Royal Belgian Institute for Natural Sciences (RBINS), Vautierstraat 29, 1000 Brussels, Belgium; everheyen@naturalsciences.be (E.V.); mvansteenberge@naturalsciences.be (M.V.S.)
4 Department Biology, Evolutionary Ecology, University of Antwerp, Universiteitsplein 1, 2610 Antwerpen, Belgium
* Correspondence: peleberodrigue@yahoo.fr or rodrigue.pelebe@ucc.edu.gh; Tel.: +229-96-97-57-62 or +233-(0)-055-242-8692

Citation: Pèlèbè, R.O.E.; Imorou Toko, I.; Verheyen, E.; Van Steenberge, M. Molecular Identification of an Invasive *Sarotherodon* Species from the Atchakpa Freshwater Reservoir (Ouémé River Basin, Benin) and Comparison within *S. melanotheron* Using COI Markers. *Diversity* **2021**, *13*, 297. https://doi.org/10.3390/d13070297

Academic Editors: Stephan Koblmüller and Michael Wink

Received: 14 May 2021
Accepted: 29 June 2021
Published: 30 June 2021

Abstract: The Atchakpa freshwater reservoir (Ouémé Basin, Benin) was found to harbour an unexpected population of a cichlid species that was presumed to be *Sarotherodon melanotheron*. This species became dominant in the reservoir and became the main fisheries target species. We applied DNA barcoding to identify this population. Besides specimens from the reservoir, we also sequenced *S. melanotheron* from its native range in Benin at the lower Ouémé and Sô Rivers, and from Lake Nokoué, and Porto-Novo Lagoon. High sequence similarity indicated that all specimens were conspecific. Hence, we cannot exclude that a natural range extension led to the presence of the species in the reservoir. A comparison with sequences from NCBI GenBank confirmed that all samples belonged to the subspecies *S. m. melanotheron*, which is native to Benin. This comparison also showed that this subspecies was previously introduced in the Philippines. We call for further studies to investigate the socioeconomic, ecological and environmental impacts of the species in the Atchakpa reservoir.

Keywords: cytochrome c oxidase subunit I; black-chin tilapia; DNA barcoding; Savè; species introductions

1. Introduction

The black-chin tilapia, *Sarotherodon melanotheron* (Rüppell, 1852), is an estuarine species endemic to West and Central Africa [1–3] that is well known from the brackish waters of southern Benin. In the Ouémé River basin, its limit of frequent occurrence is the village of Agonlin-Lowé (Commune of Adjohoun, department of Ouémé) [4]. This locality is considered the boundary between the continental and the coastal domains in Benin [4] (Figure 1). However, rare specimens of *S. melanotheron* have been found upstream of Agonlin-Lowé in southern Benin (rare occurrences at the Ouémé River at Toué, Commune of Covè, department of Zou [4], at the Hlan River at Lokoli, Commune of Zogbodomey, department of Zou [5] and in the Lake Hlan at Kpomè, Commune of Toffo, department of Atlantic [6]) (Figure 1).

A large population of presumed *S. melanotheron* was discovered prior to 2001 in the Atchakpa freshwater reservoir (SUCOBE dam built in 1982), which is located in central Benin, about 230 km north of Agonlin-Lowé [4,7]. This population, which was only morphologically identified, thrived in the landlocked reservoir. Despite its proximity to the main course of the Ouémé River, the reservoir has no direct connection with the river (Figure 2) although, during severe dry seasons, seasonal contact with river water can occur through pumping. *S. melanotheron* is quasi-inexistent at the Ouémé River near the reservoir as only one individual was captured during three years of intensive sampling from 1998 to 2001 [4]. Moreover, the Atchakpa fisheries management committee has, in recent years, not reported any catches of this species in the Ouémé River near the reservoir. If the morphological identifications of [4,7] are correct, this population is isolated from the rest of its natural geographic distribution in Benin.

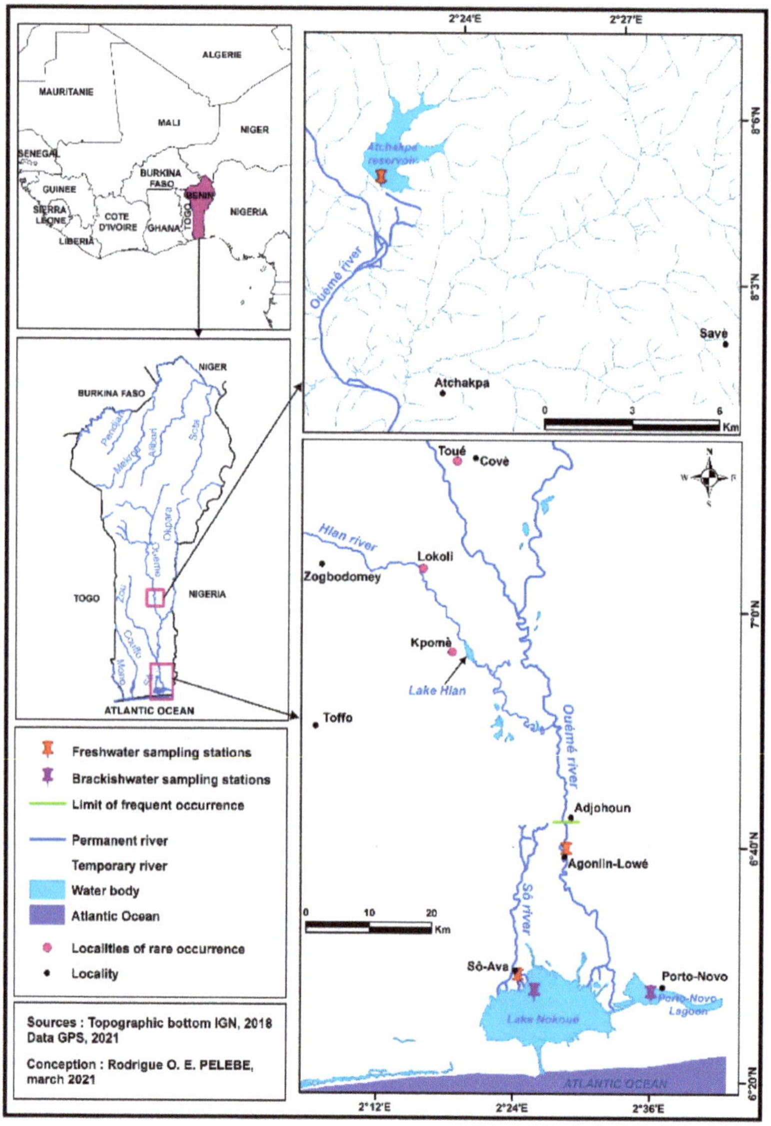

Figure 1. Hydrographic map of Benin showing the Ouémé River basin, with the location of the Atchakpa reservoir, the sampling stations, the limit of frequent occurrence and the localities of rare occurrences of *S. melanotheron*.

Figure 2. Satellite image highlighting the absence of direct communication between the main course of the Ouémé River and the Atchakpa freshwater reservoir (source: Google maps).

The fisheries in the Atchakpa reservoir ensure the supply of fish for the city of Savè. Currently, the population morphologically identified as *S. melanotheron* dominates the catches of local fishermen in terms of abundance. According to [8], *S. melanotheron* represents 72% (numerical abundance) of the fish caught at the Atchakpa reservoir and sold at the fish market in Savè. It thus surpasses the Nile tilapia *Oreochromis niloticus* (Linnaeus, 1758), which was introduced into the reservoir through aquaculture (direct restocking in 1985 and cage culture in 2014). Nile tilapia is currently rare whereas it was very common in the 1990s.

Sarotherodon melanotheron is known to be invasive, i.e., it has the potential to spread widely, reproduce rapidly and persist in a new environment with a large range of impacts [9,10]. Knowing the correct identity of an invasive species is an important prerequisite for its efficient management. The identification of invasive cichlids using only metric and meristic parameters requires caution due to their mega-diversity and the widespread introduction of alien species [11,12]. Therefore, molecular evidence is often required to provide additional verification. Hence, we chose to, for the first time in Beninese fisheries, apply a DNA barcoding approach. DNA barcoding is a molecular technique for fast and accurate determination of species identity [13]. In animals, DNA barcoding is usually performed by sequencing a fragment of the mitochondrial cytochrome c oxidase subunit I gene, COI (barcode), of the specimen under investigation and its comparison with a reference library of barcode sequences [14,15].

Sarotherodon melanotheron contains substantial intra-specific geographic structure, with three subspecies: *S. melanotheron melanotheron* Ruppell, 1852, *S. m. heudelotii* (Dumeril, 1859), *S. m. leonensis* (Thys van den Audenaerde, 1971) currently recognised [16]. The latter subspecies contains the synonym *S. m. paludinosus* (Trewavas, 1983) [16,17]. Additionally, *S. melanotheron* is highly similar to its sister species *S. nigripinnis* (Guichenot, 1861). Although the subspecies are difficult to distinguish on morphological grounds, they can be delineated with mitochondrial markers [16,17]. A molecular approach could hence allow us to confirm the subspecific nature of the *Sarotherodon* population, along with the implied geographic affinities. According to the management committee of the Atchakpa fisheries, no species of *Sarotherodon* has been stocked in the reservoir. However, it remains to be verified whether the population in the reservoir belongs to the nearest strain from the Ouémé, or whether it has been introduced from elsewhere.

Tilapia species, including *S. melanotheron*, have been extensively introduced outside of their native range and often established feral populations. This also holds for *S. melanotheron*, which is, among others, found in the Philippines [9,18], Florida [19] and Hawaii [20]. As most tilapia introductions have been poorly documented, the correct identity of feral populations often remains unknown. DNA barcoding can aid in discovering the origin of introduced populations. However, as African freshwater fishes are underrepresented in barcoding libraries [21], there is a need to barcode specimens from the species' native range.

The aims of this study are twofold: (i) the molecular identification of the invasive cichlid species from the Atchakpa reservoir and (ii) constructing a databank of DNA barcodes of *S. melanotheron* from its range in the Ouémé River in Benin, and comparing these sequences with previously published sequences of the same species in GenBank.

2. Materials and Methods

Sampling took place at five sites in the Ouémé River in Benin (Figure 1). Besides the Atchakpa reservoir, we sampled in two brackish waters: Lake Nokoué and Porto-Novo Lagoon, and two freshwaters: Agonlin-Lowé (Ouémé River) and Sô-Ava (Sô River), where *S. melanotheron* is native. Sampling took place in January 2020 and specimens were identified using the identification key of [22]. At each site, twenty-two specimens of *S. melanotheron* were collected with the help of local fishermen. We also sampled at the localities of rare occurrence (Toué, Lokoli and Kpomè, see Figure 1), but failed to collect any specimens of *S. melanotheron* there.

Caudal fin clips were sampled and put into 1.5 mL sterile Eppendorf tubes containing pure ethanol as a preservation medium. Four specimens per site were fixed in formalin and deposited at the Royal Belgian Institute for Natural Sciences (RBINS) under collection numbers: Atchakpa water reservoir (25907, 25908, 25909, and 25910), Sô-Ava (Sô River) (25923, 25924, 25925, and 25926), Agonlin-Lowé (Ouémé River) (25911, 25912, 25913, and 25914), Porto-Novo Lagoon (25915, 25916, 25917, and 25918) and Lake Nokoué (25919, 25920, 25921, and 25922). Fin clips were stored at -20°C at the Molecular Systematics Laboratory of the RBINS.

Total genomic DNA was extracted using NucleoSpin® Tissue Kit following the instructions of the manufacturer (Macherey-Nagel, Düren, Germany). The mitochondrial COI gene region for each of five specimens per locality, except only four for Lake Nokoué, was amplified by polymerase chain reaction (PCR) using the 'Fish Cocktail': an M13 tailed primer combination of VF2_t1, FishF2_t1, FishR2_t1 and VR1d_t1 [23]. A standard 25.2 µL PCR mix consisted of 2.5 µL PCR buffer (10×); 2.5 µL dNTP (2 mM); 1.25 µL 'Fish Cocktail' (2 µM); 0.2 µL Taq DNA Polymerase (5 units per µL); 16.75 µL mQ-H2O and 2.0 µL of the extracted DNA sample. The PCR profile was 3 min at 94°C, followed by 35 cycles of 40 s at 94 °C, 40 s at 52 °C, and 1 min at 72 °C, plus a final extension of 10 min at 72 °C. Amplicons were examined with 1% agarose gel electrophoresis. Products were purified with ExoSAP-IT™ PCR Product Cleanup Reagent and bidirectionally sequenced using BigDye Terminator v.3.1. Fragments were analysed on an ABI 3130 capillary sequencer. Forward and reverse sequences were edited, used to build consensus sequences, visually checked in CodonCode Aligner 5.1.4 (CodonCode Corporation) and aligned in MEGA 5.2 using Muscle Alignment [24]. Sequences were compared to the database of NCBI GenBank using nucleotide BLAST (Basic Local Alignment Search Tool). Based on earlier studies, the sequence similarity of at least 98% was used to separate conspecific from heterospecifics [25,26].

We downloaded all 39 COI sequences of *S. melanotheron* available on GenBank (Table 1) and analysed them together with our 17 newly generated sequences (Dataset S1). We constructed a minimum spanning network using PopART [27]. Additionally, a neighbour-joining (NJ) tree [28] based on Kimura-2-Parameter (K2P) distances and using *S. galileaus*

(KY84676) as an outgroup was constructed using MEGA 5.2. Nodal support was assessed with bootstrapping, using 1000 replicates. Kimura-2-Parameter (K2P) distances were calculated using unique haplotypes, obtained using FaBox 1.5.

Table 1. Information on newly generated COI sequences of *S. melanotheron* and those retrieved from GenBank.

Aquatic Ecosystems	Number of COI Sequences	Accessions Numbers GenBank	Codes for Our Samples	Country	Status	Reference
Atchakpa reservoir	05	MT180099 (1) MT180100 (1) MT180101 (1) MT180102 (2)	ATCH2 ATCH3 ATCH4 ATCH5 and ATCHF-H01	Benin		This study
Agonlin-Lowé (Ouémé River)	04	MT180103 (1) MT180106 (3)	FOUE3 FOUE2, FOUE4F-H03 and FOUE5F-G03	Benin	Native	This study
Sô-Ava (Sô River)	01	MT180104 (1)	RISO1	Benin	Native	This study
Porto-Novo Lagoon	04	MT180102 (4)	LAPO2F-E03, LAPO1F-F03, LAPO3F-D03 and LAPO4F-C03	Benin	Native	This study
Lake Nokoué	03	MT180107 (1) MT180108 (2)	LANO2 LANO3 and LANO1F-A03	Benin	Native	This study
Manila Bay Lake	04	KM212014-KM212016; KM212018	-	Philippines	Feral	[9]
Taal Lake	05	HQ654753-HQ654757	-	Philippines	Feral	[18]
Laguna de Bay Lake	05	HQ682721-HQ682725	-	Philippines	Feral	[29]
Not specified	01	*MT666031*	-	Hawaii	Feral	Unpublished
Odooba River	02	KX231781-KX231782	-	Nigeria	Native	[30]
Banc d'Arguin National Park (PNBA)	16	KJ938183-**KJ938191**; KJ938208- KJ938214	-	Mauritania	Native	[31]
Guiers Lake	04	KJ938215-KJ938218	-	Senegal	Native	[31]
State Zhongji Tilapia Farm, Hebei Province	02	NC_015611 JF894132	-	China	Farmed	[32]

In the column "Accessions numbers GenBank", the accession number in bold was mislabelled as *S. melanotheron* but instead belongs to *Coptodon guineensis*. Hence, it was not included in the phylogenetic analysis; the accession number in italics was shorter (313 bp vs. 443 bp) and was only directly compared to other sequences; numbers in brackets denote how many specimens are contained in a single sequence.

3. Results

3.1. Molecular Identification of the Specimens

A total of 22 partial mitochondrial COI sequences (five for each of the following ecosystems: Atchakpa water reservoir, Ouémé River and Porto-Novo Lagoon, and three and four for Lake Nokoué and Sô River, respectively) were successfully obtained from 24 samples. For 17 of these, a complete length of 443 bp was obtained (Table 1). The dataset consisted of three unique haplotypes that differed by only one or two nucleotides. The first contained 14 samples from all locations. The other haplotypes contained two (one from the Atchakpa Reservoir, one from Porto Novo Lagoon), and one sample (from Agonlin-Lowé). BLAST revealed a similarity of 99.8–100% with sequences from *Sarotherodon melanotheron* deposited by [18] (HQ654754 and HQ654753) and [29] (HQ682724 and HQ682721). This confirmed the identification of our samples as *S. melanotheron*.

3.2. Comparison with Available Sequences of S. melanotheron

Thirty-nine COI sequences identified as *S. melanotheron* were retrieved from NCBI GenBank (Table 1). One of these (KJ938191) was considered mislabelled and was removed from further analyses. This sequence was coded as *Coptodon guineensis* in [31]. The COI sequence of this specimen matched that of *C. guineensis* and differed by at least 15% from all *S. melanotheron* sequences. We constructed a minimum spanning network (Figure 3) and an NJ tree (Figure 4) with all available sequences of length 443 bp. This revealed three clusters in *S. melanotheron*. The first cluster included all sequences from Benin (present study), as well as sequences from Nigeria, and some sequences from the Philippines (Taal Lake and Laguna de Bay). The sequence from Hawaii was, although shorter, identical to those of the main haplotype of cluster 1 (not shown). The second, central, cluster contained sequences from Mauritania and Senegal. The third cluster contained the remaining sequences from the Philippines as well as those from China. Maximum within-cluster divergence ranged from 0.23 to 0.45%, minimum between cluster divergence ranged from 0.68% (1 vs. 2), 0.91%(2 vs. 3) and 1.14% (1 vs. 3).

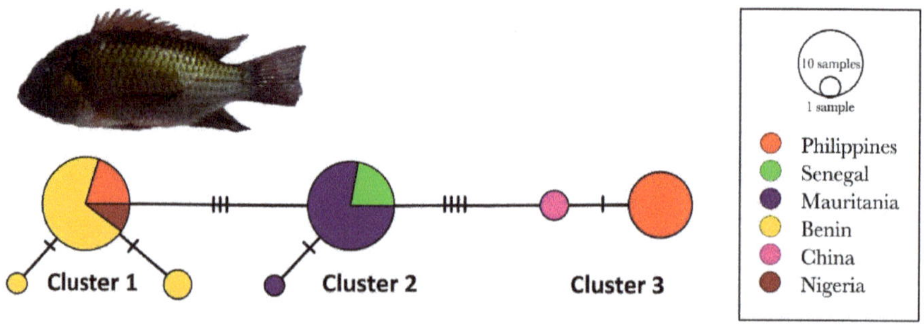

Figure 3. Minimum spanning network constructed from COI sequences of *S. melanotheron* obtained in our study and from GenBank. Colours denote countries where specimens were sampled, node size the number of sequences in a haplotype. The photograph represents a specimen of *S. melanotheron* from the Atchakpa reservoir (Photo credit: R.O.E. Pèlèbè).

Figure 4. NJ tree constructed from the COI sequences of *S. melanotheron* obtained in our study and from GenBank.

4. Discussion

In this study, the use of DNA barcoding allowed us to identify the cichlid population that became landlocked in the Atchakpa freshwater reservoir as *Sarotherodon melanotheron*. Although we could not unambiguously determine the origin of the population, we found no indication that the population in the Atchakpa reservoir belongs to a different strain than the rest of the specimens from Benin. Hence, it could have arrived in the reservoir naturally, or via pumping of river water into the reservoir. The former explanation raises doubt on whether the reservoir has always remained physically separated from the river. The mechanisms by which *S. melanotheron* replaced *O. niloticus* remain unknown. As *S. melanotheron* has a broad ecological tolerance, its success could be linked to environmental parameters in the reservoir. This, however, remains to be investigated.

The three clusters in *S. melanotheron* found using COI correspond to those found by [33] using the mitochondrial control region, which corresponds to three of the four subspecies. These are *S. melanotheron heudelotti* found from the coasts of Mauritania to Guinea (cluster 2), *S. m. leonensis* occurring from Sierra Leone to western Liberia (most likely cluster 3), and *S. m. melanotheron* known from Ivory Coast to Nigeria (cluster 1) [16,17,30,31,33]. Although we lack sequences from native populations in cluster 3, it can be assumed that this cluster corresponds to *S. m. leonensis*. All sequences from Benin fell in cluster 1, which corresponds to *S. melanotheron melanotheron*. The comparison of COI sequences revealed interesting patterns about the introduced populations of *S. melanotheron*. The presence of a cluster 1 haplotype in the specimen from Hawaii is in accordance with the recorded introduction of the species on the islands in 1962 [20]. This strain derived via Europe and New York [34] from the area of the subspecies *S. m. melanotheron* in Africa [35]. As the farmed population in the State Zhongji Tilapia Farm in China [36] has a cluster 3 haplotype, it most likely stems from the range of *S. m. leonensis*. Finally, the populations present in the Philippines contain haplotypes of both clusters 1 and 3, which suggests that at least two introductions took place on the islands.

Although we were able to identify the invasive population in the Atchakpa freshwater reservoir, further studies on its socioeconomic, ecological and environmental impacts should be done. This study represents the first successful usage of DNA barcoding in Beninese fisheries research. Its success highlights that investigations must be undertaken to assembly a national fish DNA barcodes reference library in Benin, useful for reliable monitoring and effective management and conservation of fish biodiversity.

Supplementary Materials: The following is available online at https://www.mdpi.com/article/10.3390/d13070297/s1, Dataset S1: COI sequences used for constructing the NJ phylogenetic tree.

Author Contributions: Conceptualization, R.O.E.P., M.V.S. and E.V.; methodology, R.O.E.P., M.V.S. and E.V.; formal analysis, R.O.E.P. and M.V.S.; data curation, R.O.E.P., and M.V.S.; investigation, R.O.E.P. and M.V.S.; writing—original draft preparation, R.O.E.P.; writing—review and editing, R.O.E.P., M.V.S. and E.V.; visualization, R.O.E.P.; supervision, I.I.T., M.V.S. and E.V.; funding acquisition, R.O.E.P. All authors have read and agreed to the published version of the manuscript.

Funding: This research was funded by the Belgian National Focal Point to the Global Taxonomy Initiative (GTI) that runs under the Capacities for Biodiversity and Sustainable Development (CEBioS) programme with financial support from the Belgian Directorate-General for Development Cooperation (DGD).

Institutional Review Board Statement: Not applicable.

Informed Consent Statement: Not applicable.

Data Availability Statement: The data used in this study are openly available in NCBI GenBank (https://www.ncbi.nlm.nih.gov/nuccore/?term=PELEBE) and the dataset on COI sequences is in the Supplementary Materials.

Acknowledgments: We would like to thank Heleen Maetens and Charlotte Willems for their precious help in the lab during the training internship for taxonomy capacity building of R.O.E.P in Belgium, Olivier Pauwels for curatorial services, and Guillaume Koussovi for his assistance during the sampling of fishes in Benin. The first author acknowledges the contribution of the World Bank Africa Centre of Excellence in Coastal Resilience, University of Cape Coast, Ghana for providing him with a good work environment as postdoctoral fellow in fisheries sciences, which was favourable to the finalization of this article.

Conflicts of Interest: The authors declare no conflict of interest.

References

1. Trewavas, E.; Teugels, G.G. Sarotherodon. In *Check-List of the Freshwater Fishes of Africa (CLOFFA)*; Daget, J., Gosse, J.-P., Teugels, G.G., Van Den Audenaerde, D.F.E.T., Eds.; Royal Museum for Central Africa: Tervuren, Belgium, 1991; pp. 425–437.
2. Pauly, D. The biology, fishery and potential for aquaculture of Tilapia melanotheron in a small West African lagoon. *Aquaculture* **1976**, *7*, 33–49. [CrossRef]
3. Adepo-Gourene, B.; Gourene, G. Différentiation morphologique des populations naturelles d'une sous espèce de tilapia Sarotherodon melanotheron melanotheron Rüppell, 1852 (Teleostei; Cichlidae) de Côte d'Ivoire. *Sci. Nat.* **2008**, *5*, 15–27.
4. Laleye, P.; Chikou, A.; Philippart, J.-C.; Teugels, G.; Vandewalle, P. Etude de la diversité ichtyologique du bassin du fleuve Ouémé au Bénin (Afrique de l'ouest). *Cybium* **2004**, *228*, 329–339.
5. Montchowui, E.; Niyonkuru, C.; Ahouansou Montcho, S.; Chikou, A.; Laleye, P. L'ichtyofaune de la rivière Hlan au Bénin (Afrique de l'Ouest). *Cybium* **2007**, *31*, 173–176.
6. Montchowui, E.; Chikou, A.; Kogbeto, M.-J.; Laleye, P. Biodiversité et structure des communautés de poissons du lac Hlan au Bénin. *Int. J. Biol. Chem. Sci.* **2008**, *2*, 196–206. [CrossRef]
7. Fagnon, S.M.; Chikou, A.; Youssao, I.; Laleye, P. Caractérisation morphologique des populations de Sarotherodon melanotheron (Pisces, Cichlidae) en eaux douces et saumâtres au Sud Bénin. *Int. J. Biol. Chem. Sci.* **2013**, *7*, 619–630. [CrossRef]
8. Pèlèbè, E.O.R.; Dohounkpan, G.A.; Agbohessi, T.P.; Imorou Toko, I.; Hounhoedo, A.O.; Lederoun, D. Inventaire partiel des espèces de poisson et évaluation du degré de fraîcheur de Sarotherodon melanotheron vendu au marché des produits de pêche de la commune de Savè (Bénin). *Rev. Togol. Sci.* **2016**, *10*, 187–198.
9. Ordoñez, J.; Asis, A.; Catacutan, B.; Pena, J.D.; Santos, M. First report on the occurrence of invasive black-chin tilapia Sarotherodon melanotheron (Ruppell, 1852) in Manila Bay and of Mayan cichlid Cichlasoma urophthalmus (Gunther, 1862) in the Philippines. *Bioinvasions Rec.* **2015**, *4*, 115–124. [CrossRef]
10. Cassemiro, F.A.S.; Bailly, D.; Da Graça, W.J.; Agostinho, A.A. The invasive potential of tilapias (Osteichthyes, Cichlidae) in the Americas. *Hydrobiologia* **2017**, *817*, 133–154. [CrossRef]
11. Nico, L. Discovery of the invasive Mayan Cichlid fish "Cichlasoma" urophthalmus (Günther 1862) in Thailand, with comments on other introductions and potential impacts. *Aquat. Invasions* **2007**, *2*, 197–214. [CrossRef]
12. Miller, R.R.; Minckley, W.L.; Norris, S.M. *Freshwater Fishes of Mexico*; University of Chicago Press: Chicago, IL, USA, 2005.
13. Trivedi, S.; Aloufi, A.A.; Ansari, A.A.; Ghosh, S.K. Role of DNA barcoding in marine biodiversity assessment and conservation: An update. *Saudi J. Biol. Sci.* **2016**, *23*, 161–171. [CrossRef]
14. Hebert, P.D.N.; Cywinska, A.; Ball, S.L.; deWaard, J.R. Biological identifications through DNA barcodes. *Proc. Biol. Sci.* **2003**, *270*, 313–321. [CrossRef] [PubMed]
15. Hebert, P.D.; Ratnasingham, S.; De Waard, J.R. Barcoding animal life: Cytochrome c oxidase subunit 1 divergences among closely related species. *Proc. R. Soc. B Boil. Sci.* **2003**, *270*, S96–S99. [CrossRef]
16. Falk, T.M.; Teugels, G.G.; Abban, E.K. Revision and biogeographical analysis of the black-chinned tilapia Sarotherodon melanotheron (Teleostei, Cichlidae): Results of morphometric, allozyme, globin chain and mtDNA studies. *J. Nat. Hist.* **2003**, *37*, 2191–2212. [CrossRef]
17. Falk, T.M.; Teugels, G.G.; Abban, E.K.; Villwock, W.; Renwrantz, L. Morphometric and allozyme variation in the black-chinned tilapiaSarotherodon melanotheron (Perciformes, Cichlidae), with a revision of the subspecies complex. *J. Nat. Hist.* **2000**, *34*, 1849–1863. [CrossRef]
18. Aquilino, S.V.L.; Tango, J.M.; Fontanilla, I.K.C.; Pagulayan, R.C.; Basiao, Z.U.; Ong, P.S.; Quilang, J.P. DNA barcoding of the ichthyofauna of Taal Lake, Philippines. *Mol. Ecol. Resour.* **2011**, *11*, 612–619. [CrossRef]
19. Wohlfarth, G.W.; Hulata, G. *Applied Genetics of Tilapias*, 2nd ed.; International Center for Living Aquatic Resources Management (ICLARM): Manila, Philippines, 1983.
20. Randall, J.E. Introductions of marine fishes to the Hawaiian Islands. *Bull. Mar. Sci.* **1987**, *41*, 490–502.
21. Porter, T.M.; Hajibabaei, M. Over 2.5 million COI sequences in GenBank and growing. *PLoS ONE* **2018**, *13*, e0200177. [CrossRef] [PubMed]
22. Paugy, D.; Leveque, C.; Teugels, G.G. *The Fresh and Brackish Water Fishes of West Africa*; Royal Museum for Central Africa: Tervuren, Belgium, 2003.
23. Ivanova, N.V.; Zemlak, T.S.; Hanner, R.H.; Hebert, P. Universal primer cocktails for fish DNA barcoding. *Mol. Ecol. Notes* **2007**, *7*, 544–548. [CrossRef]
24. Tamura, K.; Peterson, D.; Peterson, N.; Stecher, G.; Nei, M.; Kumar, S. MEGA5: Molecular Evolutionary Genetics Analysis Using Maximum Likelihood, Evolutionary Distance, and Maximum Parsimony Methods. *Mol. Biol. Evol.* **2011**, *28*, 2731–2739. [CrossRef]
25. Pereira, L.H.G.; Hanner, R.; Foresti, F.; Oliveira, C. Can DNA barcoding accurately discriminate megadiverse Neotropical freshwater fish fauna? *BMC Genet.* **2013**, *14*, 20. [CrossRef]
26. Ward, R.D. DNA barcode divergence among species and genera of birds and fishes. *Mol. Ecol. Resour.* **2009**, *9*, 1077–1085. [CrossRef]

27. Leigh, J.W.; Bryant, D. POPART: Full-feature software for haplotype network construction. *Methods Ecol. Evol.* **2015**, *6*, 1110–1116. [CrossRef]
28. Saitou, N.; Nei, M. The neighbor-joining method: A new method for reconstructing phylogenetic trees. *Mol. Biol. Evol.* **1987**, *4*, 406–425. [CrossRef] [PubMed]
29. Aquino, L.M.G.; Tango, J.M.; Canoy, R.J.C.; Fontanilla, I.K.C.; Basiao, Z.U.; Ong, P.S.; Quilang, J.P. DNA barcoding of fishes of Laguna de Bay, Philippines. *Mitochondrial DNA* **2011**, *22*, 143–153. [CrossRef]
30. Falade, M.O.; Opene, A.J.; Benson, O. DNA barcoding of Clarias gariepinus, Coptodon zillii and Sarotherodon melanotheron from Southwestern Nigeria. *F1000Research* **2016**, *5*, 1268. [CrossRef]
31. Kide, N.G.; Dunz, A.; Agnèse, J.-F.; Dilyte, J.; Pariselle, A.; Carneiro, C.; Correia, E.; Brito, J.; Yarba, L.O.; Kone, Y.; et al. Cichlids of the Banc d'Arguin National Park, Mauritania: Insight into the diversity of the genus Coptodon. *J. Fish Biol.* **2016**, *88*, 1369–1393. [CrossRef] [PubMed]
32. He, A.Y.; Tang, S.J.; Jiang, Y.T.; Li, S.F.; Wang, C.H. Complete mitochondrial genome of blackchin tilapiaSarotherodon melanotheron (Perciformes, Cichlidae). *Mitochondrial DNA* **2011**, *22*, 171–173. [CrossRef] [PubMed]
33. Falk, T.M.; Teugels, G.G.; Abban, E.K.; Villwock, W.; Renwrantz, L. Phylogeographic patterns in populations of the black-chinned tilapia complex (Teleostei, Cichlidae) from coastal areas in West Africa: Support for the refuge zone theory. *Mol. Phylogenet. Evol.* **2003**, *27*, 81–92. [CrossRef]
34. Trewavas, E. Tilapias: Taxonomy and Speciation. In Proceedings of the International Conference on the Biology and Culture of Tilapias, Bellagio, Italy, 2–5 September 1980; Pullin, R.S.V., Lowe-McConnell, R.H., Eds.; International Center for Living Aquatic Resources Management (ICLARM): Manila, Philippines, 1982; pp. 3–13.
35. Englund, R.A. The loss of native biodiversity and continuing nonindigenous species introductions in freshwater, estuarine, and wetland communities of Pearl Harbor, Oahu, Hawaiian Islands. *Estuaries* **2002**, *25*, 418–430. [CrossRef]
36. Li, S.F.; Yan, B.; Cai, W.Q.; Li, T.Y.; Jia, J.H.; Zhang, Y.H. Evaluation of growth, salt tolerance and parents' heterosis contribution in reciprocal hybrids F2 between Oreochromis niloticus and Sarotherodon melanotheron. *J. Fish. China* **2008**, *32*, 335–341.

diversity

MDPI

Article

DNA Barcoding of the Market Samples of Single-Drug Herbal Powders Reveals Adulteration with Taxonomically Unrelated Plant Species

Raju Balaji and Madasamy Parani *

Department of Genetic Engineering, Faculty of Engineering and Technology, SRM Institute of Science and Technology, College of Engineering and Technology, SRM Nagar, Kattankulathur, Kanchipuram, Chennai 603203, India; balajiramasrinivasan@gmail.com
* Correspondence: paranim@srmist.edu.in

Abstract: Herbal drugs are increasingly becoming a viable alternative to allopathic medicine. Since powdered herbal drugs are more prone to adulteration than intact plant parts, their authentication becomes essential to ensure the safety and efficacy of herbal drugs. This study authenticated 107 single-drug herbal powders, representing 65 species from 60 genera and 35 families, collected from the markets in Tamil Nadu, India. DNA barcoding using the *rbcL* marker revealed that 58 samples (54%) were authentic, and 49 (46%) were adulterant. About 41% of the adulterant samples were a mixture of more than one species, possibly due to unintentional cross-contamination during processing. In 59% of the adulterant samples, the authentic species was entirely substituted with taxonomically and medicinally unrelated species, 72% of which belonged to different orders and families, while 28% were from other genera. Despite the taxonomic diversity, 20% of adulterant species had a morphological resemblance to the authentic species. It is not known whether their use as adulterants is intentional. In a detailed study on DNA barcoding of 17 powder samples from *Ocimum tenuiflorum*, 88% of the samples were authentic. These results indicate that the extent of adulteration is not high in all the species. Approximately, 95% of the samples collected for this study were produced by companies with limited resources and expertise in the unorganized sector. Hence, training them on species identification and providing simple and cost-effective authentication tools will likely reduce adulteration in the market samples.

Keywords: single-drug herbal powder; *rbcL*; molecular authentication; Tulsi powder; adulteration

Citation: Balaji, R.; Parani, M. DNA Barcoding of the Market Samples of Single-Drug Herbal Powders Reveals Adulteration with Taxonomically Unrelated Plant Species. *Diversity* **2022**, *14*, 495. https://doi.org/10.3390/d14060495

Academic Editor: Stephan Koblmüller

Received: 9 May 2022
Accepted: 16 June 2022
Published: 17 June 2022

1. Introduction

India is one of the 17 countries in the world with mega-biodiversity, and 954 species of medicinal plants are actively traded in the markets [1]. About 40% to 90% of the people in different countries use traditional medicine for their primary healthcare needs [2]. These plant-based medications are less expensive and more accessible in rural areas of developing countries [3]. The expected benefits of any medicine can be realized only if authentic materials are used. It is also essential from the safety aspect of the patients. Unlike allopathic medicine, herbal medicine remains largely unregulated. This paves the way for intentional and unintentional adulterations and admixtures raising concerns about the efficacy and safety of the herbal drugs [4,5]. Increasing demand for herbal medicine is expected to boost the trade of medicinal drugs from USD 120 billion to USD 7 trillion by 2050 [6]. Such vast business and employment opportunities will be lost if the consumers do not trust the authenticity of the herbal drugs traded in the markets. Therefore, it becomes essential for consumers and traders to be interested in only authentic plant materials being traded as herbal drugs.

Several techniques such as morpho-taxonomic keys, anatomy, pharmacognosy, chemical fingerprinting, and DNA barcoding are used to differentiate authentic and non-authentic plant materials. Morpho-anatomical studies of the leaves and stems were used to identify the *Tinospora* species in dietary supplements, and it was further supported by HPTLC fingerprinting [7]. Chemical characterization was used to authenticate the presence of *Salvia* species in traditional herbal preparations in Spain [8]. DNA barcoding was used to authenticate the nut species in milk beverages [9]. Each method has its own merits and demerits [10]. While the morpho-taxonomic approach helps collect authentic plant material, it will be challenging to use the same for processed and powdered materials. Chemical fingerprinting suffers from the fact that it is difficult to establish species-specific chemical markers, and the markers are sensitive to the age, season, and place of collection of the plant material [11]. With technological advancements, drastic reduction in sequencing cost, and increasing richness of reference sequences, DNA barcoding has emerged as a more versatile and robust method for authenticating herbal products and raw drugs through molecular species identification [12,13]. DNA barcoding uses the markers that are conserved within species but divergent between species so that species-specific sequences can be retrieved using a single pair of universal primers. DNA barcoding can be used for species delimitation, identification of cryptic species, and understanding species composition in biodiversity hotspots, which are useful in taxonomy, biodiversity assessment, conservation, and environmental protection [14–16]. Several studies have employed DNA barcoding techniques to detect adulterations, product substitution, contamination, mislabeling, and admixture in herbal products and raw drugs [17–21]. In the market samples of Ashwagandha, we found that 88% of the adulterant samples were in the form of powders [22]. Therefore, we initiated a larger study to authenticate a diverse set of herbal drugs that are traded in the form of powders. A reference DNA herbal drug barcode library was assembled, and 117 single-drug herbal powders collected from the markets were authenticated. Additionally, a detailed study was conducted by analyzing 17 market samples of one herbal powder.

2. Materials and Methods

2.1. Collection of Single-Drug Herbal Powders

Single-drug herbal powders were selected based on their therapeutic value and trade volume [1,23–28]. The binomial and vernacular names of the herbal powders were obtained from the Traded Medicinal Plants Database (http://envis.frlht.org/botanical_search.php, accessed on 2 February 2022). This database maintains the data on the 960 medicinal plants that are traded from India. It provides vernacular names used in different parts of India and the botanical names for medicinal plants. It also provides the details regarding plant parts traded. The same vernacular names were used, and 117 single-drug herbal powders were collected from the markets in Tamil Nadu, India. Of these, 106 powders were from known plant parts (whole plant, root, rhizome, stem, bark, leaf, flower, and seed), and 11 were from an unknown origin. The samples collected for this study were traded by three registered companies and seven unregistered companies from the unorganized sector. Details of the samples collected for this study are given in Table S1. For detailed analysis, 17 powder samples of Tulsi (*Ocimum tenuiflorum*) were obtained from 17 manufacturers (Table S2).

2.2. Genomic DNA Extraction, PCR Amplification, and DNA Sequencing

Genomic DNA was extracted using the Cetyl Trimethyl Ammonium Bromide (CTAB) method, with minor modifications [29,30]. About 100 mg herbal powder samples were thoroughly suspended in 0.5 mL DNA extraction buffer, and the suspension was incubated for 16 h at room temperature. The suspension was mixed briefly by vortexing and incubated at 55 °C for 30 min. DNA was isolated as described before [31] and dissolved in TE buffer. Universal primers for *rbcL* (*rbc*La-F and *rbc*Lajf634-R) [32,33], *trnH-psbA* [34], and *ITS2* [35] were used for PCR amplification of the DNA barcode markers. The PCR reaction mixture contained 1X buffer with 1.5 mM MgCl2, 0.2 mM dNTPs, 5.0 pmol primers, 1 unit Taq DNA polymerase (GenetBio Inc., Nonsan-si, Korea), and 20–50 ng genomic DNA. PCR amplification was started with an initial denaturation at 95 °C for 5 min, followed by 30 cycles of denaturation at 95 °C for 30 s, annealing at 55 °C for 30 s, and extension at 72 °C for 1 min, final extension at 72 °C for 5 min, and held at 16 °C. The PCR amplified products were purified using the EZ-10 Spin Column PCR Purification Kit (Bio Basic Inc., Markham, ON, Canada). The sequencing of PCR products was carried out with BigDye Terminator v3.1 chemistry in SeqStudio, following the standard manufacturer's protocol. The quality of sequences was analyzed in Sequence Scanner Software v1.0 (Applied Biosystems, Waltham, MA, USA).

2.3. Reference DNA Barcode Library

We assembled a reference DNA barcode library consisting of 1325 accessions from 656 species (Table S3) from our previous DNA barcoding projects [12,18,30,31,36,37]. It included all species of the herbal powders collected for this study.

2.4. Data Analysis and Species Authentication

DNA barcode sequences from the single-drug herbal powders were compared with the sequences in the reference DNA barcode library. Authentic samples were identified based on the clustering pattern in the phylogenetic tree constructed using the neighbor-joining (NJ) method in MEGA version 7 [38] with Kimura 2 parameter distance model [39,40] and bootstrap analysis, with 1000 replications. The sequences from the non-authentic samples were searched against the non-redundant nucleotide database of NCBI (http://blast.ncbi.nlm.nih.gov/Blast.cgi, accessed on 12 February 2022) and the BOLD database (https://www.boldsystems.org/index.php/IDS_OpenIdEngine, accessed on 12 February 2022) using the BLAST algorithm for species identification.

3. Results and Discussion

3.1. DNA Isolation, PCR, and Sequencing

Total genomic DNA was isolated from 107 out of the 117 powder samples. The ten samples in which the DNA isolation failed were from Kottai karanthai (HRD023), Lavangappattai (HRD062), Koraikizhangu (HRD065), Boomi sakkarai kilangu (HRD066), Maathulai (HRD074), Kadukkai (HRD083), Aduthinnaipalai (HRD089), Thanrikkai (HRD117), Naval (HRD139), and Poonnankanni (HRD142). Incubation of the powders in the DNA extraction buffer overnight at room temperature was essential for better extraction of genomic DNA. In general, compared with the powders from leaves and flowers, the powders from root, rhizome, stem, and bark yielded much less DNA. However, the quantity of DNA obtained was more than sufficient to PCR amplify the DNA barcode markers. Often 10 to 30 times dilution of the DNA gives better results in terms of PCR amplification, likely due to dilution of the co-precipitated PCR inhibitors [31]. We obtained a 100% success rate for PCR amplification and sequencing of the *rbcL* marker. Therefore, we subjected all of the 107 samples for authentication. As per the label, these samples were derived from 65 species, which belong to 60 genera and 35 families. Agarose gel electrophoresis of DNA and *rbcL* marker amplified from ten samples are given in Figure 1. The chromatograms of the *rbcL* sequences were manually edited before using the data for further analysis.

3.2. Non-Authentic Mixed Samples

Chromatograms of 20 samples (19%) were completely not readable or contained several overlapping peaks originating from more than one DNA fragment in the same sample (Figure 2). These samples included tissues from more than one divergent species and, therefore, are called mixed samples. Since we collected single-drug herbal powders, all of the mixed samples were considered non-authentic (Table S4). The significant number of mixed samples found among the single-drug herbal powders samples is a concern. This may be due to unintentional cross-contamination during sample processing, though economically motivated intentional adulteration cannot be ruled out. As PCR amplification is very sensitive, DNA barcoding can detect even a minute quantity of contamination and identify it as a mixed sample. Although PCR was used to detect adulteration of up to 0.5% Chili in pepper [41], its sensitivity could be much higher considering the extraordinarily high copy number of the chloroplast genome, which can reach as high as 10,000 copies per cell [42]. Therefore, the proportion of the adulterant species needs to be determined to evaluate the clinical consequences of using the mixed samples for treatment purposes. Additionally, it would require determining the species composition of the mixed samples. The DNA sequence from mixed samples cannot be used for any DNA-sequence-based analysis, and therefore, the adulterant species present in them cannot be determined by DNA barcoding. The species composition of the mixed samples can be determined by meta-DNA barcoding [43,44] or DNA barcoding after cloning [45].

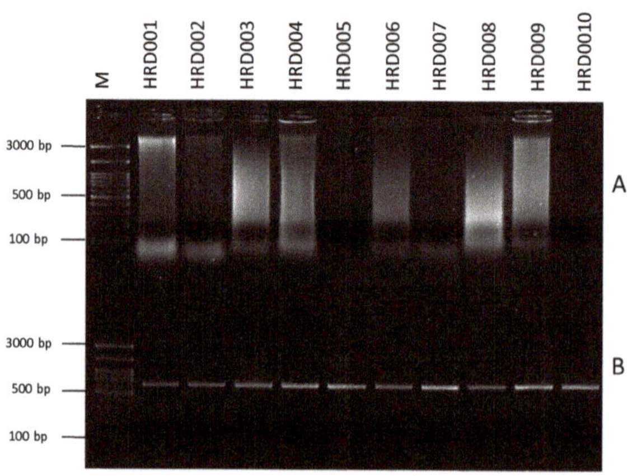

Figure 1. Agarose gel electrophoresis of the genomic DNA from ten single-drug herbal powder samples (**A**) and the PCR amplified *rbcL* DNA barcode markers (**B**) from the respective samples along with 100 bp DNA ladder (M).

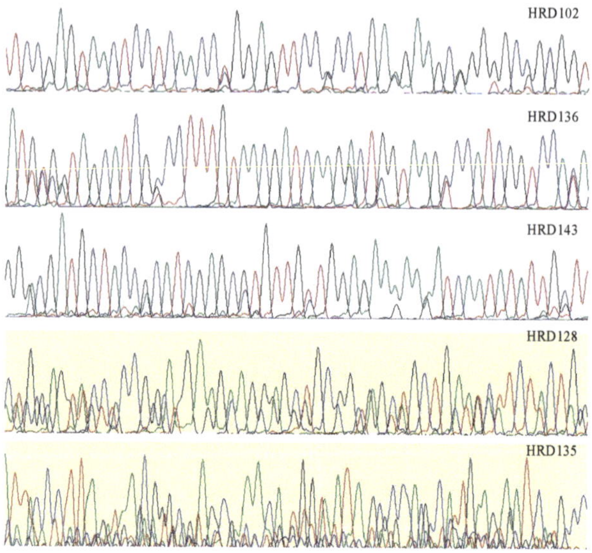

Figure 2. Chromatograms of the mixed samples showing a few overlapping peaks (HRD102, HRD136, and HRD143) to a completely unreadable DNA sequence (HRD128 and HRD135).

3.3. Non-Authentic Samples with Complete Substitution

All of the 87 samples that yielded readable DNA sequences were subjected to further analysis to identify authentic samples. In the phylogenetic tree based on the sequences from the reference DNA barcode library and market samples, 58 samples (~54%) clustered with the expected species. These samples were considered as authentic (Table S5, Figure S1). The remaining 29 samples (~27%) did not cluster with the expected species. In these samples, authentic species were completely substituted with a different species and, therefore, non-authentic samples (Table S6, Figure S1). About 72% of them clustered with species from other orders and families. The remaining 28% of the non-authentic samples clustered with species from different genera, and none of them clustered with congeneric species. These results demonstrate that the adulterant species are not closely related to the authentic species. In our earlier study on authentication of traded medicinal plants not specific to powder samples, about 13% and 7% of samples were adulterated with species from different families and genera [12]. This observation has two implications. First, the adulterant species is not likely to have the same medicinal property as the authentic species. Second, the identification of authentic samples may not need species-specific markers.

3.4. Identification of the Adulterant Species in Non-Authentic Samples

The DNA barcode sequences of the 29 non-authentic samples in which the expected species was completely substituted with a different species were subjected to BLAST analysis for species identification. The species that showed the highest identity (99.65% to 100%) are given in Table 1. The presence of adulterant species in market samples may be due to the same or similar vernacular names, morphological resemblance, mishandling, mislabeling, and species admixture [12,43,46,47]. In this study, only in one sample in which *Pavonia zeylanica* was substituted with *Sida acuta,* the adulteration may be due to a similar vernacular name ("Kurunthotti" is the vernacular name in the Tamil language for both species). Though their vernacular names are the same, the medicinal properties of these two species are different. While the *P. zeylanica* roots are used as a laxative and expectorant [48], *S. acuta* is used as an aphrodisiac and liver tonic [49]. In contrast, 35% of the completely substituted samples (10 samples from six species) likely contain a different species due to morphological resemblance (Figure 3).

We collected two powder samples of *Abutilon indicum*—one was a mixed sample, and the other was substituted with *Sida cordifolia.* Earlier, we found *A. indicum* as an adulterant in *S. cordifolia* [31]. While the aerial parts of *A. indicum* are used for treating asthma [50], the roots of *S. cordifolia* are used to prepare nervine tonic [51]. It appears that these two species are often mixed up during collection due to highly similar leaves and fruits (Figure 3A,B). It is also possible that, after harvesting the roots of *S. cordifolia,* the leftover aerial parts are used for adulteration in *A. indicum.* Flowers of *H. rosa-sinensis* are used for treating hair loss and extracting natural dyes [52]. We collected three powder samples of *H. rosa-sinensis* flowers, and all were adulterated with *Rhododendron. H. rosa-sinensis* is commonly available in the areas from where we collected the market samples. However, due to high demand, this species seems to be adulterated with *Rhododendron*, which grows in the Himalayan regions. The red-colored flowers of the *Rhododendron* highly resemble that of *H. rosa-sinensis* (Figure 3C,D). Based on morphological and powder microscopy studies, it was reported that the dried flowers of *R. arboreum* are adulterated with *H. rosa-sinensis* [53]. *Cynodon dactylon* is used as a laxative, expectorant, analgesic, and in the treatment of dropsy and diabetes [54,55]. We collected three powder samples of *C. dactylon* leaves, and all of them were adulterated with *Sporobolus helvolus.* Both are grass species with phenotypic resemblance (Figure 3E,F), and they co-occur in the same habitat. We collected three powder samples of *Senna auriculata* flowers, which are used for hair wash, as well as for treating diabetes and fever [56]. We recovered *Indigofera tinctoria* in place of *S. auriculata*, and both have similar leaf morphology (Figure 3G,H). Similarly, the morphological resemblance of the leaves could be associated with the adulteration of *Mukia maderaspatana* with *Cucumis melo* (Figure 3I,J).

Tribulus terrestris (Devil's thorn) is a highly traded medicinal plant (~3000 metric tonnes per year), and its dried fruits are used for treating urinary stones, impotence, and venereal diseases in the Indian Ayurvedic and the Chinese traditional medicine [57]. We collected four powder samples of *T. terrestris* fruits from four different manufacturers. Three samples were authentic, but one was adulterated with *Harpagophytum* (Devil's claw). Dried fruits of both *T. terrestris* and *Harpagophytum* are brown with thorns. Species of *Harpagophytum* are distributed only in southern parts of Africa [58]. It needs to be further investigated how it is found to be an adulterant in *T. terrestris* samples collected from India. The rhizomes of *H. procumbens* are used for treating arthritis, rheumatism, and labor pain [59] and are exported from Africa to Europe [60]. There is no reported medicinal use for the fruits. After harvesting the rhizomes, it is surmised that the fruits may be exported to countries such as India for adulteration with *T. terrestris.* (Figure 3K,L).

Table 1. Non-authentic single-drug powder samples in which authentic species was entirely substituted with adulterant species. Taxonomic affiliations of the authentic and substituted species are provided for comparison.

S. No	Collection ID	Species Expected as per the Label	Family	Order	Species Identified by DNA Barcoding	Family	Order
1	HRD031	*Abutilon indicum*	Malvaceae	Malvales	*Sida cordifolia*	Malvaceae	Malvales
2	HRD050	*Alpinia galanga*	Zingiberaceae	Zingiberales	*Indigofera stachyodes*	Fabaceae	Fabales
3	HRD017	*Cardiospermum halicacabum*	Sapindaceae	Sapindales	*Trigonella foenum-graecum*	Fabaceae	Fabales
4	HRD004	*Centella asiatica*	Apiaceae	Apiales	*Ipomea imperati*	Convolvulaceae	Solanales
5	HRD130	*Centella asiatica*	Apiaceae	Apiales	*Trigonella foenum-graecum*	Fabaceae	Fabales
6	HRD103	*Coscinium fenestratum*	Menispermaceae	Ranunculales	*Vigna mungo*	Fabaceae	Fabales
7	HRD084	*Curcuma aromatica*	Zingiberaceae	Zingiberales	*Cullen corylifolium*	Fabaceae	Fabales
8	HRD038	*Cynodon dactylon*	Poaceae	Poales	*Sporobolus helvolus*	Poaceae	Poales
9	HRD054	*Cynodon dactylon*	Poaceae	Poales	*Sporobolus helvolus*	Poaceae	Poales
10	HRD107	*Cynodon dactylon*	Poaceae	Poales	*Sporobolus helvolus*	Poaceae	Poales
11	HRD093	*Ficus benghalensis*	Moraceae	Rosales	*Thespesia populnea*	Malvaceae	Malvales
12	HRD085	*Ficus racemosa*	Moraceae	Rosales	*Abutilon indicum*	Malvaceae	Malvales
13	HRD138	*Ficus racemosa*	Moraceae	Rosales	*Abutilon grandiflorum*	Malvaceae	Malvales
14	HRD078	*Ficus religiosa*	Moraceae	Rosales	*Indigofera tinctoria*	Fabaceae	Fabales
15	HRD052	*Glycyrrhiza glabra*	Fabaceae	Fabales	*Canavalia* sp.	Fabaceae	Fabales
16	HRD039	*Gymnema sylvestre*	Apocynaceae	Gentianales	*Trigonella foenum-graecum*	Fabaceae	Fabales
17	HRD068	*Hibiscus rosa-sinensis*	Malvaceae	Malvales	*Rhododendron* sp.	Ericaceae	Ericales
18	HRD108	*Hibiscus rosa-sinensis*	Malvaceae	Malvales	*Rhododendron* sp.	Ericaceae	Ericales
19	HRD110	*Hibiscus rosa-sinensis*	Malvaceae	Malvales	*Rhododendron* sp.	Ericaceae	Ericales
20	HRD127	*Hybanthus enneaspermus*	Violaceae	Malpighiales	*Cardiospermum halicacabum*	Sapindaceae	Sapindales
21	HRD040	*Mangifera indica*	Anacardiaceae	Sapindales	*Mollugo cerviana*	Molluginaceae	Caryophyllales
22	HRD016	*Melia azedarach*	Meliaceae	Sapindales	*Justicia adhatoda*	Acanthaceae	Lamiales
23	HRD081	*Moringa oleifera*	Moringaceae	Brassicales	*Cassia senna*	Fabaceae	Fabales
24	HRD095	*Mukia maderaspatana*	Cucurbitaceae	Cucurbitales	*Cucumis melo*	Cucurbitaceae	Cucurbitales
25	HRD079	*Pavonia zeylanica*	Malvaceae	Malvales	*Sida acuta*	Malvaceae	Malvales
26	HRD019	*Senna auriculata*	Fabaceae	Fabales	*Indigofera tinctoria*	Fabaceae	Fabales
27	HRD099	*Terminalia arjuna*	Combretaceae	Myrtales	*Mucuna pruriens*	Fabaceae	Fabales
28	HRD098	*Tribulus terrestris*	Zygophyllaceae	Zygophyllales	*Harpagophytum* sp.	Ericaceae	Ericales
29	HRD126	*Zingiber officinale*	Zingiberaceae	Zingiberales	*Cajanus cajan*	Fabaceae	Fabales

Figure 3. Morphological resemblance between the expected authentic species and non-authentic species identified by DNA barcoding. *Abutilon indicum* versus *Sida cordifolia* (**A**,**B**); *Cynadon dactylon* versus *Sporobolus helvolus* (**C**,**D**); *Hibiscus rosa-sinensis* versus *Rhododenron delavayi* (**E**,**F**); *Senna auriculata* versus *Indigofera tinctoria* (**G**,**H**); *Mukia maderaspatana* versus *Cucumis melo* (**I**,**J**); and *Tribulus terrestris* versus *Harpaophytum procumbens* (**K**,**L**).

3.5. Authentication of Tulsi (Ocimum tenuiflorum)

To investigate a single-drug herbal powder in more detail, we collected 17 samples of Tulsi (*O. tenuiflorum*) from 17 different manufacturers. Tulsi is an important medicinal species, and its leaves are used for treating bronchitis, rheumatism, pyrexia, asthma, and tooth pain [61,62]. DNA barcoding of *O. americanum*, *O. basilicum*, *O. filamentosum*, *O. gratissimum*, *O. kilimandscharicum*, *O. tenuiflorum*, and *O.* x *citriodorum*, using *rbcL*, *matK*, and *trnH-psbA* markers, showed that *trnH-psbA* was the most suitable marker for species differentiation; however, it did not differentiate all of the species [63]. In a similar study, *O. filamentosum* was replaced with *O. carnosum*, and DNA barcoding was performed using the same markers. None of the markers was species-specific [64]. In both studies, *ITS2* was not included, probably due to the problems in PCR amplification. We found that the *ITS2* marker of *O. basilicum* has GC rich sequence, and it could be PCR amplified only in the presence of 5% DMSO as a PCR additive [65]. In the current study, we used DMSO and obtained perfect amplification of *ITS2* from *O. americanum*, *O. basilicum*, *O. gratissimum*,

O. kilimandscharicum, and *O. tenuiflorum* (Figure 4). However, a complete *ITS2* sequence (451 bp) could be obtained only for *O. tenuiflorum*, and its GC content was 67%. Supplementing the sequencing reaction with DMSO indeed improved the sequence quality. Higher GC content and specific stretches of G and C nucleotides may affect strand separation or form strong secondary structures that affect the binding and extension of the sequencing primer. However, we obtained a good-quality sequence of 115 bp in the 5′ ends and 132 bp in the 3′ ends of the *ITS2* marker. Diagnostic nucleotides in these regions were used for authentication. Out of the 17 samples tested, 15 were authentic, and 2 were mixed samples. Since the *trnH-psbA* marker often shows length variations, we PCR-amplified this marker from the two mixed samples. Two differentially sized *trnH-psbA* markers were amplified from both samples (Figure 5). Sequencing and BLAST analyses of those two markers revealed that both samples contained *O. tenuiflorum* but were mixed with *Indigofera tinctoria* or *Trigonella foenum-graecum*. Compared with the overall adulteration in herbal powders, adulteration in Tulsi was relatively much less (~12%). This may be because Tulsi is abundantly available in the study area, and being a sacred species used in temples, people are familiar with this species.

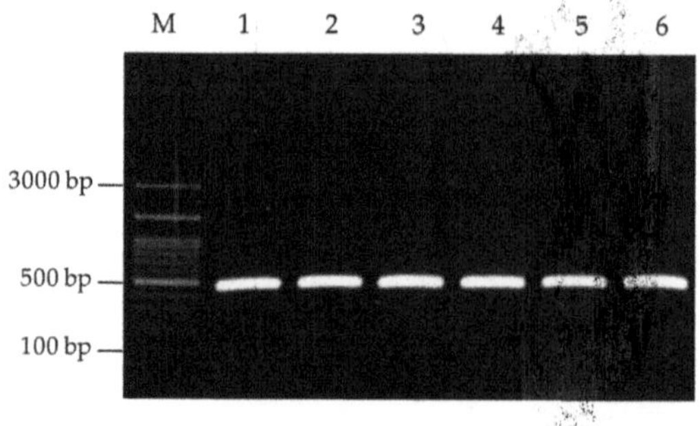

Figure 4. Agarose gel electrophoresis of PCR amplified *ITS2* DNA barcode markers from *O. tenuiflorum*—Rama type (1), *O. tenuiflorum*—Krishna type (2), *O. americanum* (3), *O. basilicum* (4), *O. gratissimum* (5), and *O. kilimandscharicum* (6). Lane M was loaded with 100 bp DNA.

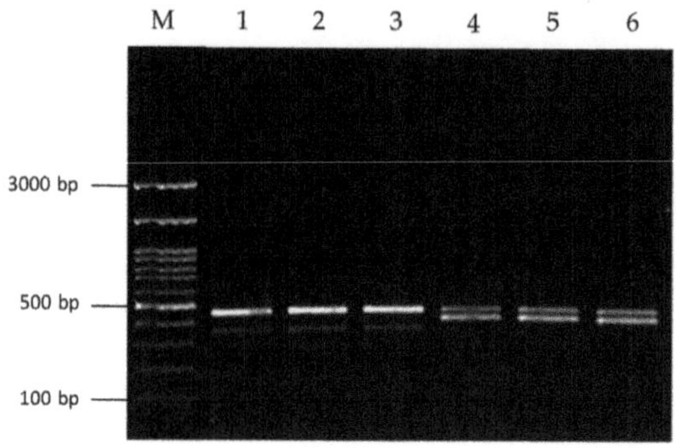

Figure 5. Agarose gel electrophoresis of PCR amplified *trnH-psbA* DNA barcode markers from the two mixed samples of Tulsi, TUL008 (lanes 1–3) and TUL015 (lanes 4–6), along with 100 bp DNA ladder (M).

4. Conclusions

The present study supports the applicability of DNA barcoding to authenticate the market samples of single-drug herbal powders by successfully identifying adulterated samples. It was not necessary to use species-specific markers because the adulterant species were not taxonomically closely related to the authentic species. Mixed samples constituted a significant percentage of adulterated samples; however, information on species composition and the proportion of the adulterant species is needed to assess the clinical significance of such adulterations. Though the authentic and adulterant species were morphologically similar in quite a few cases, it is unknown if adulteration was due to a lack of knowledge or intentionally carried out by taking advantage of the morphological resemblance. The adulteration range varies, and the authentic and adulterated species often have unrelated medicinal properties. It is worth noting that 95% of the market samples authenticated in this study belonged to unregistered companies from unorganized sectors. These companies typically sell their products through local herbal shops. Large numbers of rural and urban populations from low-income groups depend on these herbal shops for their herbal drug requirements. However, their products may reach a wider population directly through online sales or indirectly through large herbal manufacturers. Since adulteration is not high in all cases, proper training on species identification for people who collect samples and the development of simple and cost-effective technologies to verify the collected samples' taxonomic identity are likely to decrease adulteration. Therefore, it is necessary to develop methods to authenticate the market samples so that only the correct species appropriate for the particular treatment or formulation are used.

Supplementary Materials: The following supporting information can be downloaded at: https://www.mdpi.com/article/10.3390/d14060495/s1, Figure S1: Phylogenetic tree based on the sequences from the reference DNA barcode library and the market samples. Authentic and non-authentic market samples are highlighted in green and red, respectively, File S1: DNA barcode sequences from the market samples collected for this study, Table S1: Details of the single-drug herbal powder samples collected for the current study, Table S2: Details of the single-drug herbal powder samples of Tulsi collected for the current study, Table S3: Details of the species included in the reference DNA barcode library, Table S4: Details of the single-drug herbal powders identified as non-authentic mixed samples, Table S5: Details of the single-drug herbal powders identified as authentic samples, Table S6: Details of the single-drug herbal powders identified as non-authentic samples with complete substitution.

Author Contributions: Conceptualization, M.P.; data curation, R.B.; formal analysis, R.B. and M.P.; methodology, M.P.; writing—original draft preparation, R.B.; writing—review and editing, M.P. and R.B. All authors have read and agreed to the published version of the manuscript.

Funding: This research was funded by SRM-DBT Partnership Platform for Contemporary Research Services and Skill Development in Advanced Life Sciences Technologies (Grant Number BT/PR12987/INF/22/205/2015).

Institutional Review Board Statement: Not applicable.

Informed Consent Statement: Not applicable.

Data Availability Statement: Not applicable.

Acknowledgments: The authors thank D. Narasimhan (Madras Christian College, India), K. Ravikumar (I-AIM, Bengaluru, India), and G. Gnanasekaran (Madras Christian College, India) for their valuable suggestions. The authors acknowledge the SRM Institute of Science and Technology and financial support from the SRM-DBT Partnership Platform for Contemporary Research Services and Skill Development in Advanced Life Sciences Technologies (Order No. BT/PR12987/INF/22/205/2015).

Conflicts of Interest: The authors declare no conflict of interest.

References

1. Ravikumar, K.; Begum, S.N.; Ved, D.K.; Bhatt, J.R.; Goraya, G.S. *Compendium of Traded Medicinal Plants*; Foundation for Revitalization of Local Health Traditions (FRLHT): Bengaluru, India, 2018; pp. 1–338.
2. World Health Organization. Programme on Traditional Medicine. In *WHO Traditional Medicine Strategy 2002–2005*; World Health Organization: Geneva, Switzerland, 2002; Available online: https://apps.who.int/iris/handle/10665/67163 (accessed on 12 February 2022).
3. Meena, A.K.; Bansal, P.; Kumar, S. Plants-herbal wealth as a potential source of ayurvedic drugs. *Asian J. Tradit. Med.* **2009**, 4, 152–170.

4. Newmaster, S.G.; Grguric, M.; Shanmughanandhan, D.; Ramalingam, S.; Ragupathy, S. DNA barcoding detects contamination and substitution in North American herbal products. *BMC Med.* **2013**, *11*, 222. [CrossRef] [PubMed]
5. Simmler, C.; Graham, J.G.; Chen, S.N.; Pauli, G.F. Integrated analytical assets aid botanical authenticity and adulteration management. *Fitoterapia* **2018**, *129*, 401–414. [CrossRef] [PubMed]
6. Marichamy, K.; Kumar, N.Y.; Ganesan, A. Sustainable development in exports of herbals and Ayurveda, Siddha, Unani and Homeopathy (Ayush) in India. *Sci. Park Res. J.* **2014**, *27*, 1–6.
7. Parveen, A.; Adams, J.S.; Budel, J.M.; Zhao, J.; Babu, G.N.M.; Ali, Z.; Khan, I.A. Comparative morpho-anatomical and HPTLC profiling of *Tinospora* species and dietary supplements. *Planta Med.* **2020**, *86*, 470–481. [CrossRef]
8. Martinez-Frances, V.; Hahn, E.; Rios, S.; Rivera, D.; Rreich, E.; Vila, R.; Canigueral, S. Ethnopharmacological and chemical characterization of *Salvia* species used in Valencian traditional herbal preparations. *Front. Pharmacol.* **2017**, *8*, 467. [CrossRef]
9. Ding, Y.; Jiang, G.; Huang, L.; Chen, C.; Sun, J.; Zhu, C. DNA barcoding coupled with high-resolution melting analysis for nut species and walnut milk beverage authentication. *J. Sci. Food Agric.* **2020**, *100*, 2372–2379. [CrossRef]
10. Han, J.; Pang, X.; Liao, B.; Yao, H.; Song, J.; Chen, S. An authenticity survey of herbal medicines from markets in China using DNA barcoding. *Sci. Rep.* **2016**, *6*, 18723. [CrossRef]
11. Li, S.; Han, Q.; Qiao, C.; Song, J.; Cheng, L.C.; Xu, H. Chemical markers for the quality control of herbal medicines: An overview. *Chin. Med.* **2008**, *3*, 7. [CrossRef]
12. Nithaniyal, S.; Vassou, S.L.; Poovitha, S.; Raju, B.; Parani, M. Identification of species adulteration in traded medicinal plant raw drugs using DNA barcoding. *Genome* **2017**, *60*, 139–146. [CrossRef]
13. Pandit, R.; Travadi, T.; Sharma, S.; Joshi, C.; Joshi, M. DNA meta-barcoding using *rbcL* based mini-barcode revealed presence of unspecified plant species in Ayurvedic polyherbal formulations. *Phytochem. Anal.* **2021**, *32*, 804–810. [CrossRef] [PubMed]
14. Gill, B.A.; Musili, P.M.; Kurukura, S.; Hassan, A.A.; Goheen, J.R.; Kress, W.J.; Kuzmina, M.; Pringle, R.M.; Kartzinel, T.R. Plant DNA barcode library and community phylogeny for a semi-arid East African savanna. *Mol. Ecol. Resour.* **2019**, *19*, 838–846. [CrossRef] [PubMed]
15. Slipiko, M.; Myszczynski, K.; Buczkowska, K.; Baczkiewicz, A.; Szczecinska, M.; Sawicki, J. Molecular delimitation of European leafy liverworts of the genus *Calypogeia* based on plastid super-barcodes. *BMC Plant Biol.* **2022**, *20*, 243. [CrossRef] [PubMed]
16. Nitta, J.H.; Chambers, S.M. Identifying cryptic fern gametophytes using DNA barcoding: A review. *Appl. Plant Sci.* **2022**, *10*, e11465. [CrossRef] [PubMed]
17. Kool, A.; de Boer, H.J.; Kruger, A.; Rydberg, A.; Abbad, A.; Bjork, L.; Martin, G. Molecular identification of commercialized medicinal plants in southern Morocco. *PLoS ONE* **2012**, *7*, e39459. [CrossRef]
18. Vassou, S.L.; Nithaniyal, S.; Raju, B.; Parani, M. Creation of reference DNA barcode library and authentication of medicinal plant raw drugs used in Ayurvedic medicine. *BMC Complement. Altern. Med.* **2016**, *16* (Suppl. 1), 186. [CrossRef]
19. Shanmughanandhan, D.; Ragupathy, S.; Newmaster, S.G.; Mohanasundaram, S.; Sathishkumar, R. Estimating Herbal Product Authentication and Adulteration in India Using a Vouchered, DNA-Based Biological Reference Material Library. *Drug Saf.* **2016**, *39*, 1211–1227. [CrossRef]
20. Urumarudappa, S.K.J.; Tungphatthong, C.; Sukrong, S. Mitigating the Impact of Admixtures in Thai Herbal Products. *Front. Pharmacol.* **2019**, *10*, 1205. [CrossRef]
21. Senapati, A.; Basak, S.; Rangan, L. A Review on Application of DNA Barcoding Technology for Rapid Molecular Diagnostics of Adulterants in Herbal Medicine. *Drug Saf.* **2022**, *45*, 193–213. [CrossRef]
22. Amritha, N.; Bhooma, V.; Parani, M. Authentication of the market samples of Ashwagandha by DNA barcoding reveals that powders are significantly more adulterated than roots. *J. Ethnopharmacol.* **2020**, *256*, 112725. [CrossRef]
23. Department of AYUSH, Ministry of health and Family Welfare, Government of India. *The Ayurvedic Pharmacopoeia of India*, 1st ed.; Department of AYUSH, Ministry of Health and Family Welfare, Government of India: New Delhi, India, 1990; Part 1; Volume I.
24. Department of AYUSH, Ministry of health and Family Welfare, Government of India. *The Ayurvedic Pharmacopoeia of India*, 1st ed.; Department of AYUSH, Ministry of Health and Family Welfare, Government of India: New Delhi, India, 1999; Part 1; Volume II.
25. Department of AYUSH, Ministry of health and Family Welfare, Government of India. *The Ayurvedic Pharmacopoeia of India*, 1st ed.; Department of AYUSH, Ministry of Health and Family Welfare, Government of India: New Delhi, India, 2001; Part 1; Volume III.
26. Department of AYUSH, Ministry of health and Family Welfare, Government of India. *The Ayurvedic Pharmacopoeia of India*, 1st ed.; Department of AYUSH, Ministry of Health and Family Welfare, Government of India: New Delhi, India, 2004; Part 1; Volume IV.
27. Department of AYUSH, Ministry of health and Family Welfare, Government of India. *The Ayurvedic Pharmacopoeia of India*, 1st ed.; Department of AYUSH, Ministry of Health and Family Welfare, Government of India: New Delhi, India, 2006; Part 1; Volume V.
28. Department of AYUSH, Ministry of health and Family Welfare, Government of India. *The Ayurvedic Pharmacopoeia of India*, 1st ed.; Department of AYUSH, Ministry of Health and Family Welfare, Government of India: New Delhi, India, 2008; Part 1; Volume VI.
29. Doyle, J.J.; Doyle, J.L. A rapid DNA isolation procedure for small quantities of fresh leaf tissue. *Phytochem Bull.* **1987**, *19*, 11–15.
30. Poovitha, S.; Stalin, N.; Balaji, R.; Parani, M. Multi-locus DNA barcoding identifies matK as a suitable marker for species identification in *Hibiscus* L. *Genome* **2016**, *59*, 1150–1156. [CrossRef] [PubMed]
31. Vassou, S.L.; Kusuma, G.; Parani, M. DNA barcoding for species identification from dried and powdered plant parts: A case study with authentication of the raw drug market samples of *Sida cordifolia*. *Gene* **2015**, *559*, 86–93. [CrossRef] [PubMed]
32. Levin, R.A.; Wagner, W.L.; Hoch, P.C.; Nepokroeff, M.; Pires, J.C.; Zimmer, E.A.; Sytsma, K.J. Family-level relationships of Onagraceae based on chloroplast *rbcL* and *ndhF* data. *Am. J. Bot.* **2003**, *90*, 107–115. [CrossRef]

33. Fazekas, A.J.; Burgess, K.S.; Kesanakurti, P.R.; Graham, S.W.; Newmaster, S.G.; Husband, B.C.; Percy, D.M.; Hajibabaei, M.; Barrett, S.C. Multiple multilocus DNA barcodes from the plastid genome discriminate plant species equally well. *PLoS ONE* **2008**, *3*, e2802. [CrossRef]
34. Kress, W.J.; Wurdack, K.J.; Zimmer, E.A.; Weigt, L.A.; Janzen, D.H. Use of DNA barcodes to identify flowering plants. *Proc. Natl. Acad. Sci. USA* **2005**, *102*, 8369–8374. [CrossRef]
35. Chen, S.; Yao, H.; Han, J.; Liu, C.; Song, J.; Shi, L.; Zhu, Y.; Ma, X.; Gao, T.; Pang, X.; et al. Validation of the *ITS2* region as a novel DNA barcode for identifying medicinal plant species. *PLoS ONE* **2010**, *5*, e8613. [CrossRef]
36. Purushothaman, N.; Newmaster, S.G.; Ragupathy, S.; Stalin, N.; Suresh, D.; Arunraj, D.R.; Gnanasekaran, G.; Vassou, S.L.; Narasimhan, D.; Parani, M. A tiered barcode authentication tool to differentiate medicinal *Cassia* species in India. *Genet Mol. Res.* **2014**, *13*, 2959–2968. [CrossRef]
37. Nithaniyal, S.; Parani, M. Evaluation of chloroplast and nuclear DNA barcodes for species identification in *Terminalia* L. *Biochem. Syst. Ecol.* **2016**, *68*, 223–229. [CrossRef]
38. Kumar, S.; Stecher, G.; Tamura, K. MEGA7: Molecular Evolutionary Genetics Analysis Version 7.0 for Bigger Datasets. *Mol. Biol. Evol.* **2016**, *33*, 1870–1874. [CrossRef]
39. Kimura, M. A simple method for estimating evolutionary rates of base substitutions through comparative studies of nucleotide sequences. *J. Mol. Evol.* **1980**, *16*, 111–120. [CrossRef] [PubMed]
40. Tamura, K.; Stecher, G.; Kumar, S. MEGA11: Molecular Evolutionary Genetics Analysis Version 11. *Mol. Biol. Evol.* **2021**, *38*, 3022–3027. [CrossRef] [PubMed]
41. Parvathy, V.A.; Swetha, V.P.; Sheeja, T.E.; Leela, N.K.; Chempakam, B.; Sasikumar, B. DNA Barcoding to Detect Chilli Adulteration in Traded Black Pepper Powder. *Food Biotechnol.* **2014**, *28*, 25–40. [CrossRef]
42. Zoschke, R.; Liere, K.; Börner, T. From seedling to mature plant: *Arabidopsis* plastidial genome copy number, RNA accumulation and transcription are differentially regulated during leaf development. *Plant J.* **2007**, *50*, 710–722. [CrossRef]
43. Seethapathy, G.S.; Raclariu-Manolica, A.C.; Anmarkrud, J.A.; Wangensteen, H.; de Boer, H.J. DNA Metabarcoding Authentication of Ayurvedic Herbal Products on the European Market Raises Concerns of Quality and Fidelity. *Front. Plant Sci.* **2019**, *10*, 68.
44. Urumarudappa, S.K.J.; Tungphatthong, C.; Prombutara, P.; Sukrong, S. DNA metabarcoding to unravel plant species composition in selected herbal medicines on the National List of Essential Medicines (NLEM) of Thailand. *Sci. Rep.* **2020**, *10*, 18259. [CrossRef]
45. Bhooma, V.; Parani, M. DNA barcoding after cloning identifies taxonomically diverse botanical adulterants in the market samples of saffron. *Res. J. Biotechnol.* **2018**, *13*, 29–33.
46. Srirama, R.; Senthilkumar, U.; Sreejayan, N.; Ravikanth, G.; Gurumurthy, B.R.; Shivanna, M.B.; Sanjappa, M.; Ganeshiah, K.N.; Shaanker, U.R. Assessing species admixtures in raw drug trade of Phyllanthus, a hepato- protective plant using molecular tools. *J. Ethnopharmacol.* **2010**, *130*, 208–215. [CrossRef]
47. Kumar, S.J.U.; Krishna, V.; Seethapathy, G.S.; Senthilkumar, U.; Ragupathy, S.; Ganeshaiah, K.N.; Ganesan, R.; Newmaster, S.; Ravikanth, G.; Shaanker, R.U. DNA barcoding to assess species adulteration in raw drug trade of "Bala" (genus: *Sida* L.) herbal products in South India. *Biochem. Syst. Ecol.* **2015**, *61*, 501–509. [CrossRef]
48. Rana, A.C. Plumbao zeylanica: A phytopharmacological review. *IJPSR* **2011**, *2*, 247–255.
49. Shankar, R.; Rawat, M.S. Medicinal plants used in traditional medicine in lohit and Dibang valley districts of Arunachal Pradesh. *IJTK* **2008**, *7*, 288–295.
50. Nelluri, N.R.; Kumar, P.; Agarwal, N.K.; Gouda, T.S.; Setty, S.R. Phytochemical and pharmacological evaluation of leaves of *Abutilon indicum*. *IJTK* **2003**, *2*, 79–83.
51. Sivarajan, V.V.; Balachandran, I. *Ayurvedic Drugs and Their Plant Sources*; Oxford and IBH Publishing Cooperation Private Limited: New Delhi, India, 1994; p. 71.
52. Bose, S.; Nag, S. Isolation of natural dyes from the flower of *Hibiscus rosa-sinensis*. *Am. J. PharmTech Res.* **2012**, *2*, 761–770.
53. Nartunai, G.; Susikumar, S.; Narayanan, K.; Ilavarasan, R. Macro-Microscopic Identification of Dried Flowers of *Hibiscus rosa-sinensis* L. and its Differentiation from *Rhododendron arboreum* Arboreum Sm. *Pharmacog. J.* **2019**, *11*, 613–616.
54. Kesari, A.N.; Gupta, R.K.; Singh, S.K.; Diwakar, S.; Watal, G. Hypoglycemic and antihyperglycemic activity of *Aegle marmelos* seed extract in normal and diabetic rats. *J. Ethnopharmacol.* **2006**, *107*, 374–379. [CrossRef]
55. Singh, S.K.; Rai, P.K.; Jaiswal, D.; Watal, G. Evidence-based Critical Evaluation of Glycemic Potential of Cynodon dactylon. *Evid. Based Complementary Altern. Med.* **2008**, *5*, 415–420. [CrossRef]
56. Nille, G.C.; Mishra, S.K.; Chaudhary, A.K.; Reddy, K.R.C. Ethnopharmacological, Phytochemical, Pharmacological, and Toxicological Review on *Senna auriculata* (L.) Roxb.: A Special Insight to Antidiabetic Property. *Front. Pharmacol.* **2021**, *12*, 647887. [CrossRef]
57. Chhatre, S.; Nesari, T.; Somani, G.; Kanchan, D.; Sathaye, S. Phytopharmacological overview of *Tribulus terrestris*. *Phcog. Rev.* **2014**, *8*, 45–51. [CrossRef]
58. Mowa, E.; Maas, E. Influence of resting period on fruits and secondary tubers of *Harpagophytum procumbens* in Namibia. *ISTJN* **2016**, *8*, 73–90.
59. Mundy, P.J.; Ncube, S.F. Devil's claw-A natural substitute for diclofenac? *Vulture News* **2014**, *67*, 43–47. [CrossRef]
60. Marshall, N.T. *Searching for a Cure: Conservation of Medicinal Wildlife Resources in East and Southern Africa*; TRAFFIC-International: Cambridge, UK, 1998.

61. Hebbar, S.S.; Harsha, V.H.; Shripathi, V.; Hegde, G.R. Ethnomedicine of Dharwad district in Karnataka, India-plants used in oral health care. *J. Ethnopharmacol.* **2004**, *94*, 261–266. [CrossRef]
62. Nadkarni, K.M. *Indian Materia Medica*; reprinted; Bombay Popular Prakashan: Mumbai, India, 2009; Volume 1.
63. Christina, V.L.; Annamalai, A. Nucleotide based validation of *Ocimum* species by evaluating three candidate barcodes of the chloroplast region. *Mol. Ecol. Resour.* **2014**, *14*, 60–68. [CrossRef]
64. Jürges, G.; Sahi, V.; Rodriguez, D.R.; Reich, E.; Bhamra, S.; Howard, C.; Slater, A.; Nick, P. Product authenticity versus globalisation-The Tulsi case. *PLoS ONE* **2018**, *13*, e0207763. [CrossRef]
65. Varadharajan, B.; Parani, M. DMSO and betaine significantly enhance the PCR amplification of ITS2 DNA barcodes from plants. *Genome* **2021**, *64*, 165–171. [CrossRef]

diversity

MDPI

Communication

First Eastern Mediterranean Record of *Xenoligophoroides cobitis*, the Only Dactylogyrid Monogenean Infecting Mediterranean Gobies: Just Arrived or Missed the Boat?

Maarten P. M. Vanhove [1,2,3,4,*], Sofia Giakoumi [2,5], Dimitris Zogaris [6], Marcelo Kovačić [7] and Tine Huyse [3,8]

1 Centre for Environmental Sciences, Research Group Zoology: Biodiversity and Toxicology, Hasselt University, Agoralaan Gebouw D, B-3590 Diepenbeek, Belgium
2 Institute of Marine Biological Resources and Inland Waters, Hellenic Centre for Marine Research, 46.7 km Athens-Sounio Avenue, 190 13 Anavyssos, Greece; sgiakoum@hcmr.gr
3 Laboratory of Biodiversity and Evolutionary Genomics, Department of Biology, KU Leuven, Charles Deberiotstraat 32, B-3000 Leuven, Belgium; tine.huyse@kuleuven.be
4 Department of Botany and Zoology, Faculty of Science, Masaryk University, Kotlářská 2, 611 37 Brno, Czech Republic
5 Department of Biology, National and Kapodistrian University of Athens, Panepistimiopolis Athens, 15784 Athens, Greece
6 Hydrobiology—Ichthyology Lab., Department of Ichthyology and Aquatic Environment (DIAE), University of Thessaly, 38 445 Volos, Greece; zogarisd@gmail.com
7 Natural History Museum Rijeka, Lorenzov Prolaz 1, 51000 Rijeka, Croatia; marcelo@prirodoslovni.com
8 Biology Department, Royal Museum for Central Africa, Leuvensesteenweg 13, 3080 Tervuren, Belgium
* Correspondence: maarten.vanhove@uhasselt.be

Citation: Vanhove, M.P.M.; Giakoumi, S.; Zogaris, D.; Kovačić, M.; Huyse, T. First Eastern Mediterranean Record of *Xenoligophoroides cobitis*, the Only Dactylogyrid Monogenean Infecting Mediterranean Gobies: Just Arrived or Missed the Boat?. *Diversity* **2022**, *14*, 580. https://doi.org/10.3390/d14080580

Academic Editors: Mark C. Belk and Simon Blanchet

Received: 31 May 2022
Accepted: 14 July 2022
Published: 22 July 2022

Abstract: Gobies and their ectoparasitic monogenean flatworms are promising models for species diversification because of their species richness. Recent decades have seen the discovery of several new species of *Gyrodactylus* (Monogenea: Gyrodactylidae) on European gobies, mostly in the sand goby lineage and especially in the eastern Mediterranean. However, the monogenean fauna of other gobies is much less understood. Therefore, we inspected five gobiid species (34 specimens, vouchered, with some representatives sequenced), sampled in Greece, for monogenean ectoparasites. Only specimens of the giant goby, *Gobius cobitis*, were infected; they harbored *Xenoligophoroides cobitis* (Monogenea: Dactylogyridae) on their gills. Here, we provide the first record from Greece, and the first ITS rDNA and COI sequences of the representative of this monotypic genus. Additionally, 28S rDNA was sequenced and compared with published data from across its known distribution, suggesting clinal variation. No sister-group for *Xenoligophoroides* could be proposed, nor could we explain the presence of a single known member of this genus on gobies, due to a lack of sequence data of closely related dactylogyrid monogeneans in public databases. Possible hypotheses include either the ancestral long-term presence on gobiids but "missing the boat" of the diversification events in the "*Gobius*-lineage", or a recent host switch from a non-gobiid host.

Keywords: barcoding; Dactylogyridae; Dactylogyrinae; ectoparasites; giant goby; Gobiidae; *Gobius cobitis*; Greece; Monogenea; Platyhelminthes

1. Introduction

Gobiidae is the most species-rich fish family worldwide, with 1964 valid species as of 28 May 2022 [1], and the most species-rich group of European marine fishes [2]. There are 76 Mediterranean species [3]. Their diversity makes them prime models in evolutionary biology. For example, the occurrence of representatives in a wide variety of salinity conditions, and the often high levels of endemicity of gobies, render them conducive to study biogeographical patterns in aquatic ecosystems [2]. Furthermore, the radiation and local adaptation events in gobies have been fruitfully exploited in speciation research [4–7].

In addition, several goby species are successful invaders, rendering them useful models for the genomics of colonization and invasion [8] or as proof-of-principle of the use of fish parasites to elucidate introduction pathways [9]. Indeed, not only are the gobies valuable targets for biodiversity research, the same goes for their parasites. European sand gobies, with the gyrodactylid monogeneans that infect them [10,11], constitute one of the best studied fish–*Gyrodactylus* host–parasite systems. Since flatworms belonging to *Gyrodactylus* von Nordmann, 1832 are considered "the drosophilids of the parasitic world", sand gobies and their gyrodactylids are, therefore, a promising model in ecological and evolutionary parasitology [12]. Indeed, the parasites of assemblages of closely related host species may reveal important insights in parasite speciation [13]. Goby parasites also hold a lot of potential for biodiversity discovery. The recent description of seven species of *Gyrodactylus* infecting freshwater sand gobies from the Balkan region, the center of endemism of these hosts, underscores that even European species diversity is far from fully inventoried [14]. Despite the recent focus on sand gobies, the discovery of *Gyrodactylus quadratidigitus* Longshaw, Pursglove et Shinn, 2003 on British *Thorogobius ephippiatus* (Lowe, 1839) by Longshaw et al. [15] illustrates that non-sand gobies also hold the promise of undiscovered gyrodactylids. Next to Gyrodactylidae, another species-rich family of Monogenea is represented among goby parasites: Dactylogyridae. In Europe, only one dactylogyrid species is reported from marine gobies: *Xenoligophoroides cobitis* (Ergens, 1963), with its only known host the giant goby, *Gobius cobitis* Pallas, 1814 [16,17]. In general, the discovery of a plethora of marine monogenean species is to be expected: Appeltans et al. [18] estimated that among marine flatworms, Monogenea is the group of which the lowest percentage (7 to 13%) of existing species has been formally described.

Here, we expand our survey of the monogenean parasite diversity of eastern Mediterranean gobies, specifically towards marine gobies outside of the sand goby lineage, with the expectation of retrieving representatives of *Gyrodactylus* and *Xenoligophoroides*.

2. Materials and Methods

2.1. Sampling and Morphological Characterization of Parasites

Gobies outside of the sand goby lineage were collected as bycatch by Vanhove et al. [19]. They were diagnosed by a minimum combination of characters that positively identified the collected specimens among species of the family Gobiidae in the CLOFNAM area ([20,21] and references therein). We focused on larger gobiid species that often may also occur in brackish water. The fish were inspected for monogeneans on their gills, body, and fins; also, the vial and medium were checked.

About half of the monogeneans recovered were transferred to a water droplet using a dissection needle, fixed in Hoyer's medium, and mounted between slide and coverslip for morphological characterization of haptoral and genital hard parts. Measurements and micrographs were taken under phase contrast with an Olympus BX61 microscope fitted with a DP71 camera and Olympus Stream Motion software. Since only dactylogyrid monogeneans were found (Figure 1a–e), measurements followed Sasal et al. [16], whose study was the most recent publication on dactylogyrid parasites of European gobies at the onset of this work.

The remaining specimens were stored in absolute ethanol for subsequent molecular work. To allow taxonomic identification of these animals, photographic vouchers were made prior to DNA extraction. To this end, flatworms were temporarily mounted in water and photographed under a 100× (oil immersion) phase contrast objective using a Leica DM5000B microscope equipped with a Leica DFC420C camera and LAS imaging software (Figure 1f,g). Host vouchers were deposited in the Natural History Museum Rijeka (Rijeka, Croatia) (PMR), and parasite vouchers in the general invertebrate collection of the Department of Zoology, Swedish Museum of Natural History (Stockholm, Sweden) (SMNH).

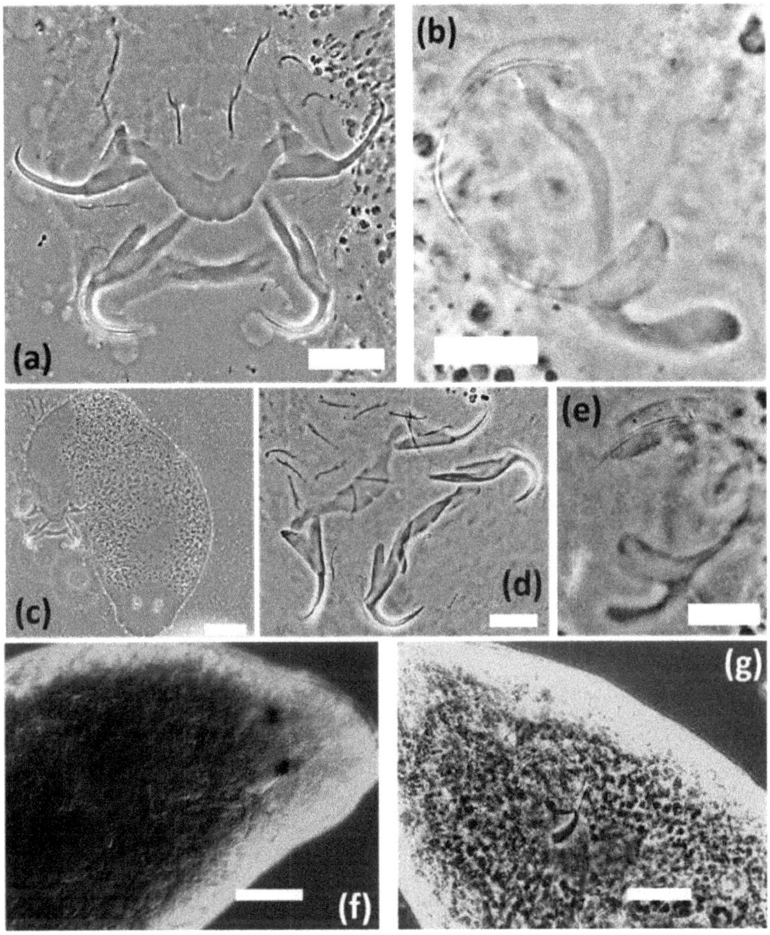

Figure 1. Micrographs of *Xenoligophoroides cobitis*. (**a**) Haptoral hard parts, specimen from Acheloos Delta. (**b**) Male copulatory organ, specimen from Acheloos Delta. (**c**) Whole mount, specimen from Acheloos Delta. (**d**) Haptoral hard parts, specimen from Kryoneri Estuary. (**e**) Male copulatory organ, specimen from Kryoneri Estuary. (**f**) Photo voucher of temporarily water-mounted specimen from Acheloos Delta. (**g**) Photo voucher of temporarily water-mounted specimen from Kryoneri Estuary. Both photo vouchers show the characteristic "bilobed" (sensu Sasal et al. [16]) or "two-chambered" base (sensu Dmitrieva et al. [17]) of the male copulatory organ of *X. cobitis*. Scale bars: 50 µm (**c**), 20 µm (**a**,**d**,**f**,**g**), 10 µm (**b**,**e**).

2.2. Molecular and Genetic Analysis

For the host specimens, DNA extraction, PCR amplification of mitochondrial 12S and 16S rDNA, subsequent purification of the PCR product, and Sanger sequencing were performed following the procedures described by Vanhove et al. [19]. Host sequences were deposited in the NCBI GenBank under accession numbers ON847338-45 (16S rDNA) and ON853912-19 (12S rDNA). Parasite DNA was extracted with the DNeasy Blood and Tissue Kit (Qiagen) according to the manufacturer's instructions. Three potential monogenean barcoding markers [22] with different mutation rates were amplified for the parasites: a fragment of the large nuclear ribosomal subunit gene (28S rDNA); the region spanning internal transcribed spacer (ITS) 1 and 2 including the 5.8S rDNA; and a fragment of the mitochondrial cytochrome *c* oxidase subunit 1 (COI). Although the former is widely used in monogenean molecular systematics and in somewhat deeper phylogenetic reconstruction (e.g., [23]), the combination of the two latter markers has recently proven valuable for phylogenetics of closely related monogenean species in the context of host (and parasite) radiation [24]. Sequences of the ITS region are well-established in the molecular taxonomy of gyrodactylid monogeneans and, hence, also in the species assemblage of *Gyrodactylus* on European gobies [10]; this includes the first-ever western Mediterranean record of members of *Gyrodactylus* [25]. Conversely, the broad applicability of COI as a marker for monogeneans and other flatworms has been questioned [22], although recent work on representatives of *Cichlidogyrus* Paperna, 1960 (Monogenea, Dactylogyridae) highlights its potential for barcoding [26]. Hence, taken together, sequence data of COI and of 28S rDNA and ITS rDNA served as versatile genetic resources for work on the monogenean goby parasites retrieved in this study. Primer combinations were C1 (5′-ACCCGCTGAATTTAAGCAT-3′) and D2 (5′-TGGTCCGTGTTTCAAGAC-3′) [27] for 28S rDNA, ITS1A (5′-GTAACAAGGTTTCCGTAGGTG-3′) and ITS2 (5′-TCCTCCGCTTAGTGATA-3′) [28] for ITS rDNA, and ASmit1 (5′–TTTTTTGGGCATCCTGAGGTTTAT–3′) [29] and Schisto3 (5′–TAATGCATMGGAAAAAAACA–3′) [30] for COI. In the latter case, the first polymerase chain reaction (PCR) was followed by a nested PCR, replacing the Schisto3 primer with ASmit2 (5′–TAAAGAAAGAACATAATGAAAATG–3′) [29]. We performed PCR using Illustra PuReTaq Ready-To-Go PCR Beads (GE Healthcare), adding 1 µL of each primer (20 µM) (Sigma Aldrich), 2 µL of template DNA, and 21 µL of double distilled, autoclaved, and filter-sterilized water. A GeneAmp PCR System 9700 (Applied Biosystems) thermocycler was used. Cycling conditions are outlined in Table 1.

Table 1. Polymerase chain reaction protocols for the genetic markers of monogenean parasites; the expected amplicon size is mentioned for each marker. The number of cycles was 39 for the fragment of large subunit rDNA, and 40 for internal transcribed spacer rDNA and the partial cytochrome *c* oxidase subunit 1 gene.

Protocol	Large Subunit 28S rDNA (*ca.* 700–900 bp)	Internal Transcribed Spacer rDNA (*ca.* 900–1200 bp)	Cytochrome *c* Oxidase Subunit 1 (*ca.* 445 bp)
initial denaturation	2 min/94 °C	3 min/96 °C	5 min/95 °C
cycle: denaturation	20 s/94 °C	50 s/95 °C	1 min/94 °C
annealing	30 s/56 °C	50 s/52 °C	1 min/50 °C
elongation	1 min 30 s/72 °C	50 s/72 °C	1 min/72 °C
final elongation	10 min/72 °C	7 min/72 °C	7 min/72 °C
cooling	4 °C	4 °C	4 °C

We purified the PCR product using the QIAquick PCR Purification Kit (Qiagen), following the manufacturer's guidelines. Bidirectional sequencing was carried out in an Applied Biosystems 3730 DNA analyzer using the BigDye protocol v.1.1. Sequences were validated by eye in MEGA v.7 [31] and aligned in the same software using ClustalW [32]. Pairwise distances were also calculated in MEGA. Sequences were subject to a BLAST search [33] on NCBI GenBank, and deposited there under accession numbers ON847354 (COI), ON853990-96 (28S rDNA) and ON854080-83 (ITS rDNA). In case published sequences of the same markers for conspecifics were found, a median-joining network [34] was inferred in PopART (http://popart.otago.ac.nz/index.shtml, accessed on 5 December 2021). Maps were rendered in QGIS [35].

3. Results

3.1. Host Records

A total of 34 specimens belonging to five goby species were checked for monogenean ectoparasites; species identities and sampling data are provided in Figure 2 and Table 2, rendering the host–parasite relationships traceable (see [36]). Interestingly, for *Gobius couchi* Miller & El-Tawil, 1974, our sample contains the first record in the Corinthian Gulf, quite geographically distant from surrounding records of this species from Corfu, Crete, and the northeastern Aegean Sea coast [37].

Figure 2. Species and localities sampled for gobies and parasites; see Table 2 for more sampling information.

Table 2. Non-sand gobies inspected for monogenean ectoparasites and hosts sequenced for host identification in this study.

Species	Locality	Sampling Date	Number of Host Specimens Inspected for Parasites/Sequenced	Voucher Specimens	GenBank Accession Numbers
Gobius cobitis Pallas, 1814	Acheloos Delta 38°20′17.6″ N 21°07′39.0″ E	10 June 2008	1/-	PMR VP 3175	/
	Kryoneri Estuary 38°22′23.4″ N 21°51′55.0″ E	7 June 2008	1/1	PMR VP 3215	ON847338 (16S rDNA), ON853912 (12S rDNA)
Gobius couchi [1] Miller & El-Tawil 1974	Lake Heraion 38°01′31.8″ N 22°52′34.6″ E	8 September 2008	1/-	PMR VP 3208	/
Gobius niger Linnaeus, 1758	Drepano Beach 39°30′54.3″ N 20°12′39.4″ E	8 June 2008	16/2	PMR VP 3179 to PMR VP 3194	ON847339-40 (16S rDNA), ON853913-14 (12S rDNA)
	Lake Heraion 38°01′31.8″ N 22°52′34.6″ E	8 September 2008	11/2	PMR VP 3195 to PMR VP 3206, PMR VP 3207, and PMR VP 3209	ON847341-42 (16S rDNA), ON853915-16 (12S rDNA)
Gobius ophiocephalus Pallas, 1814	Acheloos Delta 38°20′17.6″ N 21°07′39.0″ E	10 June 2008	2/1	PMR VP 3176, PMR VP 3178	ON847343 (16S rDNA), ON853917 (12S rDNA)
Gobius paganellus Linnaeus, 1758	Acheloos Delta 38°20′17.6″ N 21°07′39.0″ E	10 June 2008	1/1	PMR VP 3177	ON847344 (16S rDNA), ON853918 (12S rDNA)
	Euboea Island (Livadaki, Karystos) 38°00′15.8″ N 24°23′30.3″ E	3 June 2008	1/1	PMR VP 3210	ON847345 (16S rDNA), ON853919 (12S rDNA)

[1] The specimen PMR VP 3208 was identified as *G. couchi* based on the following diagnosis: (1) suborbital papillae of lateral-line system without longitudinal row *a*; (2) all three head canals of lateral-line system present; (3) predorsal area scaled; (4) six suborbital transversal papillae rows; (5) anterior oculoscapular head canal with pore α at rear of orbit; (6) oculoscapular papillae row *x1* not extending forwards to head canal pore β; (7) scales in lateral series 35–45 (present specimen had 35 and 36); (7) suborbital row *d* divided below between suborbital rows 2 and 3; (8) pelvic disc complete or no more than 1/8 emarginate (pelvic disc complete in present specimen); (9) pectoral fin count 15–18 (present specimen, for both sides, had 17).

3.2. Parasite Identification

Among the five goby species studied, only representatives of *G. cobitis* were infected by monogenean ectoparasites. A specimen caught in the Acheloos Delta was infected with nine monogenean gill parasites, five of which were used for genetic analyses and four of which were prepared as whole mounts (SMNH 207585-86). For an individual from Kryoneri Estuary, the infection intensity was seven. Three of these worms were subject to molecular analyses and the four others mounted on a slide (SMNH 207587). The two infected individuals were the only representatives of *Gobius cobitis* studied, leading to a prevalence of 100%. The results of measurements performed on the whole mounts are provided in Table 3. Based on Sasal et al. [16] and Dmitrieva et al. [17], all specimens belong to *X. cobitis* (Figure 1).

Table 3. Morphometric data of the haptoral and genital hard parts of specimens of *Xenoligophoroides cobitis* sampled in Greece. Measurements and their symbols follow Sasal et al. [16] with terminology adapted from Řehulková et al. [38]; measurements, all in µm, are presented as the range, followed by the average and the number of measured structures (n) in parentheses.

Parameter	Acheloos Delta Population	Kryoneri Estuary Population
Body		
Total length	387.6–476.2 (425.7, n = 4)	337.3–408.3 (369.3, n = 4)
Total width	140.1–181.4 (160.0, n = 4)	134.5–203.7 (157.0, n = 4)
Dorsal anchor		
Total length (a)	43.0–46.4 (44.5, n = 4)	49.6–53.7 (51.2, n = 3)
Length to notch (b)	32.3–34.1 (33.2, n = 4)	36.4–40.7 (38.6, n = 3)
Inner root length (c)	18.3–22.9 (20.3, n = 4)	23.7–24.6 (24.2, n = 3)
Outer root length (d)	5.1–9.6 (8.0, n = 4)	9.5–11.5 (10.5, n = 3)
Point length (e)	11.8–16.4 (14.4, n = 4)	13.5–20.0 (17.0, n = 3)
Ventral anchor		
Total length (a)	34.0–40.1 (37.4, n = 4)	36.8–40.1 (38.0, n = 3)
Length to notch (b)	40.4–44.2 (42.0, n = 4)	42.8–46.2 (45.0, n = 3)
Inner root length (c)	10.3–10.9 (10.6, n = 2)	9.8–10.9 (10.4, n = 3)
Outer root length (d)	5.4–6.3 (6.0, n = 3)	6.4–9.3 (8.2, n = 3)
Point length (e)	4.3–5.5 (4.7, n = 4)	5.6–6.3 (6.1, n = 3)
Dorsal bar		
Branch length (h)	35.2–38.4 (36.2, n = 4)	40.1–43.3 (41.7, n = 3)
Thickness at mid-length (w)	8.7–13.5 (11.7, n = 4)	12.8–19.3 (15.4, n = 3)
Total straight width (x)	37.2–59.3 (50.7, n = 4)	56.6–72.4 (67.0, n = 3)
Ventral bar		
Thickness at mid-length (w)	4.6–6.2 (5.5, n = 4)	7.7–9.2 (8.4, n = 3)
Total straight width (x)	41.1–45.0 (43.0, n = 4)	50.6–54.0 (52.3, n = 3)
Hook		
Pair I: total length (o)	15.5–18.1 (16.9, n = 4)	16.9–19.2 (17.7, n = 3)
Pair I: shank length (p)	9.1–12.2 (10.9, n = 4)	10.7–12.7 (11.5, n = 3)
Pair V: total length (o)	13.8–16.0 (14.9, n = 4)	14.6–15.5 (15.1, n = 3)
Pair V: shank length (p)	8.6–9.5 (9.1, n = 4)	8.3–10.2 (9.2, n = 3)
Other pairs: total length (o)	14.2–18.4 (16.4, n = 20)	14.1–20.7 (17.5, n = 18)
Other pairs: shank length (p)	8.7–11.7 (10.4, n = 20)	8.7–14.4 (11.5, n = 18)
Male copulatory organ		
Copulatory tube total straight length (q)	23.3–31.6 (26.7, n = 4)	14.7–35.2 (24.5, n = 4)
Total straight length of the base of the copulatory tube (r)	15.3–19.2 (16.5, n = 4)	13.9–21.0 (17.5, n = 4)
Copulatory tube total curved length (s)	46.7–53.7 (50.4, n = 4)	53.3–60.9 (56.2, n = 4)

3.3. Sequence Analyses

After trimming, sequence fragments of a maximal length of 858 bp (28S rDNA), 880 bp (ITS rDNA), and 171 bp (COI) were retained. Although only a single COI sequence was retrieved (from the Acheloos Delta), the four obtained ITS sequences (two from each locality) yielded a maximal uncorrected pairwise distance of 0.9% between specimens collected from the two different sites.

After searching nucleotide BLAST for the ITS fragment of *X. cobitis*, the highest identity score (93.2%) was found for two species of *Dactylogyrus* Diesing, 1850 (unpublished sequences KX369215 and KX369219), followed by a score of 91.2% for several representatives of *Cichlidogyrus* (sequences of [24,39]). These high scores were only found for a fragment that covered 23–25% of the total query (ca. 220 bp), or more specifically, in the region spanning 5.8S rDNA and internal transcribed spacer 2. For the COI fragment, the sequences with the highest pairwise similarity score belonged to *Kapentagyrus tanganicanus* Kmentová, Gelnar et Vanhove, 2018 (between 79.5% and 80.2% identity, sequences of [40]), followed by *Euryhaliotrema pirulum* (Plaisance et Kritsky, 2004) (identity 84.5%, sequence of [41]) and by species of *Cichlidogyrus* (maximal similarity of 82.0%, sequences of [24]) and *Sciadicleithrum* Kritsky, Thatcher et Boeger, 1989 (maximal similarity of 81.4%, sequences of [42]).

Only a single 28S rDNA genotype was found for all seven specimens successfully sequenced for this marker. Since other 28S rDNA sequences of *X. cobitis* are available [17], we only carried out intraspecific comparisons for this marker. The Greek genotype differed 0.7 to 3.1% (uncorrected p-distance) from those from the Black Sea, and between 0.6 and 1.3% from Sardinian conspecifics. The median-joining haplotype network, including all other published sequences of *X. cobitis* for this marker, situated the Greek population in between the Sardinian and Black Sea ones (Figure 3). In contrast to the Greek specimens all sharing a genotype, all genotypes from the Sardinian and Black Sea populations were unique.

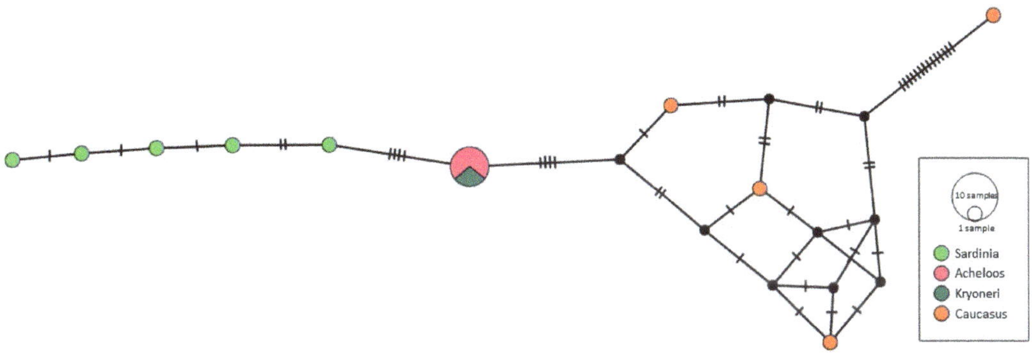

Figure 3. Median-joining haplotype network based on 702 bp of 28S rDNA from the newly sequenced individuals of *Xenoligophoroides cobitis* from Greece, and the sequences from Dmitrieva et al. [17]. Genotypes are represented by circles with the size of the circle correlating with the number of specimens displaying the respective genotype. Colors denote sampling localities; genotypes are connected with lines, indicating the number of mutations between them. Colors correspond to the sampling localities in Figure 4.

Figure 4. Distribution of *Xenoligophoroides cobitis*. Star: type locality, Gulf of Vlorë, Albania. Triangles: previously published records without accompanying genetic data (overview: see [17] and references therein). Green: population from northwestern Sardinia, Italy; orange: population from Gelendzhik, Russia, both sequenced by Dmitrieva et al. [17]. Pink: population from the Acheloos Delta, Greece; turquoise: population from Kryoneri Estuary, Greece, both sequenced in the present study. Colors correspond to those in the haplotype network (Figure 3).

4. Discussion

To further our understanding of the monogenean fauna of eastern Mediterranean gobies, we screened a number of Greek gobies belonging to five species for monogenean ectoparasites. The 2 individuals of the giant goby *G. cobitis* harbored a total of 16 dactylogyrid flatworms on their gills, morphologically and genetically identified as *X. cobitis*; no other monogeneans were found in any of the studied individuals. Given these infection parameters, and since *X. cobitis* is, here and in other studies (e.g., [16,17,43]), consistently reported on *G. cobitis* from different localities, an accidental infection is unlikely.

Ergens [43] described this parasite species as *Ancyrocephalus cobitis*. It was synonymized with *Haliotrema cupensis* Sasal, Pages et Euzet, 1998 and assigned to *Haliotrema* Johnston et Tiegs, 1922 by Merella et al. [44] as *Haliotrema cobitis* (Ergens, 1963). Dmitrieva et al. [17] erected the new monospecific genus *Xenoligophoroides* Dmitrieva, Sanna, Piras, Garippa et Merella, 2018 for it. With its type locality in the Adriatic Sea, and earlier observations from the western Mediterranean and the Black Sea, we here report *X. cobitis* (and hence any representative of *Xenoligophoroides*) for the first time in Greece, in the Ionian Sea and the eastern Mediterranean (Figure 4).

Dmitrieva et al. [17] provided an overview of morphometric data of the various populations of *X. cobitis* hitherto studied. In addition to the similarity these authors observed in overall body morphology, in shape of the hard parts, and in soft-part anatomy, they also mentioned considerable size ranges. For example, the (inner) length of the copulatory tube varies from minimally 25 μm in France to maximally 63 μm in the Black Sea; of the dorsal anchor, from 37 μm to 57 μm, and of the ventral anchor, from 25 μm to 45 μm (each time the minimal size in the French population and the maximal size in that of the Black Sea). Dmitrieva et al. [17] mentioned the geographical origin of the parasite populations, and the different sizes of the hosts studied from the Mediterranean compared to the Black Sea localities, as potential explanations for the size difference in the parasites. Measurements performed on the specimens from Greece fit within the ranges mentioned by Dmitrieva et al. [17]. It is noteworthy that the average value for most hard-part measurements is higher for the parasites collected at Kryoneri Estuary in comparison to their conspecifics in the Acheloos Delta (Table 3). The Kryoneri host specimen (total length = 88.9 mm, standard length = 72.1 mm + caudal fin length 16.8 mm) was larger than the one from the Acheloos (total length = 35.6 mm, standard length = 28.1 mm + caudal fin length 7.5 mm). Despite the limitations of our sample size, given the fact that these two Greek localities are only about 65 km apart, this could suggest an influence of host size on the size of the hard parts of *X. cobitis*. Morphological differences in various haptoral structures correlating with host size have been reported in other dactylogyrid monogeneans (e.g., [40]) though other studies did not find such correlations (e.g., [45] for a gill-infecting polyopisthocotylean monogenean). An increasing size of the gill lamellae in larger host specimens, potentially rendering larger haptoral hard parts an advantage for attachment to bigger-sized hosts, has been cited as a potential explanation for the link between host size and haptor morphology in monogeneans ([46] and references therein).

We consider the variation found in the ITS rDNA sequences of our specimens of *X. cobitis* to be intraspecific, as it remains below the threshold of 1% divergence, for this marker often associated with a difference between species (in *Gyrodactylus*: [47]). In contrast to the Sardinian and Black Sea populations sequenced by Dmitrieva et al. [17], all parasites in our sample sequenced for 28S rDNA yielded a single identical genotype. This contrasts with the diversity found by Dmitrieva et al. [17], where all five specimens from a single site in Sardinia and all four specimens from a single site in the Black Sea had a unique genotype (Figure 3).

Based on phylogenetic analyses by Dmitrieva et al. [17] using 28S rDNA, the species most similar to *X. cobitis* all seemed to share a marine lifestyle, belonging to *Ergenstrema* Paperna, 1964, *Ligophorus* Euzet et Suriano, 1977, *Euryhaliotrema* Kritsky et Boeger, 2002, and *Haliotrematoides* Kritsky, Yang et Sun, 2009. Our BLAST analyses based on the ITS and COI markers pointed towards similarities with both marine (e.g., *Euryhaliotrema*) and freshwater (e.g., *Kapentagyrus* Kmentová, Gelnar et Vanhove, 2018) genera. We consider our BLAST results a consequence of the scarcity of barcoding data for monogenean flatworms. Hopefully, the advent of mitochondrial genomes of an ever-increasing phylogenetic range of monogeneans (e.g., [48,49]) will help alleviate this important data gap in the near future.

The most recent and much more comprehensive phylogeny of dactylogyrids [23] placed all above-mentioned genera under Dactylogyrinae, but did not confirm a close relationship between any of them and *X. cobitis*. This dactylogyrid phylogeny also positions the members of *Gobioecetes* Ogawa et Itoh, 2017, infecting freshwater and diadromous gobies, in the Palearctic Far East [50], in Dactylogyrinae. *Gobioecetes*, with its sister taxon *Ancyrocephalus mogurndae* (Yamaguti, 1940) that is known to infect the Asian freshwater goby *Gymnogobius urotaenia* (Hilgendorf, 1879) among other hosts (see [51]), belonged to an entirely different clade than *X. cobitis*, indicating that gobies have been colonized by members of Dactylogyrinae at least twice independently.

Kmentová et al. [23] could not suggest a lineage closely related to *Xenoligophoroides*. Hence, it is impossible to propose a scenario of how (and when) this monogenean, currently the only known dactylogyrid from European gobies, colonized its host. Additionally, the host phylogeny adds little information: as the "*Gobius*-lineage" also includes eastern Atlantic genera from Norway to South Africa, Ponto–Caspian freshwater representatives, and even has affinities with tropical Pacific and neotropical gobies [4,52], we cannot reliably hypothesize where this lineage first infected gobies, and whether this happened in a marine or freshwater environment.

In contrast to the species-rich assemblage of *Gyrodactylus* on European (sand) gobies, only a single species of *Xenoligophoroides* is known. Although one has to caution against overinterpretation as many more goby species should be studied for parasites, this difference in species richness between goby-infecting monogenean lineages could be due to, for example, a recent colonization of the "*Gobius*-lineage" (long after it diversified) from another host lineage, or to "missing the boat" of diversification of these gobies, i.e., the absence on the founder populations of diverging goby lineages (see [53]). In order to favor any of these scenarios, identifying the sister group relationships of *Xenoligophoroides* would be key, as would inspecting fish species occurring in sympatry with *G. cobitis*. It would perhaps allow the inference of host-switching events such as those proposed by Huyse et al. [54], who studied sand goby-infecting members of *Gyrodactylus*. These authors suggested recent host-switches between sand gobies and sticklebacks, and between sand gobies and eels. In this respect, it may be useful to reflect on the former taxonomic affinities of *X. cobitis*. This species was previously assigned to *Haliotrema* and to *Ancyrocephalus* Creplin, 1839. Although no other Mediterranean monogeneans are classified under *Haliotrema*, other monogeneans occurring in the Mediterranean (e.g., *Ancyrocephalus salinus* Paperna, 1964 infecting *Aphaniops dispar* (Rüppell, 1829)) currently belong to *Ancyrocephalus*, which can safely be assumed to be a catch-all taxon (see [51,55]). Therefore, we concur with Dmitrieva et al. [17] that it will be fruitful to verify their systematic position in general, and their potential affinity to *X. cobitis* in particular. Apart from, e.g., the killifish host of *A. salinus*, further screening of gobiids and fishes occurring sympatrically with gobiids seems a fruitful approach to increase our understanding of the Mediterranean gyrodactylid and dactylogyrid fauna.

Author Contributions: Conceptualization, M.P.M.V.; methodology, M.P.M.V.; formal analysis, M.P.M.V.; investigation, M.P.M.V. and M.K.; resources, S.G. and T.H.; data Curation, M.P.M.V. and M.K.; writing—original draft preparation, M.P.M.V.; writing—review and editing, S.G., M.K. and T.H.; visualization, D.Z.; supervision, T.H.; funding Acquisition, M.P.M.V., S.G. and T.H. All authors have read and agreed to the published version of the manuscript.

Funding: This research received support from the Special Research Fund of Hasselt University (BOF20TT06) and research grant 1513419N of the Research Foundation–Flanders (FWO–Vlaanderen). At the time of sampling, MPMV was the recipient of a PhD fellowship and TH of a postdoctoral fellowship from the Research Foundation–Flanders (FWO–Vlaanderen). MPMV was supported for fieldwork by a travel grant from the Belgian Network for Coastal Research (BeNCoRe: ENCORA coordination action contract number FP6-2004-Global-3–518120), and for lab work by the Synthesis of Systematic Resources (SYNTHESYS) project (GB-TAF-2984) (http://www.synthesys.info/), which is financed by the European Community Research Infrastructure Action under the FP7 Integrating Activities Programme, and by the European Centre for IchthyoParasitology (Centre of excellence financed by the Czech Science Foundation, project no. P505/12/G112).

Institutional Review Board Statement: This study was carried out in accordance with Research Permit 85404/130/9-1-2008 from the Greek Ministry of Environment, Energy and Climate Change, Directorate of Development and Protection of Forests and Natural Environment.

Data Availability Statement: Voucher specimens were deposited in the Natural History Museum Rijeka (PMR VP 3175 to PMR VP 3210, and PMR VP 3215), and in the Swedish Museum of Natural History (SMNH 207585-87). Sequences were deposited in NCBI GenBank, accession numbers ON847338-45, ON847354, ON853990-96, ON853912-19, and ON854080-83. Observations of *X. cobitis* were posted on iNaturalist under https://www.inaturalist.org/observations/119560694 to https://www.inaturalist.org/observations/119560698, https://www.inaturalist.org/observations/124777745, https://www.inaturalist.org/observations/124777746, https://www.inaturalist.org/observations/124774873, https://www.inaturalist.org/observations/124774874, https://www.inaturalist.org/observations/124774877, https://www.inaturalist.org/observations/124775414, https://www.inaturalist.org/observations/124775939, and https://www.inaturalist.org/observations/124777025 to https://www.inaturalist.org/observations/124777027.

Acknowledgments: Andrea Waeschenbach, D. Tim J. Littlewood, and the staff of the Wolfson Wellcome Biomedical Laboratories (Natural History Museum) are thanked for hosting and supporting the molecular work; Alcibiades N. Economou and Stamatis Zogaris (Hellenic Centre for Marine Research) for co-organizing this study and for hosting MPMV as a post-doctoral fellow; Walter A. Boeger (Universidade Federal do Paraná) and Eva Řehulková (Masaryk University) for sharing insights on dactylogyrid morphology and identification; Chahinez Bouguerche and Ulf Jondelius (Swedish Museum of Natural History) for curatorial services; Evgenija V. Dmitrieva (A.O. Kovalevsky Institute of Marine Biological Research), Daria Sanna (Università di Sassari), and Armando J. Cruz

Laufer (Hasselt University) for interesting discussions on the current study; and Filip A.M. Volckaert (KU Leuven) for PhD supervision of MPMV and for constructive comments on this manuscript.

Conflicts of Interest: The authors declare no conflict of interest.

References

1. Fricke, R.; Eschmeyer, W.N.; Fong, J.D. Eschmeyer's Catalog of Fishes: Genera/Species by Family/Subfamily. Available online: http://researcharchive.calacademy.org/research/ichthyology/catalog/SpeciesByFamily.asp (accessed on 28 May 2022).
2. Schwarzhans, W.; Agiadi, K.; Carnevale, G. Late Miocene–Early Pliocene evolution of Mediterranean gobies and their environmental and biogeographic significance. *Riv. Ital. Paleontol. Stratigr.* **2020**, *126*, 657–724.
3. Kovačić, M.; Renaoult, J.P.; Pillon, R.; Svensen, R.; Bogorodsky, S.; Engin, S.; Louisy, P. Identification of Mediterranean marine gobies (Actinopterygii: Gobiidae) of the continental shelf from photographs of "in situ" individuals. *Zootaxa* **2022**, *5144*. [CrossRef]
4. McCraney, W.T.; Thacker, C.E.; Alfaro, M.E. Supermatrix phylogeny resolves goby lineages and reveals unstable root of Gobiaria. *Mol. Phylogenet. Evol.* **2020**, *151*, 106862. [CrossRef] [PubMed]
5. Yamasaki, Y.Y.; Takeshima, H.; Kano, Y.; Oseko, N.; Suzuki, T.; Nishida, M.; Watanabe, K. Ecosystem size predicts the probability of speciation in migratory freshwater fish. *Mol. Ecol.* **2020**, *29*, 3071–3083. [CrossRef] [PubMed]
6. Bernardi, G. Inter-island local adaptation in the Galápagos Archipelago: Genomics of the Galápagos blue-banded goby, *Lythrypnus gilberti*. *Coral Reefs* **2021**, *41*, 625–633. [CrossRef]
7. Ito, R.K.; Harada, S.; Tabata, R.; Watanabe, K. Molecular evolution and convergence of the rhodopsin gene in *Gymnogobius*, a goby group having diverged into coastal to freshwater habitats. *J. Evol. Biol.* **2022**, *35*, 333–346. [CrossRef]
8. Adrian-Kalchhauser, I.; Blomberg, A.; Larsson, T.; Musilova, Z.; Peart, C.R.; Pippel, M.; Solbakken, M.H.; Suurväli, J.; Walser, J.-C.; Wilson, J.Y.; et al. The round goby genome provides insights into mechanisms that may facilitate biological invasions. *BMC Biol.* **2020**, *18*, 11. [CrossRef]
9. Huyse, T.; Vanhove, M.P.M.; Mombaerts, M.; Volckaert, F.A.M.; Verreycken, H. Parasite introduction with an invasive goby in Belgium: Double trouble? *Parasitol. Res.* **2015**, *114*, 2789–2793. [CrossRef]
10. Huyse, T.; Volckaert, F.A.M. Identification of a host associated species complex using molecular and morphometric analyses, with the description of *Gyrodactylus rugiensioides* n. sp. (Gyrodactylidae, Monogenea). *Int. J. Parasitol.* **2002**, *32*, 907–919. [CrossRef]
11. Huyse, T.; Volckaert, F.A.M. Comparing host and parasite phylogenies: *Gyrodactylus* flatworms jumping from goby to goby. *Syst. Biol.* **2005**, *54*, 710–718. [CrossRef]
12. Bakke, T.A.; Cable, J.; Harris, P.D. The biology of gyrodactylid monogeneans: The "Russian-doll killers". *Adv. Parasitol.* **2007**, *64*, 161–460. [CrossRef] [PubMed]
13. Pariselle, A.; Morand, S.; Deveney, M.; Pouyaud, L. Parasite species richness of closely related hosts: Historical scenario and "genetic" hypothesis. In *Hommage à Louis Euzet—Taxonomie, Écologie et Évolution des Métazoaires Parasites. Taxonomy, Ecology and Evolution of Metazoan Parasites*; Combes, C., Jourdane, J., Eds.; Presses Universitaires de Perpignan: Perpignan, France, 2003; pp. 147–166.
14. Vanhove, M.P.M.; Economou, A.N.; Zogaris, S.; Giakoumi, S.; Zanella, D.; Volckaert, F.A.M.; Huyse, T. The *Gyrodactylus* (Monogenea, Gyrodactylidae) parasite fauna of freshwater sand gobies (Teleostei, Gobioidei) in their centre of endemism, with description of seven new species. *Parasitol. Res.* **2014**, *113*, 653–668. [CrossRef] [PubMed]
15. Longshaw, M.; Pursglove, M.; Shinn, A.P. *Gyrodactylus quadratidigitus* n. sp. (Monogenea: Gyrodactylidae), a parasite of the leopard-spotted goby *Thorogobius ephippiatus* (Lowe) from the south-western coast of the UK. *Syst. Parasitol.* **2003**, *55*, 151–157. [CrossRef] [PubMed]
16. Sasal, P.; Pagès, J.R.; Euzet, L. *Haliotrema cupensis* n. sp. (Monogenea, Ancyrocephalidae) from a marine gobiid (Teleostei, Perciformes) of the Mediterranean coast. *Syst. Parasitol.* **1998**, *39*, 107–112. [CrossRef]
17. Dmitrieva, E.V.; Sanna, D.; Piras, M.C.; Garippa, G.; Merella, P. *Xenoligophoroides cobitis* (Ergens, 1963) n. g., n. comb. (Monogenea: Ancyrocephalidae), a parasite of *Gobius cobitis* Pallas (Perciformes: Gobiidae) from the Mediterranean and Black seas. *Syst. Parasitol.* **2018**, *95*, 625–643. [CrossRef]
18. Appeltans, W.; Ahyong, S.T.; Anderson, G.; Angel, M.V.; Artois, T.; Bailly, N.; Bamber, R.; Barber, A.; Bartsch, I.; Berta, A.; et al. The magnitude of global marine species diversity. *Curr. Biol.* **2012**, *22*, 2189–2202. [CrossRef]
19. Vanhove, M.P.M.; Economou, A.N.; Zogaris, S.; Larmuseau, M.H.D.; Giakoumi, S.; Kalogianni, E.; Volckaert, F.A.M.; Huyse, T. Phylogenetics and biogeography of the Balkan "sand gobies" (Teleostei, Gobiidae): Vulnerable species in need of taxonomic revision. *Biol. J. Linn. Soc.* **2012**, *105*, 73–91. [CrossRef]
20. Miller, P.J. Gobiidae. In *Fishes of the North-Eastern Atlantic and the Mediterranean* 3; Whitehead, P.J.P., Bauchot, M.-L., Hureau, J.-C., Nielsen, J., Tortonese, E., Eds.; UNESCO: Paris, France, 1986; pp. 1019–1085.
21. Kovačić, M. Checklist of gobies (Actinopterii: Gobiidae) of the Mediterranean Sea and a key for species identification. *Zootaxa* **2020**, *4877*, 75–101. [CrossRef]
22. Vanhove, M.P.M.; Tessens, B.; Schoelinck, C.; Jondelius, U.; Littlewood, D.T.J.; Artois, T.; Huyse, T. Problematic barcoding in flatworms: A case-study on monogeneans and rhabdocoels (Platyhelminthes). *ZooKeys* **2013**, *365*, 355–379. [CrossRef]

23. Kmentová, N.; Cruz-Laufer, A.J.; Pariselle, A.; Smeets, K.; Artois, T.; Vanhove, M.P.M. Dactylogyridae 2022: A meta-analysis of phylogenetic studies and generic diagnoses of parasitic flatworms using published genetic and morphological data. *Int. J. Parasitol.* **2022**, *52*, 427–457. [CrossRef]
24. Vanhove, M.P.M.; Pariselle, A.; Van Steenberge, M.; Raeymaekers, J.A.M.; Hablützel, P.I.; Gillardin, C.; Hellemans, B.; Breman, F.C.; Koblmüller, S.; Sturmbauer, C.; et al. Hidden biodiversity in an ancient lake: Phylogenetic congruence between Lake Tanganyika tropheine cichlids and their monogenean flatworm parasites. *Sci. Rep.* **2015**, *5*, 13669. [CrossRef] [PubMed]
25. Huyse, T.; Pampoulie, C.; Audenaert, V.; Volckaert, F.A.M. First report of *Gyrodactylus* spp. (Platyhelminthes: Monogenea) in the western Mediterranean sea: Molecular and morphological descriptions. *J. Parasitol.* **2006**, *92*, 682–690. [CrossRef] [PubMed]
26. Jorissen, M.W.P.; Vanhove, M.P.M.; Pariselle, A.; Snoeks, J.; Vreven, E.; Šimková, A.; Wamuini Lunkayilakio, S.; Chocha Manda, A.; Kapepula Kasembele, G.; Muterezi Bukinga, F.; et al. Molecular footprint of parasite co-introduction with Nile tilapia in the Congo Basin. *Org. Divers. Evol.* **2022**. [CrossRef]
27. Hassouna, N.; Michot, B.; Bachellerie, J.P. The complete nucleotide sequence of mouse 28S rRNA gene. Implications for the process of size increase of the large subunit rRNA in higher eukaryotes. *Nucleic Acids Res.* **1984**, *12*, 3563–3583. [CrossRef]
28. Matějusová, I.; Gelnar, M.; McBeath, A.J.A.; Collins, C.M.; Cunningham, C.O. Molecular markers for gyrodactylids (Gyrodactylidae: Monogenea) from five fish families (Teleostei). *Int. J. Parasitol.* **2001**, *31*, 738–745. [CrossRef]
29. Littlewood, D.T.J.; Rohde, K.; Clough, K.A. Parasite speciation within or between host species? Phylogenetic evidence from site-specific polystome monogeneans. *Int. J. Parasitol.* **1997**, *27*, 1289–1297. [CrossRef]
30. Lockyer, A.E.; Olson, P.D.; Østergaard, P.; Rollinson, D.; Johnston, D.A.; Attwood, S.W.; Southgate, V.R.; Horak, P.; Snyder, S.D.; Le, T.H.; et al. The phylogeny of the Schistosomatidae based on three genes with emphasis on the interrelationships of *Schistosoma* Weinland, 1858. *Parasitology* **2003**, *126*, 203–224. [CrossRef]
31. Kumar, S.; Stecher, G.; Tamura, K. MEGA7: Molecular evolutionary genetics analysis version 7.0 for bigger datasets. *Mol. Biol. Evol.* **2016**, *33*, 1870–1874. [CrossRef]
32. Higgins, D.; Thompson, J.; Gibson, T.; Thompson, J.D.; Higgins, D.G.; Gibson, T.J. CLUSTAL W: Improving the sensitivity of progressive multiple sequence alignment through sequence weighting, position-specific gap penalties and weight matrix choice. *Nucleic Acids Res.* **1994**, *22*, 4673–4680. [CrossRef]
33. Altschul, S.F.; Gish, W.; Miller, W.; Myers, E.W.; Lipman, D.J. Basic local alignment search tool. *J. Mol. Biol.* **1990**, *215*, 403–410. [CrossRef]
34. Bandelt, H.; Forster, P.; Röhl, A. Median-joining networks for inferring intraspecific phylogenies. *Mol. Biol. Evol.* **1999**, *16*, 37–48. [CrossRef] [PubMed]
35. QGIS Association. QGIS Geographic Information System. 2021. Available online: http://www.qgis.org (accessed on 30 May 2022).
36. Justine, J.L.; Rahmouni, C.; Gey, D.; Schoelinck, C.; Hoberg, E.P. The monogenean which lost its clamps. *PLoS ONE* **2013**, *8*, e79155. [CrossRef] [PubMed]
37. Renoult, J.P.; Pillon, R.; Kovačić, M.; Louisy, P. Frontiers in Fishwatching Series—Gobies of the North-eastern Atlantic and the Mediterranean: *Gobius* and *Thorogobius*. In *Les Cahiers de la Fondation Biotope*; Fondation Biotope: Cayenne, France, 2022; Volume 37, pp. 1–237.
38. Řehulková, E.; Mendlová, M.; Šimková, A. Two new species of *Cichlidogyrus* (Monogenea: Dactylogyridae) parasitizing the gills of African cichlid fishes (Perciformes) from Senegal: Morphometric and molecular characterization. *Parasitol. Res.* **2013**, *112*, 1399–1410. [CrossRef] [PubMed]
39. Vanhove, M.P.M.; Briscoe, A.G.; Jorissen, M.W.P.; Littlewood, D.T.J.; Huyse, T. The first next-generation sequencing approach to the mitochondrial phylogeny of African monogenean parasites (Platyhelminthes: Gyrodactylidae and Dactylogyridae). *BMC Genom.* **2018**, *19*, 520. [CrossRef]
40. Kmentová, N.; Koblmüller, S.; Van Steenberge, M.; Raeymaekers, J.A.M.; Artois, T.; De Keyzer, E.L.R.; Milec, L.; Muterezi Bukinga, F.; Mulimbwa N'sibula, T.; Masilya Mulungula, P.; et al. Weak population structure and expansive demographic history of the monogenean parasite *Kapentagyrus* spp. infecting clupeid fishes of Lake Tanganyika, East Africa. *Int. J. Parasitol.* **2020**, *50*, 471–486. [CrossRef]
41. Plaisance, L.; Rousset, V.; Morand, S.; Littlewood, D.T.J. Colonization of Pacific islands by parasites of low dispersal ability: Phylogeography of two monogenean species parasitizing butterflyfishes in the South Pacific Ocean. *J. Biogeogr.* **2008**, *35*, 76–87. [CrossRef]
42. Santacruz, A.; Barluenga, M.; Pérez-Ponce de León, G. The macroparasite fauna of cichlid fish from Nicaraguan lakes, a model system for understanding host-parasite diversification and speciation. *Sci. Rep.* **2022**, *12*, 3944. [CrossRef]
43. Ergens, R. Über *Pseudochetostoma leucisci* n. sp. (Trematoidea) und *Ancyrocephalus cobitis* n. sp. (Monogenoidea), zwei neue parasitische Würmer der fische Albaniens. *Z. Parasitenkd.* **1963**, *22*, 287–291. [CrossRef]
44. Merella, P.; Dmitrieva, E.V.; Piras, M.C.; Huyse, T.; Gerasev, P.; Garippa, G. Two monogenean species (Platyhelminthes) infecting *Gobius cobitis* Pallas, 1811 (Osteichthyes: Gobiidae) off Sardinia, western Mediterranean Sea. SOIPA XXVI Abstracts. *Parassitologia* **2010**, *52*, 359.
45. Baker, T.G.; Viricel, A.; Meraziz, L.; de Buron, I. Size variation of adult polyopisthocotylid *Metamicrocotyla macracantha* (Monogenea) in relation to host size. *Comp. Parasitol.* **2005**, *72*, 179–182. [CrossRef]
46. Lakshmi Perera, K.M. The effect of host size on large hamuli length of *Kuhnia scombri* (Monogenea: Polyopisthocotylea) from Eden, New South Wales, Australia. *Int. J. Parasitol.* **1992**, *22*, 123–124. [CrossRef]

47. Ziętara, M.S.; Lumme, J. Speciation by host-switching and adaptive radiation in a fish parasite genus *Gyrodactylus* (Monogenea, Gyrodactylidae). *Evolution* **2002**, *56*, 2445–2458. [CrossRef] [PubMed]
48. Zhang, D.; Li, W.X.; Zou, H.; Wu, S.G.; Li, M.; Jakovlić, I.; Zhang, J.; Chen, R.; Wang, G. Homoplasy or plesiomorphy? Reconstruction of the evolutionary history of mitochondrial gene order rearrangements in the subphylum Neodermata. *Int. J. Parasitol.* **2019**, *49*, 819–829. [CrossRef] [PubMed]
49. Zhang, D.; Zou, H.; Jakovlić, I.; Wu, S.G.; Li, M.; Zhang, J.; Chen, R.; Li, W.X.; Wang, G.T. Mitochondrial genomes of two *Thaparocleidus* species (Platyhelminthes: Monogenea) reveal the first rRNA gene rearrangement among the Neodermata. *Int. J. Mol. Sci.* **2019**, *20*, 4214. [CrossRef] [PubMed]
50. Nitta, M.; Nagasawa, K. *Gobioecetes longibasis* n. sp. (Monogenea: Dactylogyridae) from *Rhinogobius similis* Gill (Perciformes: Gobiidae) from Okinawa-jima Island, the Ryukyu Archipelago, southern Japan, with a new host record for *Gobioecetes biwaensis* Ogawa & Itoh, 2017. *Syst. Parasitol.* **2020**, *97*, 193–200. [CrossRef]
51. Pugachev, O.N.; Gerasev, P.I.; Gussev, A.V.; Ergens, R.; Khotenowsky, I. *Guide to Monogenoidea of Freshwater Fish of Palaeartic and Amur Regions*; Galli, P., Pugachev, O.N., Kritsky, D., Eds.; Ledizione-LediPublishing: Milan, Italy, 2009.
52. Agorreta, A.; San Mauro, D.; Schliewen, U.; Van Tassell, J.L.; Kovačić, M.; Zardoya, R.; Rüber, L. Molecular phylogenetics of Gobioidei and phylogenetic placement of European gobies. *Mol. Phylogenet. Evol.* **2013**, *69*, 619–633. [CrossRef]
53. Johnson, K.P.; Adams, R.J.; Page, R.D.M.; Clayton, D.H. When do parasites fail to speciate in response to host speciation? *Syst. Biol.* **2003**, *52*, 37–47. [CrossRef]
54. Huyse, T.; Audenaert, V.; Volckaert, F.A.M. Speciation and host-parasite relationships in the parasite genus *Gyrodactylus* (Monogenea, Platyhelminthes) infecting gobies of the genus *Pomatoschistus* (Gobiidae, Teleostei). *Int. J. Parasitol.* **2003**, *33*, 1679–1689. [CrossRef]
55. Kmentová, N.; Van Steenberge, M.; Raeymaekers, J.A.M.; Koblmüller, S.; Hablützel, P.I.; Muterezi Bukinga, F.; Mulimbwa N'sibula, T.; Masilya Mulungula, P.; Nzigidahera, B.; Ntakimazi, G.; et al. Monogenean parasites of sardines in Lake Tanganyika: Diversity, origin and intra-specific variability. *Contrib. Zool.* **2018**, *87*, 105–132. [CrossRef]

diversity

MDPI

Article

Phylogeographic and Morphological Analysis of *Botrylloides niger* Herdman, 1886 from the Northeastern Mediterranean Sea

Berivan Temiz [1,2], Esra Öztürk [2], Simon Blanchoud [3,*] and Arzu Karahan [2,*]

1 Developmental Biology and Genomics Laboratory, Department of Anatomy, Otago School of Medical Sciences, University of Otago, P.O. Box 56, Dunedin 9054, New Zealand
2 Institute of Marine Sciences, Middle East Technical University, Erdemli-Mersin 33731, Türkiye
3 Department of Biology, University of Fribourg, 1700 Fribourg, Switzerland
* Correspondence: simon.blanchoud@unifr.ch (S.B.); arzukarahan@ims.metu.edu.tr (A.K.)

Abstract: *Botrylloides niger* (class Ascidiacea) is an invasive marine filter-feeding invertebrate that is believed to originate from the West Atlantic region. This species of colonial tunicate has been observed in several locations along the coasts of Israel and around the Suez Canal, but it has not yet been reported on the coasts of the Northeastern Mediterranean Sea (NEMS), suggesting an ongoing Lessepsian migration. However, the extent of this invasion might be concealed by reports of other potentially misidentified species of *Botrylloides*, given that the strong morphological similarities within this genus renders taxonomical identification particularly challenging. In this study, we performed a phylogeographic and morphological analysis of *B. niger* in the NEMS. We collected 238 samples from 8 sampling stations covering 824 km of the coastlines of NEMS. We reported 14 different morphotypes, of which the orange-brown, orange, and brown-striped morphs were the most abundant. Using the mitochondrial cytochrome C oxidase I (COI) as a DNA barcode marker, we identified 4 haplotypes. The COI haplotypes clustered with the reference *B. niger* sequences from GenBank and differed significantly from the sister *Botrylloides* species. We confirmed our identification using three additional barcoding markers (Histone 3, 18S rRNA, and 28S rRNA), which all matched with over 99% similarity to reference sequences. In addition, we monitored a station for a year and conducted a temporal analysis of the collected colonies. The colonies were absent during the winter and spring, while new colonies were established in the summer and expanded during autumn. We performed demographic population analysis on our spatial data that identified a possible population subdivision at a sampling site, which might have been caused by local freshwater input. Herein, we present the first report on the presence of *Botrylloides niger* in the NEMS. This study represents a key step toward understanding the diversity and the propagation of this highly invasive species of colonial ascidians, both within the Mediterranean basin as well as globally.

Keywords: ascidian; DNA barcoding; COI; northeastern Mediterranean Sea; phylogeography

Citation: Temiz, B.; Öztürk, E.; Blanchoud, S.; Karahan, A. Phylogeographic and Morphological Analysis of *Botrylloides niger* Herdman, 1886 from the Northeastern Mediterranean Sea. *Diversity* **2023**, *15*, 367. https://doi.org/10.3390/d15030367

Academic Editor: Stephan Koblmüller

Received: 4 December 2022
Revised: 28 February 2023
Accepted: 1 March 2023
Published: 3 March 2023

1. Introduction

Phylogeography investigates the geographical distribution of genealogical lineages by combining spatial snapshots of a portion of a population [1]. Populations divided by long distances or geographical barriers may display a higher level of genetic differentiation, representing the accumulation of mutations acquired over long periods of isolation [2]. Geography is, thus, intrinsically connected with evolutionary relations, and this connection can be detected via molecular markers that estimate genetic variations to understand the linkage between populations [1]. Consequently, spatial analyses of genetic diversity can provide important insights into the evolution of entire populations over potentially large spatial and temporal scales.

While such analyses are reasonably accessible in terrestrial environments, in aquatic systems, the population dynamics are more turbulent and harder to estimate. Considering the development of maritime traffic and aquaculture, the establishment of non-indigenous marine organisms in new environments is inevitable [3]. In particular, the global spreading of invasive species is a major threat to native ecosystems, and their prevention is of major biological and economical interest. Thus, phylogeographic studies on marine environments are essential to estimate genetic diversity, isolation patterns, and hydrographic barriers to elucidate the dispersal of populations.

DNA barcoding is a prominent molecular tool used to identify species and catalogue biodiversity. The main marker used for barcoding is a 500–650 bp fragment of the mitochondrial cytochrome C oxidase subunit I (COI) gene. COI sequences are compared to global reference databases, such as GenBank or Barcoding of Life Database (BOLD) [4,5], to assign to the sample of interest the same species as the one of the closest sequences found among the references. COI sequences from the same population can be compared to estimate the genetic diversity of the population and identify haplotypes [6]. COI sequences can also be compared across spatial and temporal sampling locations to study the movement of genetic material and infer population dynamics [7]. Additional genetic markers include the sequences of the chromatin component Histone 3 (H3), the ribosomal small subunit 18S rRNA, and the ribosomal large subunit 28S rRNA, all of which have been used to resolve population dynamics [8].

In the Northeastern Mediterranean Sea (NEMS), encompassing the coasts of Türkiye, Syria, and Lebanon, DNA barcoding studies have increased in recent years to monitor possible invasive Lessepsian migrations from the Suez Canal into NEMS [9]. These studies have proven to be decisive for the implementation of marine conservation policies by providing the required scientific knowledge on the ecosystem's dynamics [9–21].

Botrylloides is a globally present genus of sessile marine invertebrate filter-feeders that belong to the Tunicata subphylum [22]. These colonial chordates live on hard substrata by establishing flat, hard, gelatin-encrusted colonies after the motile larva attaches and undergoes metamorphosis [23]. *Botrylloides* can also undergo asexual reproduction, which is called blastogenesis, whereby new zooids bud from the peribranchial wall of the parental zooids [24]. The blastogenic cycle culminates with the death and absorption of all parental zooids by the colony and the emergence of a new generation of zooids in a process known as the takeover [25]. Natural chimerism following the fusion of two closely related kins, hibernation, and even whole-body regeneration have commonly, but not exhaustively, been reported in *Botrylloides* [6,26–31]. The *Botrylloides* genus is composed of 21 reported species, all of which are morphologically similar, with zooids aligned in a ladder-like arrangement. Anatomical features that differentiate species can be extremely difficult to assess, such as the number of stigmata rows on the branchial basket, the presence of a pyloric caecum, or the location where the larvae incubate. Consequently, taxonomical assignments based on their anatomical features are challenging for species-level categorization [32].

Several species of *Botrylloides* have been identified as invasive and have been reported in the Mediterranean Sea [10,32–41]. *Botrylloides niger* [42] is classified within the Styelidae family and has been synonymized with reports of other colonial ascidians, including *Metrocarpa nigrum* [42], *Botryllus niger* [42], *Botryllus nigrum* [42], *Botrylloides chazaliei* [43], and *Botrylloides nigrum* [42]. *B. niger* is predicted to be native to the West Atlantic due to its frequent presence there, although it was first identified on the coasts of Bermuda, an island located in the temperate region of the North-Atlantic Ocean [6]. Peres documented the presence of *B. niger* on the Mediterranean coasts of Israel more than fifty years ago [44], a presence that was recently confirmed using DNA barcoding by Griggio et al. [45,46]. Since then, Halim and Messeih have reported its presence in the Suez Canal [41]; Sallona et al. in the Ionian Sea [47]; and Crocetta et al. at several sites in Italy [48–50].

The only record of *Botrylloides* in the NEMS is that of *Botrylloides leachii*, documented in the Mersin harbor in Türkiye [36]. However, this report was solely based on morphology and has a number of missing key anatomical insights, which suggests that the resulting taxonomical assignment at the species level could be debatable. In particular, pigmentation patterns of colonial ascidians were shown to be highly polymorphic, and, thus, not suitable for species determination [51]. For instance, Brunetti [38] and Reem et al. [10] disagree with Sheets et al. [6] on the identification of *Botrylloides* colonies on the coasts of Israel, with the former two assigning them as *B. leachii*, but the latter as *B. niger*.

In this study, we investigated the genetic and morphological diversity of *B. niger* colonies from the NEMS by combining COI barcoding with H3, 18S rRNA, and 28S rRNA to support our identification, and conducted a time-series sampling for a year to investigate the effect of seasonal changes on population genetics. Herein, we present the first report on the presence of *Botrylloides niger* in the NEMS. This study is a key step towards understanding the diversity and the propagation of this highly invasive species of colonial ascidians, both within the Mediterranean basin and globally.

2. Materials and Methods

2.1. Colonies Sampling

Spatial samples were collected from the coastal areas of the Antalya (Kemer-Side-Alanya-Tersane), Mersin (Tisan-Kızkalesi-Mezitli), and Hatay (Konacık) sites in September and October 2018 at water depths between 30–50 cm, as previously described for *Botrylloides anceps* (Figure 1A) [31]. Samples were collected from submerged stones using a single-edged razor blade. The colonies to be monitored were placed onto a microscope slide and attached with a sewing thread, while the samples for barcoding were put into 1.5 mL tubes filled with 70 % (v/v) ethanol. The sampling area's date, coordinates, salinity, and pH were documented (Table S1). In total, 238 samples were collected, 203 of them to be processed for DNA barcoding (Table S2). Of these samples, 100 were used in our spatial analysis, 65 in the temporal analysis, and 38 in both analyses. The Kızkalesi station was visited monthly for the time-series between November 2017 and October 2018, with samples collected in November 2017 and between July 2018 and October 2018 (Table S3).

2.2. Colonies Characterization

A total of 218 of our specimens were characterized via morphological examination regarding their zooid distribution, color, and habitat preferences. Pigmentation patterns for each morphotype (Figure S1) were characterized by adapting the criteria used for the sister genus *Botryllus* [51]. Living specimens that were taken to the laboratory were photographed under a stereo and a light microscope (Olympus SZX16-UC30 camera; Olympus CX43-ToupTek camera (Olympus, Tokyo, Japan; ToupTek Photonics, Hangzhou, China). Living colonies were kept at 20 °C in the IMS-METU aquaculture room. The temperature and light of the room were stable, and the salinity of the water was constant at 40 ppt. The colonies were fed and the water was replaced every two days.

Hibernation of the animals was characterized by visual inspection based on the regression of all the zooids, thus resulting in a dense vascular system. Blastogenic cycles were monitored from the ventral side of the zooids, and the budding through the atrial epithelium was monitored by visual inspection.

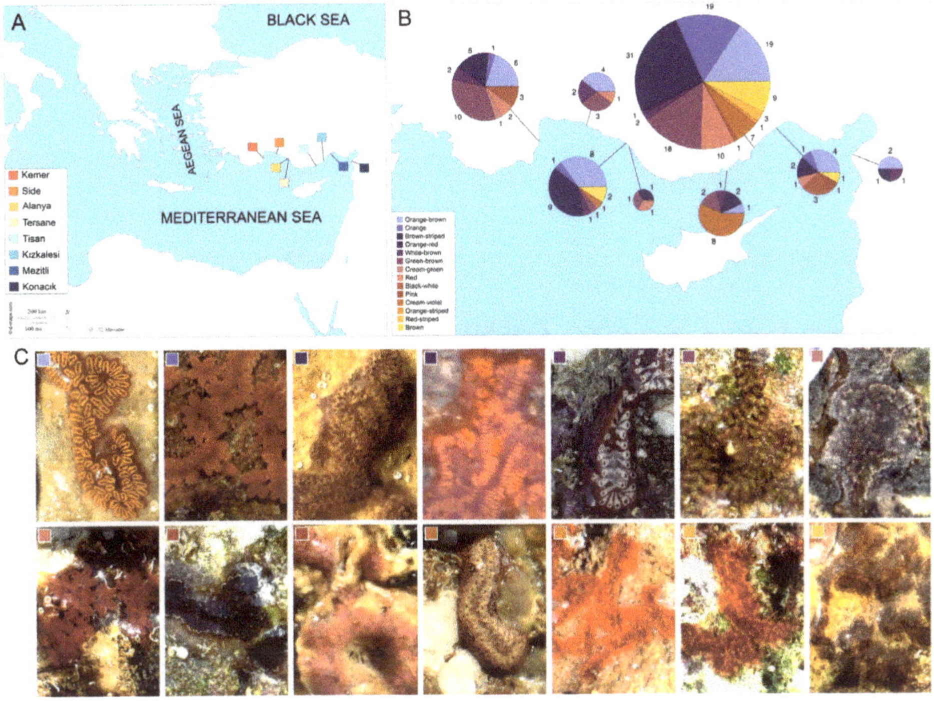

Figure 1. Morphotypes of *Botrylloides niger* from the sampling stations of the NEMS. (**A**) NEMS sampling sites, from west to east, cover the Antalya region (Kemer, Side, Alanya, Tersane), the Mersin region (Tisan, Kızkalesi, Mezitli) and the Hatay region (Konacık). (**B**) Regional diversity in the sampled morphotypes, depicted with pie charts proportional to the sample sizes. (**C**) In situ images of the corresponding morphotypes.

2.3. Molecular Analyses

DNA extractions were completed as previously described [31]. The amplification reactions were executed as previously described [10] for the COI, H3, 18S, and 28S genes (Table S4). All PCR products were purified with the NucleoSpin® DNA clean-up kit [52] and then sent to Macrogen Inc. (Seoul, Republic of Korea) for sequencing. The primers used in this study are given in Table S5. Sequences were aligned for each gene region separately using Clustal X v2 [53]. Acquired contigs were aligned on BioEdit version 7 [54], edited, and trimmed.

2.4. Haplotypes Network Analyses and Bayesian Trees

The minimum spanning network of haplotypes were calculated using Arlequin version 3.5.2.2 [55] and visualized via HapStar version 0.7 [56].

Bayesian trees of COI haplotypes and database-mined samples show the phylogenetic distances. The best model for the MrBayes v3.2 [57] was chosen via PhyML-SMS v3 software [58]. In total, four MrBayes runs (two independent for each) were conducted for the haplotypes (H) alone, as well as together, with all the database-mined samples (DM). The runs were performed based on the general time-reversible model with a proportion of invariable sites (GTR + I) for 900′000 (H) and 5′400′000 (DM) combined states with two independent runs. In total, 15 (H) and 45,002 (DM) trees were sampled after discharging a burn-in fraction of 25 % that verified the log likelihood of the cold chain (LnL) stationarity. As the convergence diagnostic, the average standard deviation of split frequencies was recorded as 0.004 (H) and 0.005 (DM), and potential scale reduction factors (PSRFs) were close to 1.0 for both [57]. The sister group of *Botrylloides*, the colonial ascidian *Symplegma*, was used as an out-group. The validity of the MCMC chains was confirmed by visual inspection of the LnL distribution to ensure stationarity. The final trees were visualized with FigTree v.1.4.4 [59].

2.5. Database Sequences

Ascidian sequences were mined from NCBI GenBank [4] in September 2022 to be used for species delimitation analyses. Sequences were selected based on their assigned genera, whether they had a voucher record, and whether they were approved by a taxonomist. All utilized reference sequences are listed in Table S6.

2.6. Species Analyses

Species delimitation analyses were carried out using the Automatic Barcode Gap Discovery method through the Assemble Species by Automatic Partitioning (ASAP, https://bioinfo.mnhn.fr/abi/public/asap/asapweb.html, accessed on 20 September 2022) [60] and the Poisson Tree Processes (PTP, http://species.h-its.org/ptp/, accessed on 20 September 2022) [61]. The hypothetical species are defined as Operational Taxonomic Units (OTUs) using these methods. ASAP clusters sequences into partitions consisting of hypothetical species based on the statistical inference of the "barcode gap", i.e., the gap in the distribution of intra-species and inter-species pairwise distances. ASAP analyses were performed using a web-based interface [60] (last accessed on 20 September 2022). Two metric options provided by ASAP for the pairwise distance calculations were used: Jukes–Cantor (JC69) [62] and Kimura 2 parameters (K80) [63]. This strategy allowed us to exclude possible biases of the selected evolutionary model for OTU delimitation. PTP analyses were conducted using the Bayesian implementation (bPTP; adds Bayesian support values to delimited species on the input tree), available on the web-based interface [61] (last accessed on 20 September 2022). MrBayes trees were generated using 100′000 Markov Chain Monte Carlo (MCMC) generations, subsampling every 100 generations, a burn-in fraction of 0.1, and 123 seed.

2.7. Population Analyses

Mean COI distances between the NEMS populations and the NCBI database sequences were calculated by a Kimura 2-parameter model [64] using MEGA version X [65]. The Blast suite of NCBI was used to find the matching substitution rates between bases [66]. The population-wide statistics, including the genetic diversity, neutrality test, and genetic differentiation, were calculated from the corresponding multiple sequence alignments of the COI locus per population with DnaSP version 6 [67].

The following four genetic diversity indices were measured: number of polymorphic sites (N_p), number of haplotypes (N_h), nucleotide diversity (π; window length: 100, step size: 25) [68], haplotype diversity (H_d, window length: 100, step size: 25) [69]. In addition, three associated neutrality test statistics were computed to test the hypothesis that all mutations were selectively neutral: Fu and Li's D* (F&LD) [68], Fu and Li's F* (F&LF) [70], and Tajima's D (TajD) [71].

To compare populations, the pairwise genetic differentiation (F_{st}, permutations number: 10,000) [72,73] and the population size changes were calculated with DnaSP. For the population size change, measured population mismatch distributions were compared to expected values for a population with a constant population size [74] using the raggedness statistic, r [75].

All figures were edited with Inkscape version 1.1 [76].

3. Results

3.1. Morphological Records

Overall, 14 different morphotypes were observed among the sampled colonies (Figure 1). Single- and double-colored morphs were recorded: orange-brown, orange, brown-striped, orange-red, white-brown, green-brown, cream-green, red, black-white, pink, cream-violet, orange-striped, red-striped, and brown (Figure S1). These 14 major morphotypes were composed of multiple sub-types with minute differences in the patterns of the colors.

All colonies presented the typical *Botrylloides* ladder-like organization, where zooids are aligned side by side while their dorsal lamina face the surrounding environment, and the ventral side is located on the attachment side (Figure S2). Four large and four small tentacles were recorded inside the buccal siphon, and eight alternative smaller protrusions were observed (Figure S2). Pigmented blood cells were recorded on the tentacles, especially on the two largest ones. Many pigmented blood cells on both sides of the endostyle were observed over the whole length of the animals.

3.2. Life History

The blastogenic cycle of 19 colonies was examined according to the Watanabe [77] four-phase (A–D) staging method. The duration of the cycle varied between 4 to 7 days (Figure S3). In stage A, two primary blastozooids were formed by budding from a single parental zooid. During stage B, blood flow was observed in the cardiac swellings of the primary buds. In stage C, secondary blastozooids were formed from the primary zooids. In stage D, known as takeover, the parental zooids were reabsorbed by the colony while the clonal primary zooids matured (Figure S3).

Fourteen of the nineteen cultured colonies hibernated during the winter season. During hibernation, there were no zooids in the dormant colony. The overall morphology thus consisted of only a carpet-like layer of ampullae (Figure S4). The blood circulation was lower and thicker than what is usually observed in healthy colonies. Termination of hibernation was not observed in any colony, even after the end of winter.

The life spans and morphologies of 19 colonies were documented (Table S7). The average life span of the *B. niger* colonies under lab conditions was ~5 months.

3.3. Network Analysis Based on COI

Our samples clustered under four haplotypes; H1, H2, H3, and H4 (Figure 2). The populations from the sampling sites of Kızkalesi, Mezitli, and Konacık were observed to contain more genetic variation, with three unique haplotypes in these regions. No correlation was identified between the genetic and morphological variations for the haplotypes. The highest divergence was in Konacık, where the main haplotype (H1) was separated by four mutation steps.

3.4. Species Delimitation Analysis

In total, five OTUs were assigned for all samples. *Botrylloides* cf. *lentus* (ON098245_1) was assigned in OTU-1, *Symplegma brakenhielmi* (LS992554_1) in OTU-2, all 27 *Botrylloides niger/nigrum/aff. leachii* reference samples together with all 11 from the present study in OTU-3, two uncharacterized *Botrylloides sp.* samples in OTU-4, and the 54 reference samples of *Botrylloides diegensis/leachii* in OTU-5 (Figure 3). All the main lineages reached 100% support, with some peripheral lineages having support as low as 60%. ASAP and PTP results supported identical species clustering (Figure S5).

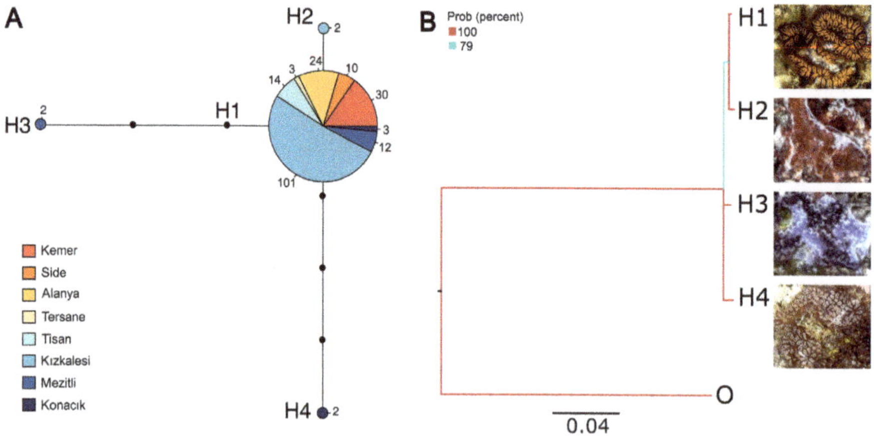

Figure 2. COI haplotype network of *B. niger* from the NEMS. (**A**) Minimum spanning network based on COI haplotypes of *B. niger* locus, overlaid with the corresponding haplotype name. Size differences of the circles indicate frequency, while colors reflect the regions where the colony was sampled. Numbers represent the number of samples in each category, and each correcting edge indicates a distance of one mutation step between the haplotypes. (**B**) Phylogenetic distance between the COI haplotypes is depicted by the best Bayesian tree, together with representative images of each haplotype. The branch colors represent bootstrap probability. The tree was rooted with the outgroup species *Symplegma brakenhielmi*.

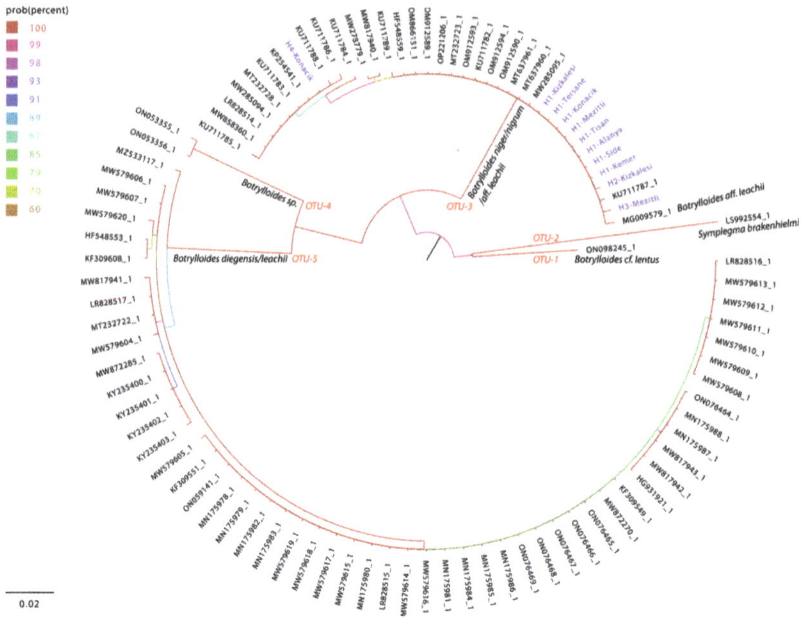

Figure 3. Bayesian majority rule consensus tree, reconstructed from the 519 bp COI sequence alignment. Support value is color-coded, as depicted on the left side of the tree. Corresponding OTUs, as determined by the ASAP and bPTP analyses, are written at the root of each group. The distance scale is given under the tree.

3.5. *Species Assignment by DNA Barcoding Analysis*

The final length of aligned and trimmed COI partial sequences was 519 bp. In total, we analyzed 203 COI sequences from the NEMS. The divergence among the studied populations was 0–0.4% (Table 1). The greatest distance was observed between the Konacık and the rest of the NEMS populations. The *B. aff. leachii* (MG009579) and *B. niger* (MW858360, LR828514, MW278779) database sequences diverged by less than 1% from the present populations. The distances between the NEMS samples and the database reference *B. leachii*/*B. diegensis* samples were ~17–23%.

Table 1. Pairwise mean COI distances between the NEMS populations and reference populations.

	1	2	3	4	5	6	7	8	9	10	11	12	13	14	15	16
1. Kızkalesi	-															
2. Mezitli	0.001	-														
3. Tershane	0.000	0.001	-													
4. Alanya	0.000	0.001	0.000	-												
5. Side	0.000	0.001	0.000	0.000	-											
6. Kemer	0.000	0.001	0.000	0.000	0.000	-										
7. Tisan	0.000	0.001	0.000	0.000	0.000	0.000	-									
8. Konacık	0.003	0.004	0.003	0.003	0.003	0.003	0.003	-								
9. BN-IL	0.000	0.001	0.000	0.000	0.000	0.000	0.000	0.003	-							
10. BL-IL	0.004	0.005	0.004	0.004	0.004	0.004	0.004	0.007	0.004	-						
11. BN-US	0.005	0.005	0.005	0.005	0.005	0.005	0.005	0.005	0.005	0.008	-					
12. BN-BR	0.008	0.008	0.008	0.008	0.008	0.008	0.008	0.005	0.008	0.011	0.005	-				
13. BL-IT	0.169	0.169	0.169	0.169	0.169	0.169	0.169	0.174	0.169	0.176	0.175	0.180	-			
14. BD-FR	0.196	0.195	0.196	0.196	0.196	0.196	0.196	0.201	0.196	0.192	0.202	0.209	0.008	-		
15. BL-ES	0.197	0.196	0.197	0.197	0.197	0.197	0.197	0.200	0.197	0.187	0.202	0.204	0.009	0.000	-	
16. BL-FR	0.230	0.230	0.230	0.230	0.230	0.230	0.230	0.232	0.230	0.234	0.232	0.234	0.196	0.203	0.199	-
17. Out-group	0.227	0.226	0.227	0.227	0.227	0.227	0.227	0.230	0.227	0.218	0.233	0.235	0.216	0.236	0.215	0.268

We sequenced 46 samples for the H3, three samples for the 18S, and two samples for the 28S genomic region. These sequences paired with at least 99% similarity to the reference sequences of *B. aff. leachii* (Figure S6) [10].

3.6. *Spatial Diversity Analysis Based on COI*

The evaluated regional population genetic diversity metrics (Table 2) showed that, despite its lower sample size (n = 5), the highest polymorphism (N_p = 4), haplotype diversity (N_h = 0.6), and nucleotide diversity (π = 0.0046) were measured in the Konacık population. No polymorphism was documented among the Tisan, Alanya, Side, Tersane, or Kemer populations.

Table 2. Genetic diversity indices and neutrality test statistics per NEMS geographical population based on COI sequences. Sequence numbers (n), number of polymorphic sites (N_p), number of haplotypes (N_h), observed haplotypes (H), nucleotide diversity (π), haplotype diversity (H_d), Fu and Li's D (F and LD), Fu and Li's F (F and LF), and Tajima's D (TajD). [a]: These groups have small population sizes (n < 10). Statistical significance: $p < 0.05$ (*).

Populations	n	N_p	N_h	H	π	H_d	F and LD	F and LF	TajD
Kemer	30	0	1	H1	0	0	0	0 [NS]	0
Side	10	0	1	H1	0	0	0	0 [NS]	0
Alanya	24	0	1	H1	0	0	0	0 [NS]	0
Tersane [a]	3	0	1	H1	0	0	0 [a]	0 [a]	0 [a]
Tisan	14	0	1	H1	0	0	0	0 [NS]	0
Kızkalesi	38	1	2	H1, H2	0.0001	0.053	−1.758	−1.823	−1.129
Mezitli	14	3	2	H1, H3	0.0015	0.264	1.070	0.757	−0.494
Konacık [a]	5	4	2	H1, H4	0.0046	0.600	1.641 [a]	1.670 [a]	1.641 [a]
All	138	7	4	H1, H2, H3, H4	0.0004	0.071	0.257	−0.557	−1.857 *

Fu and Li's D and Fu, Li's F, and Tajima's D tests were utilized to examine the neutrality of the mutations (Table 2). While no test was statistically significant at the level of geographical sub-populations, Tajima's D value for all the NEMS populations was significant negative (-1.857, $p < 0.05$).

Pairwise genetic differentiation (F_{st}) estimations (Table 3) showed that the Kemer-Mezitli pair was the only one that showed a statistically significant differentiation.

Table 3. Genetic differentiation (F_{st}) between the spatial populations. The statistical significance for F_{st} is indicated. [a]: These groups have small population sizes ($n < 10$). Statistical significance: p-value < 0.05 (*); n.c.: not calculated due to a lack of polymorphism.

F_{st}	Kemer	Side	Alanya	Tersane [a]	Tisan	Kızkalesi	Mezitli	Konacık [a]
Kemer	-	n.c.	n.c.	n.c.	n.c.	0	0.08 *	0.25 [a]
Side	-	-	n.c.	n.c.	n.c.	0	0.08	0.25 [a]
Alanya	-	-	-	n.c.	n.c.	0	0.08	0.25 [a]
Tersane [a]	-	-	-	-	n.c.	0 [a]	0.08 [a]	0.25 [a]
Tisan	-	-	-	-	-	0	0.08	0.25 [a]
Kızkalesi	-	-	-	-	-	-	0.07	0.25 [a]
Mezitli	-	-	-	-	-	-	-	0.17 [a]
Konacık [a]	-	-	-	-	-	-	-	-

Population size changes raggedness statistic (r) indicated that the only significant result was recorded for the Side population (Figure S7).

3.7. Temporal Diversity Analysis Based on COI

The genetic diversity metrics for the Kızkalesi time-series (Table 4) showed two haplotypes and one polymorphic site. The highest nucleotide and haplotype diversities were observed within the November population ($\pi = 0.0002$, $H_d = 0.111$), while no diversity was observed for the August–September period. The same polymorphism ($N_p = 1$) was recorded within the November and September populations.

Table 4. Genetic diversity indices and neutrality test statistics per Kızkalesi temporal population. Sequence numbers (n), number of polymorphic sites (N_p), number of haplotypes (N_h), nucleotide diversity (π), haplotype diversity (H_d), Fu and Li's D* (F and LD), Fu and Li's F* (F and LF) and Tajima's D (TajD). No value was measured with a statistically significant p-value < 0.05.

Populations	n	N_p	N_h	π	H_d	F and LD	F and LF	TajD
November 2017	18	1	2	0.0002	0.111	−1.450	−1.612	−1.165
July 2018	4	0	1	0	0	0	0	0
August 2018	17	0	1	0	0	0	0	0
September 2018	26	0	1	0	0	0	0	0
October 2018	38	1	2	0.0001	0.053	−1.758	−1.823	−1.129
All	103	1	2	0.0001	0.038	0.491	0.080	−0.912

The neutrality test statistics (Table 4) showed no statistically significant values for any population, but negative TajD values for the November and October populations. Pairwise genetic differentiation (F_{st}) and the population size changes raggedness statistic (r) showed no significant values for any of the temporal populations (Figure S8 and Table S8).

4. Discussion

In the present study, one primary (COI) and three additional molecular markers (H3, r18S, and r28S) were used to identify the genetic diversity of the *B. niger* colonies from the NEMS coasts of Türkiye. Taxonomical assignment based on the morphology of the samples was limited to the genus level (i.e., as a *Botrylloides* sp.) due to the high similarities between the sister species of this taxon [10,38,47,78]. Although the type of larval incubation and the structure of the pyloric cecum have been reported to support species differentiation [38], these features are very challenging to precisely assess during punctual field sampling.

The morphological characteristics of the ascidians, as a tool to distinguish species and genera, have remained cryptic, without consensus. According to Van Name [79] and Boyd et al. [80], morphological variations are accepted to have no taxonomic importance for ascidians. On the other hand, Tarjuelo et al. [81] suggested that it is possible to differentiate the color morphs of *Pseudodistoma crucigaster*, a colonial ascidian, based on COI locus. Furthermore, a recent study proposes that although there are morphological overlaps between *Botryllus* and *Botrylloides*, they possess different features [78]. While this proposition keeps the two genera separated, the morphological separation within *Botrylloides* remains untangled. Aside from the general morphological confusion of *B. niger* with its sister species *B. leachii*, *B. diegensis*, or *B. violaceus*, the color morphs that were given for *Botrylloides simodensis* and *Botrylloides praelongus* significantly resemble some morphs of the NEMS colonies, highlighting that *Botrylloides* species share highly similar morphotypes [82].

Based on our morphological examination, a ladder-like "leachii type" zooid organization was found in all NEMS colonies [38]. At least 14 major morphotypes with various types of striped pigmentation were recorded as sub-morphs within the NEMS. The morphological diversity was similar to the description of Sheets et al. [6], who indicated 8 different morphotypes for *B. niger* for the 16 worldwide locations. We observed most of the indicated morphotypes in our NEMS colonies, with the addition of a frequent green-violet morph that was not documented in their study. In this study, we assumed that *B. niger* had been introduced to the Mediterranean basin from the Atlantic Ocean, from which it has been proposed to originate [6]. However, the greater number of morphotypes that we found in the NEMS is challenging this hypothesis, since greater diversity suggests fewer bottlenecked populations and, thus, potentially fewer migrations. Similarly, a significant morph variation of *B. schlosseri* was recorded for the Mediterranean colonies, which suggested that different pigmentation patterns might result in different adaptive fitness levels [51]. Moreover, Mediterranean colonies of *B. schlosseri* indicated a mixture of native and non-native sub-species with diverse origins from the Pacific to the Atlantic [83,84]. Thus, to understand the history of *B. niger*, broader sampling is needed.

The variation in the life history characteristics of colonial ascidians is shaped by their high phenotypic and genetic plasticity to environmental changes [37,85]. Monitoring the blastogenic cycles of the different morphs showed that for all colonies, the blastogenic stages (A–D) were sequential, with a duration that varied from 4 to 7 days. These results are congruent with the previously suggested cycle duration of about one week [23,30]. The shorter cycles are probably related to hibernation [51,86]. We observed that most of the hibernating colonies in the lab were unable to recover. Hyams et al. [86] stated that ~80% of the hibernating colonies died within five months of the hibernation period. Hibernation seems to be one of the main reasons for the short life span of the colonies. To understand hibernation in their natural environment, we recorded the colonial diversity of our samplings in the winter period. We did not observe any colonies during the winter in the intertidal zone, where we collected our samples regularly.

The measured negative TajD value supports this interpretation of population expansion after a selective sweep, albeit without being statistically significant. However, Fu and Li's D* and Fu and Li's F* non-significant values suggest the opposite process, implying that the population has not undergone recent demographic or selective events. This apparent contradiction might be due to our sampling strategy. Indeed, F and LD and F and LF have notably little detection power for growing populations if the sampling rate is not focused around the period of maximal growth rate [87]. Consequently, having one single sampling period during August might prevent these two tests from detecting the selective sweep measured by TajD. Monitoring the colonies over more than one year, or at a higher frequency during summer, would help to clarify the demographic and selective processes that may affect the population.

Along with the morphological and life history characterization of the NEMS population, we also conducted genetic analyses. Molecular comparisons with the GenBank data showed that the current study sequences match 99–100 % with the *B. niger* sequences based on the COI locus; thus, we classified our species as *Botrylloides niger*. We found five OTUs based on our species delimitation analyses. OTU-3 consisted of all the *B. niger* haplotypes of the current study together with the GenBank *B. niger/nigrum/aff. leachii* sequences. The out-group (*Symplegma brakenhielmi*) and the other *Botrylloides* species, such as *Botrylloides diegensis*, were located in separate taxonomic units. We thus support COI as an adequate marker to identify and separate *Botrylloides* species, and *B. niger* in particular. Furthermore, our H3, r18S, and r28S sequences congruently matched with the same Israeli *B. aff. leachii* sequences whose COI barcode matched our COI barcode 100 % [10].

Despite covering only 8 stations along an 800 km coastline, the studied NEMS colonies demonstrated significant diversity, encompassing 14 morphotypes, 4 COI haplotypes, and 14 H3 genotypes. Furthermore, the overall haplotype diversity of COI was low (π_{COI}: 0.0004), which might stem from a recent bottleneck of the mitochondrial genome or from high introgression rates. Similarly, Sheets et al. [6] found low diversity of the *B. niger* populations on the Atlantic and Pacific coasts of COI and ANT loci.

Considering the significant negative Tajima's D values, the null hypothesis that the colonies of Türkiye's NEMS were overall under negative selection for COI was not rejected. The NEMS population seems to be under purifying selection, expanding from a restricted population or a selective sweep.

Likewise, Sheets et al. [6] found the *B. niger* populations on the Atlantic coast of Panama, the Pacific coast of Panama, and the coasts of Mexico to be under negative selection for the COI locus, which suggests a small population size, a founder effect, or a low dispersal. The selective forces acting as the environmental stressors on the mitochondrial DNA might be salinity or temperature [88–90].

The demographic population analyses based on the COI gene showed that the Side population was under significant subdivision, which could very likely result from major local freshwater input (the Manavgat stream). Freshwater inputs are known to cause the coastal salinity values to fluctuate. In support of this, Karahan et al. [91] stated that the dynamics of *B. schlosseri* populations from the California coasts were dramatically affected upon flooding events. Moreover, *B. violaceus* larvae exposed to low salinity were shown to express osmotic stress with an increased mortality rate, suggesting that colonial fitness would be reduced by seasonal storm events that cause seawater salinity to fluctuate [92]. Since the lowest salinity record (37.5 ppt) among the sampling stations belonged to Side, the population structure of the *B. niger* colonies appears to be affected by the lower salinity.

In this study, the population structures and interactions of *B. niger* from the NEMS were investigated using four molecular markers for eight spatial stations and one time-series station. Morphological results showed similar colonial characteristics regarding the blastogenic cycle and hibernation, but with a greater number of new morphotype records. Concerning the ambiguities in the classification of *Botrylloides*, previously suggested molecular markers were used to identify the populations from the Turkish coasts of the NEMS. As a result, the COI marker was observed to provide sufficient identification of the species.

Supplementary Materials: The following supporting information can be downloaded at: https://www.mdpi.com/article/10.3390/d15030367/s1. Figure S1: Description of the 14 morphotypes. Figure S2: Morphology of *Botrylloides niger*. Figure S3: The blastogenic cycle of *B. niger*. Figure S4: Hibernation of *B. niger*. Figure S5: Delimiting Operational Taxonomic Units (OTUs) and Bayesian tracer. Figure S6: Nuclear gene analyses of *B. niger*. Figure S7: Population size changes analysis based on the mismatch distribution of the COI gene for Kızkalesi, Mezitli, Konacık, and all populations. Figure S8: Population size changes of Kızkalesi time-series station based on mismatch distribution of COI. Table S1: Details of sampling stations. Table S2: Details of samplings and observed haplotypes and genotypes. Table S3: Kızkalesi site time-series samplings. Table S4: PCR programs for COI, H3, 28S rRNA and 18S rRNA genes. Table S5: Primers used in the present study. Table S6: Accession numbers for the GenBank reference sequences. Table S7: Life history of the 19 colonies cultured in the laboratory. Table S8:

Pairwise comparison of the genetic differentiation (Fst) between the temporal populations of Kızkalesi. Table S9: Correspondence between sample IDs, morphotypes and the determined haplotypes and genotypes for COI, H3, 18S and 28S.

Author Contributions: Conceptualization, B.T. and A.K.; data curation, B.T., E.Ö. and A.K.; methodology, B.T., E.Ö., S.B. and A.K.; writing—original draft, B.T., S.B. and A.K. All authors have read and agreed to the published version of the manuscript.

Funding: This study was supported by the YÖP-701-2018-2666 project (Middle East Technical University support program) and BAP-08-11-DPT2012K120880 for B.T., E.Ö and A.K.; and by the Swiss National Science Foundation (SNF) [grant number PZ00P3_173981] for S.B.

Institutional Review Board Statement: The study was conducted in accordance with the Animal testing regulations Directive 2010/63/EU and the national regulation of Türkiye.

Data Availability Statement: Sequences, trace files, image files, and the primer information for each of the four COI haplotype were uploaded to the Barcode of Life Data System [5] with the accession numbers IMS255-19, IMS256-19, FEE003-23, and FEE007-23, all grouped under the BOLD:ACI1328 BIN code. COI haplotype sequences were also uploaded to NCBI GenBank [4] with accession numbers OQ211497, OQ211498, OQ211500 and OQ211501. Similarly, the 14 Histone 3, the three 18S rRNA, and the two 28S rRNA sequences were submitted to BOLD with accession numbers FEE003-23, FEE007-23, FEE012-23 to FEE025-23 and IMS254-19 to IMS257-19, as well as to GenBank; accession numbers are pending attribution. Accession numbers of the COI sequences taken from GenBank are provided in Table S6 (data retrieved from the NCBI in September 2022). Haplotypes of the sampled colonies are provided in Table S9.

Acknowledgments: We thank Megan Wilson and Michael Hart for their valuable comments on the final version of the manuscript. We thank Baruch Rinkevich, Jacob Douek, and Guy Paz for their endless support and Kenan Murat Karahan for sampling help. We kindly thank Oscar Dalkjaer Wigant, Sonja Hummel and Aude Blanchoud for proofreading the manuscript. We thank Side and Erdemli Municipality for their kind assistance during our samplings. We also thank the anonymous reviewers for their helpful comments.

Conflicts of Interest: The authors declare no conflict of interest.

References

1. Avise, J.C. Phylogeography: Retrospect and Prospect. *J. Biogeogr.* **2009**, *36*, 3–15. [CrossRef]
2. Bowen, B.W.; Gaither, M.R.; DiBattista, J.D.; Iacchei, M.; Andrews, K.R.; Grant, W.S.; Toonen, R.J.; Briggs, J.C. Comparative Phylogeography of the Ocean Planet. *Proc. Natl. Acad. Sci. USA* **2016**, *113*, 7962–7969. [CrossRef] [PubMed]
3. Seebens, H.; Schwartz, N.; Schupp, P.J.; Blasius, B. Predicting the Spread of Marine Species Introduced by Global Shipping. *Proc. Natl. Acad. Sci. USA* **2016**, *113*, 5646–5651. [CrossRef]
4. Benson, D.A.; Karsch-Mizrachi, I.; Clark, K.; Lipman, D.J.; Ostell, J.; Sayers, E.W. GenBank. *Nucleic Acids Res.* **2012**, *40*, D48–D53. [CrossRef]
5. Ratnasingham, S.; Hebert, P.D.N. Bold: The Barcode of Life Data System (http://www.barcodinglife.org). *Mol. Ecol. Notes* **2007**, *7*, 355–364. [CrossRef] [PubMed]
6. Sheets, E.A.; Cohen, C.S.; Ruiz, G.M.; da Rocha, R.M. Investigating the Widespread Introduction of a Tropical Marine Fouling Species. *Ecol. Evol.* **2016**, *6*, 2453–2471. [CrossRef]
7. Villalobos, S.M.; Lambert, G.; Shenkar, N.; López-Legentil, S. Distribution and Population Dynamics of Key Ascidians in North Carolina Harbors and Marinas. *Aquat. Invasions* **2017**, *12*, 447–458. [CrossRef]
8. Reem, E.; Douek, J.; Rinkevich, B. A Critical Deliberation of the 'Species Complex' Status of the Globally Spread Colonial Ascidian *Botryllus schlosseri*. *J. Mar. Biol. Assoc. UK* **2021**, *101*, 1047–1060. [CrossRef]
9. Karahan, A.; Douek, J.; Paz, G.; Stern, N.; Kideys, A.E.; Shaish, L.; Goren, M.; Rinkevich, B. Employing DNA Barcoding as Taxonomy and Conservation Tools for Fish Species Censuses at the Southeastern Mediterranean, a Hot-Spot Area for Biological Invasion. *J. Nat. Conserv.* **2017**, *36*, 1–9. [CrossRef]
10. Reem, E.; Douek, J.; Rinkevich, B. Ambiguities in the Taxonomic Assignment and Species Delineation of Botryllid Ascidians from the Israeli Mediterranean and Other Coastlines. *Mitochondrial DNA A DNA Mapp. Seq. Anal.* **2018**, *29*, 1073–1080. [CrossRef]
11. Eryilmaz, L.; Dalyan, C. First Record of *Apogon queketti* Gilchrist (Osteichthyes: Apogonidae) in the Mediterranean Sea. *J. Fish Biol.* **2006**, *69*, 1251–1254. [CrossRef]
12. Smith, M.A.; Poyarkov, N.A.; Hebert, P.D.N. CO1 DNA Barcoding Amphibians: Take the Chance, Meet the Challenge. *Mol. Ecol. Resour.* **2008**, *8*, 235–246. [CrossRef]

13. Keskin, E.; Atar, H.H. DNA Barcoding Commercially Important Fish Species of Turkey. *Mol. Ecol. Resour.* **2013**, *13*, 788–797. [CrossRef]
14. Oter, K.; Gunay, F.; Tuzer, E.; Linton, Y.M.; Bellini, R.; Alten, B. First Record of *Stegomyia albopicta* in Turkey Determined By Active Ovitrap Surveillance and DNA Barcoding. *Vector-Borne Zoonotic Dis.* **2013**, *13*, 753–761. [CrossRef] [PubMed]
15. Azzurro, E.; Goren, M.; Diamant, A.; Galil, B.; Bernardi, G. Establishing the Identity and Assessing the Dynamics of Invasion in the Mediterranean Sea by the Dusky Sweeper, *Pempheris rhomboidea* Kossmann & Rauber, 1877 (Pempheridae, Perciformes). *Biol. Invasions* **2015**, *17*, 815–826. [CrossRef]
16. Bariche, M.; Torres, M.; Smith, C.; Sayar, N.; Azzurro, E.; Baker, R.; Bernardi, G. Red Sea Fishes in the Mediterranean Sea: A Preliminary Investigation of a Biological Invasion Using DNA Barcoding. *J. Biogeogr.* **2015**, *42*, 2363–2373. [CrossRef]
17. Seyhan, D.; Turan, C. DNA Barcoding of Scombrid Species in the Turkish Marine Waters. *J. Black Sea-Mediterr. Environ.* **2016**, *22*, 35–45.
18. Tuney, I. Molecular Identification of Puffer Fish *Lagocephalus sceleratus* (Gmelin, 1789) and *Lagocephalus spadiceus* (Richardson, 1845) from Eastern Mediterranean, Turkey. *Fresenius Environ. Bull.* **2016**, *25*, 1429–1437.
19. Ciftci, O.; Karahan, A.; Orek, Y.A.K.; Kideys, A.E. First Record of the Buccaneer Anchovy *Encrasicholina punctifer* (Fowler, 1938) (Clupeiformes; Engraulidae) in the Mediterranean Sea, Confirmed through DNA Barcoding. *J. Appl. Ichthyol.* **2017**, *33*, 520–523. [CrossRef]
20. Ozbek, E.O.; Mavruk, S.; Saygu, I.; Ozturk, B. Lionfish Distribution in the Eastern Mediterranean Coast of Turkey. *J. Black Sea-Mediterr. Environ.* **2017**, *23*, 1–16.
21. Golestani, H.; Crocetta, F.; Padula, V.; Camacho-Garcia, Y.; Langeneck, J.; Poursanidis, D.; Pola, M.; Yokes, M.B.; Cervera, J.L.; Jung, D.W.; et al. The Little *Aplysia* Coming of Age: From One Species to a Complex of Species Complexes in *Aplysia parvula* (Mollusca: Gastropoda: Heterobranchia). *Zool. J. Linn. Soc.* **2019**, *187*, 279–330. [CrossRef]
22. Shenkar, N.; Swalla, B.J. Global Diversity of Ascidiacea. *PLoS ONE* **2011**, *6*, e20657. [CrossRef]
23. Berrill, N.J. The Developmental Cycle of *Botrylloides*. *Q. J. Microsc. Sci.* **1947**, *88*, 393–407. [CrossRef] [PubMed]
24. Sabbadin, A. Experimental Analysis of the Development of Colonies of *Botryllus schlosseri* (Pallas) [Ascidiacea]. *Arch. Ital. Anat. Embriol.* **1958**, *63*, 178–221.
25. Manni, L.; Anselmi, C.; Cima, F.; Gasparini, F.; Voskoboynik, A.; Martini, M.; Peronato, A.; Burighel, P.; Zaniolo, G.; Ballarin, L. Sixty Years of Experimental Studies on the Blastogenesis of the Colonial Tunicate *Botryllus schlosseri*. *Dev. Biol.* **2019**, *448*, 293–308. [CrossRef] [PubMed]
26. Paz, G.; Rinkevich, B. Morphological Consequences for Multi-Partner Chimerism in *Botrylloides*, a Colonial Urochordate. *Dev. Comp. Immunol.* **2002**, *26*, 615–622. [CrossRef] [PubMed]
27. Rinkevich, B. Natural Chimerism in Colonial Urochordates. *J. Exp. Mar. Biol. Ecol.* **2005**, *322*, 93–109. [CrossRef]
28. Zondag, L.E.; Rutherford, K.; Gemmell, N.J.; Wilson, M.J. Uncovering the Pathways Underlying Whole Body Regeneration in a Chordate Model, *Botrylloides leachi* Using de Novo Transcriptome Analysis. *BMC Genom.* **2016**, *17*, 114. [CrossRef]
29. Blanchoud, S.; Zondag, L.; Lamare, M.D.; Wilson, M.J. Hematological Analysis of the Ascidian *Botrylloides leachii* (Savigny, 1816) during Whole-Body Regeneration. *Biol. Bull.* **2017**, *232*, 143–157. [CrossRef]
30. Blanchoud, S.; Rinkevich, B.; Wilson, M.J. Whole-Body Regeneration in the Colonial Tunicate *Botrylloides leachii*. *Results Probl. Cell Differ.* **2018**, *65*, 337–355. [CrossRef]
31. Karahan, A.; Öztürk, E.; Temiz, B.; Blanchoud, S. Studying Tunicata WBR Using *Botrylloides anceps*. In *Whole-Body Regeneration*; Springer Humana: New York, NY, USA, 2022; pp. 311–332.
32. Viard, F.; Roby, C.; Turon, X.; Bouchemousse, S.; Bishop, J. Cryptic Diversity and Database Errors Challenge Non-Indigenous Species Surveys: An Illustration With *Botrylloides* spp. in the English Channel and Mediterranean Sea. *Front. Mar. Sci.* **2019**, *6*, 615. [CrossRef]
33. Spallanzani, L. Storia naturale del mare. 1784. Unpublished. Available online: https://sites.google.com/site/ascidianbiologylab/clients (accessed on 28 February 2023).
34. Olivi, G. Zoologica Adriatica. *Biodivers. Herit. Libr.* 1792. Available online: https://www.marinespecies.org/aphia.php?p=taxdetails&id=542267 (accessed on 28 February 2023).
35. Savigny, J.C. *Memoires Sur Les Animaux sans Vertebres*; Deterville: Paris, France, 1816. [CrossRef]
36. Pinar, E. Fouling and Boring Organisms in Some Turkish Harbours and the Effectivity of Antifouling and Antiboring Paint against These Organisms. *Sci. Rep. Fac. Sci. Ege Univ.* **1974**, *170*, 1–67. (In Turkish)
37. Rinkevich, B.; Shlemberg, Z.; Lilker-Levav, T.; Goren, M.; Fishelson, L. Life History Characteristics of *Botrylloides* (Tunicata) Populations in Akko Bay, Mediterranean Coast of Israel. *Isr. J. Ecol. Evol.* **1993**, *39*, 197–212. [CrossRef]
38. Brunetti, R. Botryllid Species (Tunicata, Ascidiacea) from the Mediterranean Coast of Israel, with Some Considerations on the Systematics of *Botryllinae*. *Zootaxa* **2009**, *2289*, 18–32. [CrossRef]
39. Brunetti, R.; Mastrototaro, F. *Botrylloides Pizoni*, a New Species of *Botryllinae* (Ascidiacea) from the Mediterranean Sea. *Zootaxa* **2012**, *3258*, 28–36. [CrossRef]
40. Cinar, M.E. Checklist of the Phyla Platyhelminthes, Xenacoelomorpha, Nematoda, Acanthocephala, Myxozoa, Tardigrada, Cephalorhyncha, Nemertea, Echiura, Brachiopoda, Phoronida, Chaetognatha, and Chordata (Tunicata, Cephalochordata, and Hemichordata) from the Coasts of Turkey. *Turk. J. Zool.* **2014**, *38*, 698–722.

41. Halim, Y.; Messeih, M.A. Aliens in Egyptian Waters. A Checklist of Ascidians of the Suez Canal and the Adjacent Mediterranean Waters. *Egypt. J. Aquat. Res.* **2016**, *42*, 449–457. [CrossRef]
42. Herdman, W.A. Report on the Tunicata Collected during the Years 1873-1876. Part 2, Ascidiae Compositae. *Zool. Chall. Exp.* **1886**, *14*, 1–425.
43. Sluiter, C.P. Tuniciers Recueilli En 1896, Par La Chazalie, Dans La Mer Des Antilles. *Mém. Société Zool. Fr.* **1898**, *11*, 5–34.
44. Pérès, J.M. Ascidies Recoltées Sur Les Cotes Mediterraneenaires d'Israel. *Bull. Res. Counc. Isr.* **1958**, *7B*, 143–150.
45. Griggio, F.; Voskoboynik, A.; Iannelli, F.; Justy, F.; Tilak, M.-K.; Xavier, T.; Pesole, G.; Douzery, E.J.P.; Mastrototaro, F.; Gissi, C. Ascidian Mitogenomics: Comparison of Evolutionary Rates in Closely Related Taxa Provides Evidence of Ongoing Speciation Events. *Genome Biol. Evol.* **2014**, *6*, 591–605. [CrossRef]
46. Rubinstein, N.D.; Feldstein, T.; Shenkar, N.; Botero-Castro, F.; Griggio, F.; Mastrototaro, F.; Delsuc, F.; Douzery, E.J.; Gissi, C.; Huchon, D. Deep Sequencing of Mixed Total DNA without Barcodes Allows Efficient Assembly of Highly Plastic Ascidian Mitochondrial Genomes. *Genome Biol. Evol.* **2013**, *5*, 1185–1199. [CrossRef] [PubMed]
47. Salonna, M.; Gasparini, F.; Huchon, D.; Montesanto, F.; Haddas-Sasson, M.; Ekins, M.; McNamara, M.; Mastrototaro, F.; Gissi, C. An Elongated COI Fragment to Discriminate Botryllid Species and as an Improved Ascidian DNA Barcode. *Sci. Rep.* **2021**, *11*, 4078. [CrossRef] [PubMed]
48. Virgili, R.; Tanduo, V.; Katsanevakis, S.; Terlizzi, F.; Villani, G.; Fontana, A.; Crocetta, F. The Miseno Lake (Central-Western Mediterranean Sea): An Overlooked Reservoir of Non-Indigenous and Cryptogenic Ascidians in a Marine Reserve. *Front. Mar. Sci.* **2022**, *9*. [CrossRef]
49. Della Sala, G.; Coppola, D.; Virgili, R.; Vitale, G.A.; Tanduo, V.; Teta, R.; Crocetta, F.; de Pascale, D. Untargeted Metabolomics Yields Insights into the Lipidome of *Botrylloides Niger* Herdman, 1886, An Ascidian Invading the Mediterranean Sea. *Front. Mar. Sci.* **2022**, *9*, 866906. [CrossRef]
50. Micaroni, V.; Strano, F.; Crocetta, F.; Di Franco, D.; Piraino, S.; Gravili, C.; Rindi, F.; Bertolino, M.; Costa, G.; Langeneck, J.; et al. Project "Biodiversity MARE Tricase": A Species Inventory of the Coastal Area of Southeastern Salento (Ionian Sea, Italy). *Diversity* **2022**, *14*, 904. [CrossRef]
51. Cima, F.; Ballarin, L.; Caicci, F.; Franchi, N.; Gasparini, F.; Rigon, F.; Schiavon, F.; Manni, L. Life History and Ecological Genetics of the Colonial Ascidian *Botryllus schlosseri*. *Zool. Anz. -J. Comp. Zool.* **2015**, *257*, 54–70. [CrossRef]
52. Faber, K.L.; Person, E.C.; Hudlow, W.R. PCR Inhibitor Removal Using the NucleoSpin (R) DNA Clean-Up XS Kit. *Forensic Sci. Int.-Genet.* **2013**, *7*, 209–213. [CrossRef] [PubMed]
53. Larkin, M.A.; Blackshields, G.; Brown, N.P.; Chenna, R.; McGettigan, P.A.; McWilliam, H.; Valentin, F.; Wallace, I.M.; Wilm, A.; Lopez, R. Clustal W and Clustal X Version 2.0. *Bioinformatics* **2007**, *23*, 2947–2948. [CrossRef] [PubMed]
54. Hall, T.A. BioEdit: Version 7.0.0, A User-Friendly Biological Sequence Alignment Editor and Analysis Program for Windows 95/98/NT. *Nucleic Acids Symposium Series* **1999**, *41*, 95–98.
55. Excoffier, L.; Lischer, H.E. Arlequin Suite Ver 3.5: A New Series of Programs to Perform Population Genetics Analyses under Linux and Windows. *Mol. Ecol. Resour.* **2010**, *10*, 564–567. [CrossRef] [PubMed]
56. Teacher, A.G.F.; Griffiths, D.J. HapStar: Automated Haplotype Network Layout and Visualization. *Mol. Ecol. Resour.* **2011**, *11*, 151–153. [CrossRef] [PubMed]
57. Ronquist, F.; Teslenko, M.; Van Der Mark, P.; Ayres, D.L.; Darling, A.; Höhna, S.; Larget, B.; Liu, L.; Suchard, M.A.; Huelsenbeck, J.P. MrBayes 3.2: Efficient Bayesian Phylogenetic Inference and Model Choice across a Large Model Space. *Syst. Biol.* **2012**, *61*, 539–542. [CrossRef]
58. Lefort, V.; Longueville, J.-E.; Gascuel, O. SMS: Smart Model Selection in PhyML. *Mol. Biol. Evol.* **2017**, *34*, 2422–2424. [CrossRef] [PubMed]
59. Rambaut, A. FigTree v1.4.4. Available online: http://tree.bio.ed.ac.uk/software/figtree/ (accessed on 24 November 2022).
60. Puillandre, N.; Brouillet, S.; Achaz, G. ASAP: Assemble Species by Automatic Partitioning. *Mol. Ecol. Resour.* **2021**, *21*, 609–620. [CrossRef] [PubMed]
61. Zhang, J.; Kapli, P.; Pavlidis, P.; Stamatakis, A. A General Species Delimitation Method with Applications to Phylogenetic Placements. *Bioinformatics* **2013**, *29*, 2869–2876. [CrossRef] [PubMed]
62. Jukes, T.H.; Cantor, C.R. Evolution of Protein Molecules. *Mamm. Protein Metab.* **1969**, *3*, 21–132.
63. Kimura, M. A Simple Method for Estimating Evolutionary Rates of Base Substitutions through Comparative Studies of Nucleotide Sequences. *J. Mol. Evol.* **1980**, *16*, 111–120. [CrossRef] [PubMed]
64. Nei, M.; Kumar, S. *Molecular Evolution and Phylogenetics*; Oxford University Press: Oxford, UK, 2000.
65. Kumar, S.; Stecher, G.; Li, M.; Knyaz, C.; Tamura, K. MEGA X: Molecular Evolutionary Genetics Analysis across Computing Platforms. *Mol. Biol. Evol.* **2018**, *35*, 1547–1549. [CrossRef]
66. Johnson, M.; Zaretskaya, I.; Raytselis, Y.; Merezhuk, Y.; McGinnis, S.; Madden, T.L. NCBI BLAST: A Better Web Interface. *Nucleic Acids Res.* **2008**, *36*, W5–W9. [CrossRef]
67. Rozas, J.; Ferrer-Mata, A.; Sanchez-DelBarrio, J.C.; Guirao-Rico, S.; Librado, P.; Ramos-Onsins, S.E.; Sanchez-Gracia, A. DnaSP 6: DNA Sequence Polymorphism Analysis of Large Data Sets. *Mol. Biol. Evol.* **2017**, *34*, 3299–3302. [CrossRef] [PubMed]
68. Nei, M.; Li, W.-H. Mathematical Model for Studying Genetic Variation in Terms of Restriction Endonucleases. *Proc. Natl. Acad. Sci. USA* **1979**, *76*, 5269–5273. [CrossRef]
69. Nei, M. *Molecular Evolutionary Genetics*; Columbia University Press: New York, NY, USA, 1987.

70. Fu, Y.-X.; Li, W.-H. Statistical Tests of Neutrality of Mutations. *Genetics* **1993**, *133*, 693–709. [CrossRef]
71. Tajima, F. Evolutionary Relationship of DNA Sequences in Finite Populations. *Genetics* **1983**, *105*, 437–460. [CrossRef] [PubMed]
72. Weir, B.S.; Cockerham, C.C. Estimating F-Statistics for the Analysis of Population Structure. *Evolution* **1984**, *38*, 1358–1370. [CrossRef] [PubMed]
73. Hudson, R.R.; Boos, D.D.; Kaplan, N.L. A Statistical Test for Detecting Geographic Subdivision. *Mol. Biol. Evol.* **1992**, *9*, 138–151.
74. Watterson, G.A. On the Number of Segregating Sites in Genetical Models without Recombination. *Theor. Popul. Biol.* **1975**, *7*, 256–276. [CrossRef]
75. Harpending, H.C. Signature of Ancient Population Growth in a Low-Resolution Mitochondrial DNA Mismatch Distribution. *Hum. Biol.* **1994**, *66*, 591–600.
76. Bah, T. *Inkscape: Guide to a Vector Drawing Program (Digital Short Cut)*; Pearson Education: London, UK, 2009; ISBN 0-13-705171-9.
77. Watanabe, H. Studies on the Regulation in Fused Colonies in *Botryllus primigenus* (Ascidiae Compositae). *Sci. Rep. Tokyo Bunrika Daigaku* **1953**, *7*, 183–198.
78. Nydam, M.L.; Lemmon, A.R.; Cherry, J.R.; Kortyna, M.L.; Clancy, D.L.; Hernandez, C.; Cohen, C.S. Phylogenomic and Morphological Relationships among the Botryllid Ascidians (*Subphylum tunicata, Class ascidiacea,* Family Styelidae). *Sci. Rep.* **2021**, *11*, 8351. [CrossRef]
79. Van Name, W.G. The North and South American Ascidians. *Bull. Am. Mus. Nat. Hist.* **1945**, *84*, 1–462.
80. Boyd, H.C.; Weissman, I.L.; Saito, Y. Morphologic and Genetic Verification That Monterey *Botryllus* and Woods Hole *Botryllus* Are the Same Species. *Biol. Bull.* **1990**, *178*, 239–250. [CrossRef] [PubMed]
81. Tarjuelo, I.; Posada, D.; Crandall, K.A.; Pascual, M.; Turon, X. Phylogeography and Speciation of Colour Morphs in the Colonial Ascidian *Pseudodistoma crucigaster*. *Mol. Ecol.* **2004**, *13*, 3125–3136. [CrossRef] [PubMed]
82. Atsumi, M.O.; Saito, Y. Studies on Japanese Botryllid Ascidians. V. A New Species of the Genus *Botrylloides* Very Similar to *Botrylloides Simodensis* in Morphology. *Zool. Sci.* **2011**, *28*, 532–542. [CrossRef]
83. Nydam, M.L.; Giesbrecht, K.B.; Stephenson, E.E. Origin and Dispersal History of Two Colonial Ascidian Clades in the *Botryllus schlosseri* Species Complex. *PLoS ONE* **2017**, *12*, e0169944. [CrossRef]
84. Ulman, A.; Ferrario, J.; Occhpinti-Ambrogi, A.; Arvanitidis, C.; Bandi, A.; Bertolino, M.; Bogi, C.; Chatzigeorgiou, G.; Çiçek, B.A.; Deidun, A. A Massive Update of Non-Indigenous Species Records in Mediterranean Marinas. *PeerJ* **2017**, *5*, e3954. [CrossRef]
85. Blanchoud, S.; Rutherford, K.; Zondag, L.; Gemmell, N.J.; Wilson, M.J. De Novo Draft Assembly of the *Botrylloides leachii* Genome Provides Further Insight into Tunicate Evolution. *Sci. Rep.* **2018**, *8*, 5518. [CrossRef]
86. Hyams, Y.; Paz, G.; Rabinowitz, C.; Rinkevich, B. Insights into the Unique Torpor of *Botrylloides leachi*, a Colonial Urochordate. *Dev. Biol.* **2017**, *428*, 101–117. [CrossRef]
87. Fu, Y.-X. Statistical Tests of Neutrality of Mutations Against Population Growth, Hitchhiking and Background Selection. *Genetics* **1997**, *147*, 915–925. [CrossRef]
88. Galtier, N.; Nabholz, B.; Glémin, S.; Hurst, G.D.D. Mitochondrial DNA as a Marker of Molecular Diversity: A Reappraisal. *Mol. Ecol.* **2009**, *18*, 4541–4550. [CrossRef]
89. Morin, P.A.; Foote, A.D.; Baker, C.S.; Hancock-Hanser, B.L.; Kaschner, K.; Mate, B.R.; Mesnick, S.L.; Pease, V.L.; Rosel, P.E.; Alexander, A. Demography or Selection on Linked Cultural Traits or Genes? Investigating the Driver of Low MtDNA Diversity in the Sperm Whale Using Complementary Mitochondrial and Nuclear Genome Analyses. *Mol. Ecol.* **2018**, *27*, 2604–2619. [CrossRef]
90. Wei, W.; Tuna, S.; Keogh, M.J.; Smith, K.R.; Aitman, T.J.; Beales, P.L.; Bennett, D.L.; Gale, D.P.; Bitner-Glindzicz, M.A.K.; Black, G.C.; et al. Germline Selection Shapes Human Mitochondrial DNA Diversity. *Science* **2019**, *364*, 6520. [CrossRef] [PubMed]
91. Karahan, A.; Douek, J.; Paz, G.; Rinkevich, B. Population Genetics Features for Persistent, but Transient, *Botryllus schlosseri* (Urochordata) Congregations in a Central Californian Marina. *Mol. Phylogenet. Evol.* **2016**, *101*, 19–31. [CrossRef] [PubMed]
92. Lambert, W.J.; Dijkstra, J.A.; Clark, E.; Connolly, J. Larval Exposure to Low Salinity Compromises Metamorphosis and Growth in the Colonial Ascidian *Botrylloides violaceus*. *Invertebr. Biol.* **2018**, *137*, 281–288. [CrossRef]

diversity

MDPI

Article

Population Genetic Differentiation and Structure of *Maruca vitrata* (Lepidoptera: Crambidae) in India

Rashmi Manohar Mahalle [1,2], Snehel Chakravarty [1] and Chandra Prakash Srivastava [1,*]

1 Department of Entomology and Agricultural Zoology, Institute of Agricultural Sciences, Banaras Hindu University, Varanasi 221005, India; rashmimahalle27@gmail.com (R.M.M.); snehel.chak@gmail.com (S.C.)
2 Department of Applied Biology, College of Agriculture and Life Sciences, Chungnam National University, Daejeon 34134, Korea
* Correspondence: csrivastava63@gmail.com or cpsento@bhu.ac.in

Abstract: *Maruca vitrata* is one of the primary biotic constraints for pigeon pea production in India. The present study assessed the genetic variation and population structure of *M. vitrata* from diverse agro-ecologies in India using the mitochondrial *cytochrome c oxidase* I gene. A low number of segregating sites (10), haplotypes (13), nucleotide diversity (0.00136), and overall mean genetic distance (0.0013) were observed among the populations. The negative values of the neutrality tests and unimodal mismatch distribution supported its demographic expansion in the country. The analysis of molecular variance (AMOVA) revealed that the variation among populations or groups was only 13.91%, and the geographical distance did not significantly contribute to the genetic differentiation ($R^2 = 0.0024$, $p = 0.280$). The clustering of haplotypes was also independent of the geographical location. Overall, our results suggest the existence of low genetic variation and high gene flow among populations of *M. vitrata* in India.

Keywords: *COI* gene; diversity; haplotype; *Maruca vitrata*; pigeon pea; phylogeny

Citation: Mahalle, R.M.; Chakravarty, S.; Srivastava, C.P. Population Genetic Differentiation and Structure of *Maruca vitrata* (Lepidoptera: Crambidae) in India. *Diversity* **2022**, *14*, 546. https://doi.org/10.3390/d14070546

Academic Editor: Stephan Koblmüller

Received: 10 May 2022
Accepted: 27 June 2022
Published: 7 July 2022

1. Introduction

The crambid moth, *Maruca vitrata* (Fabricius), commonly known as the spotted pod borer or legume pod borer, is one of the most destructive pests of grain legumes across the subtropical and tropical regions of the world [1]. Although the most probable region of origin for *M. vitrata* is the Indo-Malaysian region, its geographical distribution range includes South and East Asia, sub-Saharan Africa, Oceania, and Central America, including the Caribbean islands [2]. It also exhibits a good degree of polyphagia and is known to attack several cultivated legumes, including *Cajanus cajan* (pigeon pea), *Vigna unguiculata* subsp. *unguiculata* (cowpea), *V. radiata* (green gram), *Phaseolus lunatus* (lima bean), and *Glycine max* (soybean) [3]. The larvae feed upon tender leaf axils, flowering inflorescence, and pods, by forming typical webbings or clusters [4]. The typical concealed feeding behavior is a severe challenge to management practices, as the webbed mass safeguards the larvae from natural enemies and diminishes insecticide efficacy. It has also been reported to pose a critical threat to the cultivation of early pigeon pea across India, inflicting an average annual yield loss of up to 84 percent [5] and a monetary loss of about USD 30 million, annually [6].

The control of *M. vitrata* relies mainly on the use of chemical insecticides, as no *Maruca*-resistant varieties are available in the major food legumes [7]. The frequent application of insecticides has resulted in an increased resistance to these insecticides. It has been observed in recent years that previously effective insecticides have acquired reduced effectiveness on *M. vitrata* in the country, thereby leading to population outbreaks [8]. For sustainable and effective control of pest insects, the adoption of insecticide resistance management (IRM) techniques is considered to be important. A sound knowledge of the biology and ecology of the target pest species is required to develop and apply sustainable IRM strategies. Additionally, it also requires a thorough understanding of the genetic, morphological, and physiological mechanisms governing the resistance development process. For *M. vitrata*, the factors that limit the development of IRM strategies include the deficiency of data related to its population genetic structure, including the availability of suitable species-specific molecular markers and DNA sequences [9].

The natural selection process governs variation in genetic traits resulting from an interaction between genetic forces and constantly changing environments. Host plants also play a considerable role in this process. Further, it is reported that the genus *Maruca* includes several species and/or subspecies that are difficult to distinguish morphologically, and even *M. vitrata* has long been thought to be a complex of several cryptic species [10]. There are also reports about *M. vitrata* populations exhibiting a differential response to pheromones in South and Southeast Asia (including India), due to the variability within the pheromone-binding protein genes [11]. So, it is quite possible that *M. vitrata* populations occurring across varied agro-ecological regions of India are also genetically diverse. The recent outbreaks and resistance development in the pest's field populations also raises a question of whether *M. vitrata* has experienced any sub-speciation or cryptic speciation in secluded areas of India, restricting gene flow. Inter-population genetic diversity studies using molecular markers are a more accurate means of identifying modifications that insect pests move through to address various survival challenges, as well as for the development and deployment of long-term strategies against them [12].

Several features of mitochondrial DNA such as maternal inheritance, its rapid evolution rate, and the negligible chances of recombination make it a useful marker in population genetics studies [13]. The high specificity of the *cytochrome c oxidase* I (*COI*) gene in species identification [14] and the availability of a wide range of primers for its amplification [15] make it an ideal mitochondrial genome study region [16]. This gene has been found to be valuable in distinguishing cryptic species [17], as well as for the assessment of intraspecific diversity in insects [6], because of its large size and high nucleotide substitution rate [18]. Information on intraspecific genetic variation and genetic differentiation in *M. vitrata* is lacking for most of the regions of India. Thus, we collected *M. vitrata* samples from 20 geographic localities, covering all four pigeon pea growing zones across India, and analyzed the *COI* region to examine the genetic diversity, haplotype diversity, historical demography, and population structure. Such studies will pave the way for understanding physiological or behavioral changes, population dynamics, and damage thresholds in different ecological regions, in order to design efficient and safe management strategies against *M. vitrata.*

2. Materials and Methods

2.1. Sample Collection and DNA Extraction

The larvae of *M. vitrata* were collected from 20 different geographical locations falling under the four pigeon pea growing zones of India (Figure 1, Table 1) during 2018 and 2019. For isolation of the genomic DNA, four late instar larvae were randomly sorted from each location (field population) and preserved in 95% ethyl alcohol at −20 °C. The extractions were carried out from the larval skin following the method outlined by Murray and Thompson [19] with some modifications. Briefly, the excised larval skin of individual third instar larvae was ground with a pestle and mortar using an extraction buffer containing 2% (*w*/*v*) CTAB. After incubation at 37 °C for 1 h, the extract was then emulsified with an equal volume of chloroform/isoamyl alcohol (24:1), and precipitated with chilled isopropanol and 3 M sodium acetate solution. The DNA pellet was washed twice with ethanol (70%), then the air-dried pellet was dissolved in TE buffer. The quality of the extracted DNA samples was examined on 0.8% agarose gel and quantified using a NanoDrop ND-1000 (NanoDrop products, Wilmington, DE, USA).

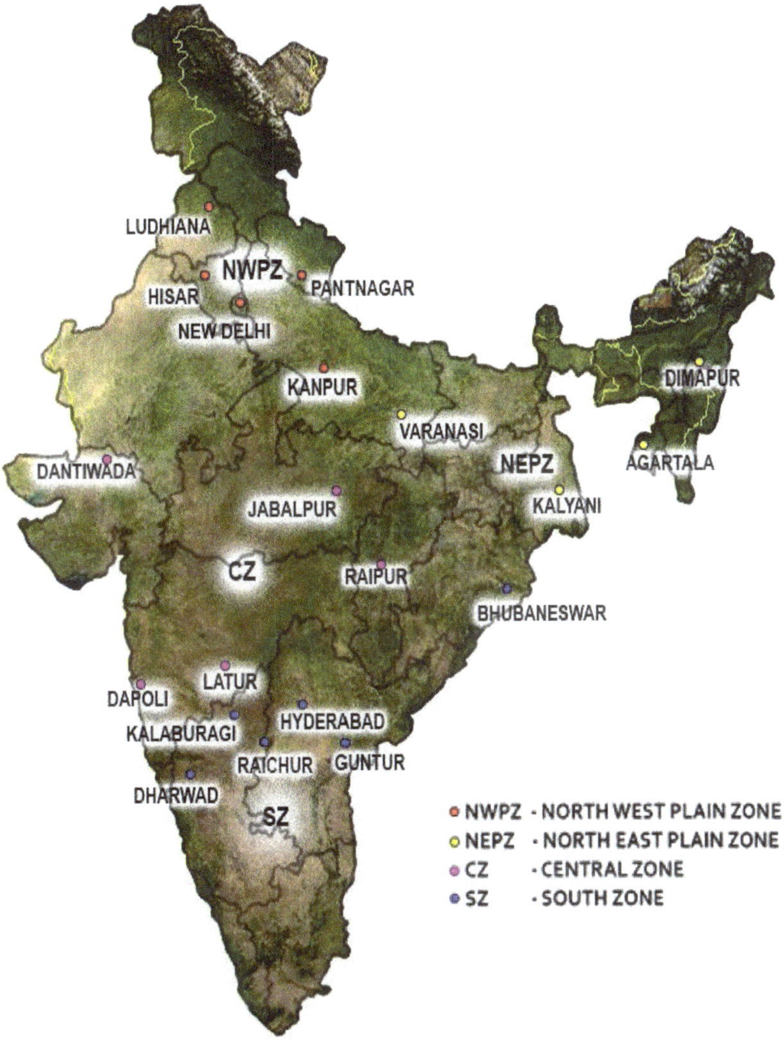

Figure 1. Sampling locations for collection of field populations of *M. vitrata* across India.

Table 1. Sampling details of *M. vitrata* populations from ecologically diverse pigeonpea growing zones of India.

Zones	Sample Code	Sampling Location (State)	Geographic Co-Ordinates	GenBank Accession Number(s) *
NWPZ (n = 20)	LDH	Ludhiana (Punjab)	30.90° N, 75.81° E	MW417880, MW417881, MW417882, MW417883
	HSR	Hissar (Haryana)	29.14° N, 75.71° E	MW417884, MW417885, MW417886, MW417887
	NDLS	New Delhi	28.64° N, 77.16° E	MW417888, MW417889, MW417890, MW417891
	PBW	Pantnagar (Uttarakhand)	29.02° N, 79.49° E	MW417892, MW417893, MW417894, MW417895
	CNB	Kanpur (Uttar Pradesh)	26.44° N, 80.33° E	MW417868, MW417869, MW417870, MW417871

Table 1. *Cont.*

Zones	Sample Code	Sampling Location (State)	Geographic Co-Ordinates	GenBank Accession Number(s) *
NEPZ (n = 16)	BSB	Varanasi (Uttar Pradesh)	25.27° N, 82.99° E	MW417872, MW417873, MW417874, MW417875
	KYI	Kalyani (West Bengal)	22.94° N, 88.53° E	MW417876, MW417877, MW417878, MW417879
	AGTL	Agartala (Tripura)	23.91° N, 91.32° E	MW417864, MW417865, MW417866, MW417867
	DMV	Dimapur (Nagaland)	25.05° N, 93.03° E	MW417860, MW417861, MW417862, MW417863
CZ (n = 20)	DWZ	Dantiwada (Gujarat)	24.32° N, 72.32° E	MW417912, MW417913, MW417914, MW417915
	JBP	Jabalpur (Madhya Pradesh)	23.21° N, 79.95° E	MW417896, MW417897, MW417898, MW417899
	R	Raipur (Chattisgarh)	21.24° N, 81.70° E	MW417900, MW417901, MW417902, MW417903
	LUR	Latur (Maharashtra)	18.42° N, 76.61° E	MW417908, MW417909, MW417910, MW417911
	DPLI	Dapoli (Maharashtra)	17.75° N, 73.18° E	MW417904, MW417905, MW417906, MW417907
SZ (n = 24)	BBS	Bhubaneswar (Odisha)	20.27° N, 85.81° E	MW417916, MW417917, MW417918, MW417919
	HYB	Hyderabad (Telangana)	17.31° N, 78.16° E	MW417920, MW417921, MW417922, MW417923
	GNT	Guntur (Andhra Pradesh)	16.36° N, 80.43° E	MW417924, MW417925, MW417926, MW417927
	KLBG	Kalaburagi (Karnataka)	17.32° N, 76.84° E	MW417936, MW417937, MW417938, MW417939
	RC	Raichur (Karnataka)	16.20° N, 77.33° E	MW417928, MW417929, MW417930, MW417931
	DWR	Dharwad (Karnataka)	15.49° N, 74.98° E	MW417932, MW417933, MW417934, MW417935

Abbreviations: NWPZ, North West Plain Zone; NEPZ, North East Plain Zone; CZ, Central Zone; SZ, South Zone; n, number of individuals sequenced for mitochondrial *COI* gene. * Listed above are the GenBank accession numbers of mitochondrial *COI* gene sequences deposited from this study.

2.2. Amplification and Sequencing of COI Gene Fragment

The polymerase chain reactions were performed for amplification of the *COI* gene (partial sequences) using LCO1490 (forward) and HCO2198 (reverse) primers [15]. The master mix (25 μL) contained 2 μL of the template DNA (100 ng), 2.5 μL of PCR buffer (10×), 0.2 μL of Taq polymerase (1U, GeNei™), 1 μL of dNTP mix (2 mM), 1μL of $MgCl_2$ (2 mM), 10 pmol of each primer, and 17.5 μL of nuclease-free water. The amplification was carried out in a thermocycler (BIO-RAD MJ Mini™) programmed for: initial denaturation for 5 min at 94 °C, followed by 30 cycles of denaturation (30 s at 94 °C), annealing (45 s at 50 °C), and extension (1 min at 72 °C). This was followed by a final extension for 10 min at 72 °C. The amplified products were examined on a 1% agarose gel, and the gel purified (GeneJET Gel Extraction Kit, Thermo Scientific, Waltham, MA, USA) PCR products were sequenced using Sanger's method in an ABI 3730 automated DNA analyzer at M/s Eurofins Analytical Services India Pvt. Ltd., Bengaluru.

2.3. Sequence Data Analyses

The alignment and trimming of the *COI* sequences was performed in Clustal W (default parameters) in the MEGA 6.0 [20]. The final alignment length was 620 bp. The obtained sequences were confirmed using nBLAST of the National Center for Biotechnology Information (https://www.ncbi.nlm.nih.gov/; accessed on 15 January 2020).) and deposited in the GenBank (Table 1). DnaSP version 5.10.1 software [21] was used to infer various diversity indices, i.e., the number of haplotypes (Hn), haplotype diversity (Hd), nucleotide diversity (π), segregating polymorphic sites (S), and an average number of nucleotide differences among haplotypes (k). Further, to ascertain the demographic history and evolutionary neutrality of the *M. vitrata COI* sequences, tests such as Fu's Fs, Tajima's D, Fu and Li's D,

Fu and Li's F, and mismatch distribution analysis were also performed using DnaSP 5.10.1. The average pairwise sequence divergences among *M. vitrata* populations were estimated using the Kimura 2-parameter distance model [22] and displayed graphically in a neighbor-joining (NJ) tree, using MEGA 6.0, with a confidence level of 1000 bootstrap replicates. The average nucleotide base composition of *COI* sequences and overall mean genetic distance were also calculated with MEGA 6.0. For understanding the genetic structure, the analysis of molecular variance (AMOVA) and pairwise F_{ST} values were computed using Arlequin 3.5 [23]. The level of significance was determined with 1000 permutation replicates. Principal coordinates analysis (PCoA) based on the genetic distance matrix and the Mantel test using the pairwise geographical distance (Ln km) against pairwise linearized genetic distance among populations were performed in GenAlEx (version 6.5), with 1000 random permutations [24]. The haplotype network was created using the median-joining algorithm in Network 4.6.1.1 software [25].

3. Results

3.1. Variability in the Mitochondrial COI Gene

The 80 sequences generated in the study showed high similarity (100 percent query coverage and 98 to 100 percent identity) to the *M. vitrata COI* gene sequences already available in the public domain database (NCBI-GenBank BLASTN search tool). The nucleotide composition in the *COI* sequences among the populations was found to be very similar, with adenine (A) = 30.93%, cytosine (C) = 15.06%, guanine (G) = 14.62% and thymine (T) = 39.39%, averaged across the multiple sequence alignment (Table 2). The sequence regions also showed comparatively much higher AT (70.32%) over GC (29.68%) content, in accordance with the general patterns in mitochondrial DNA of arthropods. A total of 10 transitions (A = G, T = C) were recorded in the studied sequences but no transversion. The number of haplotypes for the studied population sets from different zones ranged from 3 to 8, with a total of 13 haplotypes recorded from the pooled populations (Table 2). The populations from the Central Zone exhibited comparatively higher haplotype diversity (0.732) and nucleotide diversity (0.00193), as well as an average number of genetic differences among haplotypes (1.216), while the North East Plain Zone population set was lowest in all concerns (Hd = 0.242, π = 0.00060, and k = 0.375). The total nucleotide (π) and haplotype (Hd) diversity were 0.00136 and 0.554, respectively.

3.2. Demographic Inference and Population Structure

A neutrality test was employed to determine the demographic population history for all the populations across India (Table 2). Tajima's D value was statistically insignificant for all the zones and the overall population, indicating a low-frequency polymorphism or low levels of genetic variations among these populations. Most of the populations also showed insignificant negative values for Fu and Li's D and F and Fu's Fs tests, except those from the North East Plain Zone. The negative deviations from zero in the overall population indicated the occurrence of population growth or expansion in this pest species in the recent past due to an excess of rare mutations. This result was further supported by the smooth and unimodal mismatch distribution plot (Figure 2) that included all the studied populations from varied zones. The AMOVA showed that the majority of total molecular variance (86.09%) was distributed within populations, and only 13.38% and 0.54% were attributed to distribution among populations within zones and among different zones, respectively (Table 3). A non-significant degree of differentiation among groups of populations (F_{CT} = 0.00537, $p > 0.10$), among populations within groups (F_{SC} = 0.13448, $p > 0.05$), and among subpopulations within the total population (F_{ST} = 0.13912, $p > 0.05$) indicated the lack of considerable population genetic structure.

Table 2. Molecular diversity indices and neutrality test values based on *COI* gene sequences from different *M. vitrata* populations.

Index	*M. vitrata* Population				
	NWPZ	NEPZ	CZ	SZ	Pooled
N	20	16	20	24	80
Hn	4	3	8	6	13
Hd	0.558	0.242	0.732	0.550	0.554
Π	0.00139	0.00060	0.00193	0.00127	0.00136
K	0.874	0.375	1.216	0.801	0.856
S	3	3	8	4	10
Nucleotide composition (Relative values)					
A (%)	30.95	30.94	30.92	30.92	30.93
C (%)	15.05	15.08	15.06	15.05	15.06
G (%)	14.60	14.61	14.63	14.63	14.62
T (%)	39.40	39.37	39.39	39.40	39.39
A + T (%)	70.35	70.31	70.31	70.32	70.32
C + G (%)	29.65	29.69	29.69	29.68	29.68
Neutrality tests					
Fu's Fs	−0.3174	−0.8982	−4.1487	−2.6631	−9.6793
Tajima's D	−0.0087	−1.6965	−1.5532	−0.6905	−1.5224
Fu and Li's D	−1.0065	−2.2045 *	−1.4854	−0.8560	−1.3226
Fu and Li's F	−0.8685	−2.3662 *	−1.7409	−0.9355	−1.6481

Abbreviations: n, number of sequences; Hn, number of haplotypes; Hd, haplotype diversity; π, nucleotide diversity; k, average number of nucleotide differences (genetic differences) among haplotypes; S, number of polymorphic (segregating) sites; A, adenine; C, cytosine; G, guanine; T, thymine. * Significant at $0.01 < p < 0.05$, while other values are non-significant ($p > 0.10$) for neutrality tests. Here, "Pooled" denotes the combined set of populations from all four major pulse growing zones of India.

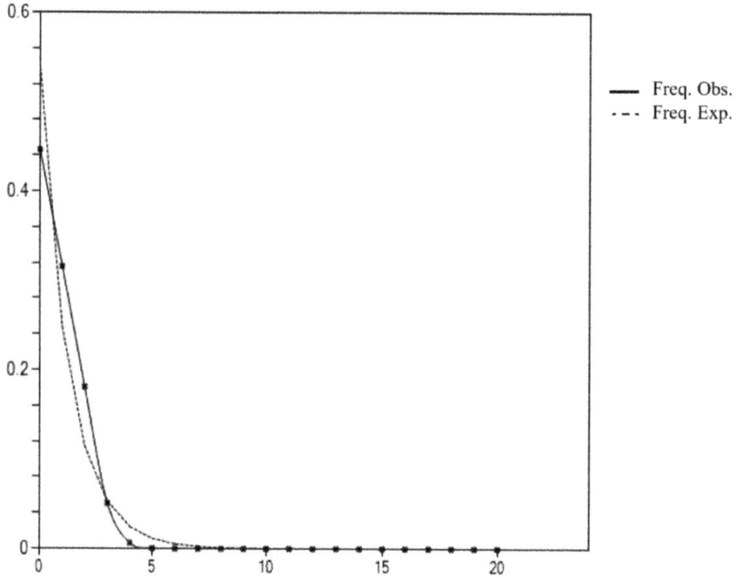

Figure 2. Frequencies of the observed and expected mismatch distribution in *M. vitrata* populations of India. X-axis (x), number of pairwise nucleotide differences; Y-axis (y), frequency of mismatches; Freq. Exp., frequency expected (dashed line); and Freq. Obs, frequency observed (solid line).

Table 3. Analysis of molecular variance (AMOVA) for the *COI* sequences of different *M. vitrata* populations.

Source of Variation	df	Sum of Squares	Variance Components	Percentage Variation	Fixation Indices
Among groups (zones)	3	1.942	0.00231	0.54	F_{CT}: 0.00537 ($p > 0.10$)
Among populations within groups	16	11.562	0.05762	13.37	F_{SC}: 0.13448 ($p > 0.05$)
Within populations	60	22.250	0.37083	86.09	F_{ST}: 0.13912 ($p > 0.05$)
Total	79	33.812	0.43026	100	

df, Degrees of freedom.

The overall mean genetic distance was also found to be very low (0.0013 ± 0.001). The principal component analysis (PCoA) showed that the first principal component accounted for 50.59% of the total variation, followed by the second component, which accounted for 21.37% of the variation, and the first three axes explained 86.14% of the cumulative variation (Figure 3). The PCoA roughly separated the populations into three main groups, but there was no clear geographical pattern in the distribution of these populations. Further, the mantel test showed a non-significant and weak correlation between the genetic distance matrix of the studied *M. vitrata* populations with the corresponding geographic distance matrix (Ln km) ($R^2 = 0.0024$, $p = 0.280$) (Figure 4) across the study area.

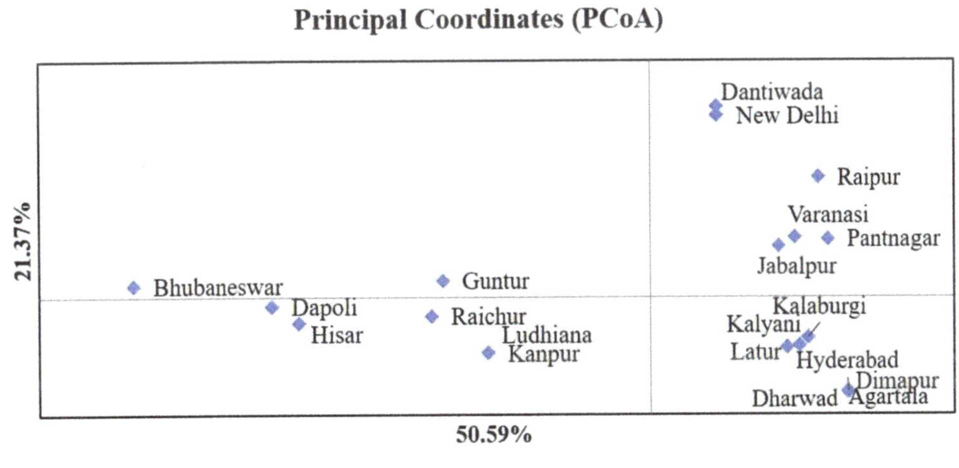

Figure 3. PCoA analysis score plot of *M. vitrata* populations from different locations across India based on *COI* gene.

3.3. Haplotype Distribution and Phylogenetic Analysis

A total of 13 *COI* haplotypes were identified (80 individuals). The haplotype network was star-like, and different localities shared haplotypes (Figure 5). Haplotype 1 contained 53 *M. vitrata* individuals and was shared by populations from all locations. It appeared to be the ancestral haplotype, as it had a central position in the network, and all other lineages arose from it. Haplotype 2 formed the second largest group (a total of six individuals) with one individual each from Kanpur, Ludhiana, Hisar, and Dapoli, and two others belonging to Bhubaneswar. Haplotype 6 occurred in individuals collected from New Delhi, Pantnagar, and Raipur. Two individuals from Dantiwada and one each from Jabalpur and Raipur shared haplotype 8. Haplotype 12 included populations only from Guntur. The remaining haplotypes were unique for a single location.

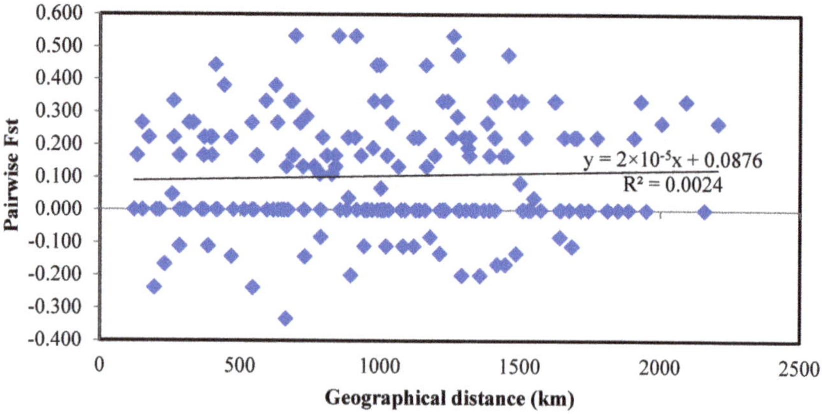

Figure 4. Correlation between pairwise genetic differentiation using *COI* gene sequences and geographical distance among the different *M. vitrata* populations sampled across India.

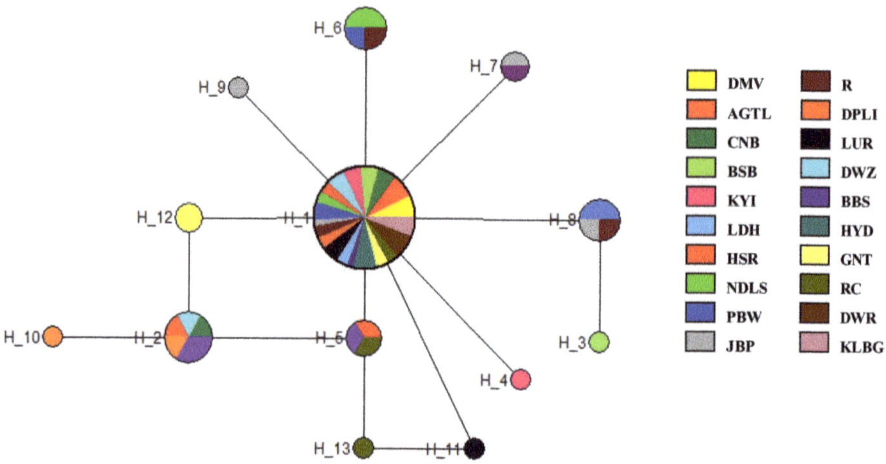

Figure 5. Median-joining haplotype network of *M. vitrata* in India based on mitochondrial *COI* gene. The circle areas are proportional to haplotype frequencies, while the color portions represent the proportions of the same haplotype occurring in each geographical region.

Further, for the phylogenetic comparison of Indian populations with global genetic assemblage, 20 additional *COI* gene sequences were mined from the NCBI database. These sequences represented populations from 18 different countries outside India, based on the spread of *M. vitrata*. Topologies of the neighbor-joining tree indicated that all the studied Indian populations belonged to a single major clade (Figure 6). *M. vitrata COI* gene sequences deposited in the GenBank database from other countries also shared similarities with studied sequences.

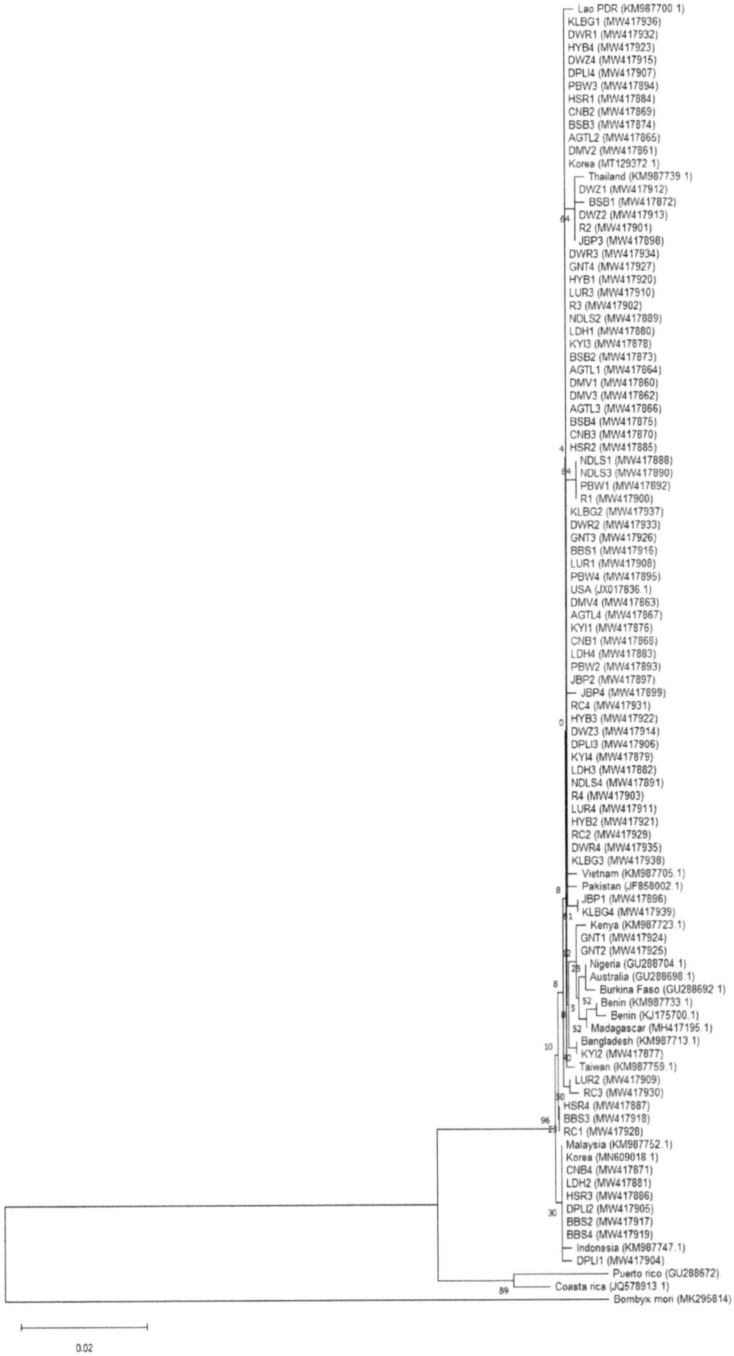

Figure 6. Neighbor-joining tree based on Kimura 2-parameter distances showing clustering of *M. vitrata* populations for *COI* gene. Numbers at branch point indicate 10,000 bootstrap values. *Bombyx mori* (accession #MK295814) was used as outlier sequence.

4. Discussion

M. vitrata is a key pest of pulses in the Indian subcontinent. The occurrence of this pest has been recorded from various agro-ecological regions of India. In this study, a total of 80 specimens from 20 locations (latitude: 15.49° N to 30.90° N and longitude 72.32° E to 85.81° E) in India were sequenced for *COI* gene fragments. In population genetic studies of insects, these markers have played an important role [26,27]. They are applicable to the assessment of population genetic structure, identification of unidentified cryptic species, and detection of an alien pest in a new area. Based on the fact that the *COI* gene has proven to be informative in population genetic studies, we examined the genetic variability using a mitochondrial marker (i.e., *COI* gene sequences) in an attempt to elucidate the population genetic structure of this pest species in India. The homology search of the *COI* sequence of each population of *M. vitrata* with the NCBI sequences confirmed the specimens' identity. Furthermore, the results showed mean A + T and G + C levels of 70.32% and 29.68%, respectively, confirming the AT-biased nature of the *COI* gene in *M. vitrata,* as in other arthropods [1,6,10].

The haplotype diversity (Hd) and nucleotide diversity (π) for all the zones, as well as the overall populations in the present investigation, suggested that the entire population exhibited a low level of genetic diversity. This was in agreement with a previous study that determined the total nucleotide diversity to be 0.00309 for the *M. vitrata* populations examined across Asia and Africa [10]. Low nucleotide diversity was also exhibited among the Indian (0.00226) and foreign populations (0.00582) [6]. The overall mean genetic distance was also found to be exceptionally low (0.0013), and this strongly supports the single species status of *M. vitrata* in the country, as a divergence of more than five percent in the *COI* gene amplicon, i.e., a genetic distance exceeding 0.05, depicts the likely occurrence of a new species in Lepidoptera [16].

The neutrality test indices and genetic differentiation values are helpful in analyzing demographic history, where negative values relate to the demographic expansion of populations after a recent sharp decline, and positive values determine populations subdivided at equilibrium [28,29]. Additionally, Fu's FS's negative values are usually associated with "an excess of singletons in a population expansion event" [29,30]. Hence, as per the present study, the Indian population of *M. vitrata* is predicted to have had recent demographic expansion events depicting an excess of low frequency polymorphism. Our results are consistent with those from the previous study by Periasamy et al. [10], where Tajima's D test values were non-significant and negative for populations of Asia and Africa. The present investigation is strongly supported by Chatterjee et al. [6] who found a moderate level of polymorphism across global and Indian populations with a positive and statistically significant Tajima's D value ($p < 0.001$). However, a statistically insignificant negative Tajima's D value was observed among the Indian populations, indicating a low-frequency polymorphism.

The genetic variability observed within populations accounted for 86.09%, and only 13.37% and 0.54% of variations were found among populations within the same zone and population sets of different zones, respectively. Low genetic variability among *M. vitrata* populations using other markers have also been reported by [2,3,9]. Further, no significant correlation was obtained between the genetic variance existing among the studied populations and their respective geographical locations. This reveals that isolation by distance is not always expected in lepidopteran species such as *M. vitrata* that have high mobility, and this could only be due to the intermingling of populations.

Among the 13 haplotypes identified, haplotype 1 (Hap1), predominantly distributed throughout the Indian populations, comprising 53 samples, is likely the ancestral haplotype among the populations sampled. According to the coalescent theory, common haplotypes at the center of a network are inferred to be ancestral, while tip haplotypes at the periphery are derived or descendant from ancestral haplotypes [31]. In a study undertaken for South and Southeast Asia and sub-Saharan Africa, 64 haplotypes were identified in 686 specimens of *M. vitrata* [10]. Except for Benin, the most common haplotype included 225 *M. vitrata* individuals collected from Asian countries. When the populations from each continent were analyzed separately, Oceania had the fewest haplotypes, while Asia had the most. The study also discovered that six out of ten *M. vitrata* individuals from Jharkhand, India, constituted a distinct haplotype. The present study is also supported by Chatterjee et al. [6], who documented six haplotypes among the populations collected from 11 locations across India. The phylogenetic tree also supported very low genetic heterogeneity in Indian *M. vitrata* and hinted at the intermingling of populations among diverse agro-ecologies.

There are two probable reasons for the very low genetic variations among Indian populations of *M. vitrata*. Firstly, it is well documented that several lepidopteran moths have high migratory potential [12]. Thus, because of the substantial gene flow, many of them very often exhibit low genetic differentiation over their wide geographical distributional ranges. In the case of *M. vitrata*, both the active migration (because of good flight capacity) and passive migration (because of long distance transport of plant materials) might be helping it to maintain a steady state of gene flow, thereby minimizing genetic variation. The previous reports on the ecology and migratory behavior of *M. vitrata* species across India suggest a gradual shift from North to South India as winter progresses, i.e., during the course of September to December. This can be inferred from the following population dynamics data of the country. The peak activity of *M. vitrata* from North India has been recorded during October [5,32,33]; in contrast, in southern parts of India, its incidence begins from the third week of November, and the peak is attained during the second fortnight of December [34–36]. Such migration can be the cause for higher gene flow among the populations, thereby decreasing the degree of genetic differentiation. The second reason can be that *M. vitrata* has a very narrow host range feeding mainly on pigeon pea and cowpea in the country. Most of the commercially grown varieties of these crops show slight variation amongst themselves for their susceptibility against this pest species. Thus, there is no selection pressure exerted on this pest due to host plants [37], preventing it from undergoing any genetic changes unlike certain other lepidopterans, mostly from the Noctuidae [38] and Tortricidae [39] families, where the existence of the host-associated genetic differences is widespread.

5. Conclusions

This study provides a clear picture regarding the homogenous genetic structure of *M. vitrata* in the country. The highly homogenous populations indicate that they are well adapted and migrate freely, which could lead to the concentration of resistant alleles in particular patches, subsequently accelerating the evolution of resistance. This information will be very helpful in designing sustainable management strategies for this pest species in an area-wide manner across India.

Author Contributions: Conceptualization: C.P.S.; methodology: R.M.M.; formal analysis: R.M.M.; investigation: R.M.M.; resources: C.P.S.; data curation: R.M.M., S.C.; writing—original draft preparation: R.M.M.; writing—review and editing: S.C. and C.P.S.; contributed intellectually to the interpretation and discussion of results: C.P.S.; and funding acquisition: C.P.S. All authors have read and agreed to the published version of the manuscript.

Funding: This research received no external funding.

Institutional Review Board Statement: Not applicable.

Informed Consent Statement: Not applicable.

Data Availability Statement: NCBI database gene bank accessions (MW417860-MW417939).

Acknowledgments: We thank Kartikeya Srivastava (Banaras Hindu University, India) for his valuable suggestions during the planning and course of the study.

Conflicts of Interest: The authors declare no conflict of interest.

References

1. Agunbiade, T.A.; Coates, B.S.; Sun, W.L.; Tsai, M.R.; Valero, M.C.; Tamò, M.; Pittendrigh, B.R. Comparison of the mitochondrial genomes of the old and new world strains of the legume pod borer, *Maruca vitrata* (Lepidoptera: Crambidae). *Int. J. Trop. Insect Sci.* **2017**, *37*, 125–136. [CrossRef]
2. Margam, V.M.; Coates, B.S.; Ba, M.N.; Sun, W.; Binso-Dabire, C.L.; Baoua, I.; Ishiyaku, M.F.; Shukle, J.T.; Hellmich, R.L.; Covas, F.G.; et al. Geographic distribution of phylogenetically distinct Legume Pod Borer, *Maruca vitrata* (Lepidoptera: Pyraloidea: Crambidae). *Mol. Biol. Rep.* **2011**, *38*, 893–903. [CrossRef] [PubMed]
3. Agunbiade, T.A.; Coates, B.S.; Datinon, B.; Djouaka, R.; Sun, W.; Tamò, M.; Pittendrigh, B.R. Genetic Differentiation among *Maruca vitrata* F. (Lepidoptera: Crambidae) populations on cultivated cowpea and wild host plants: Implications for insect resistance management and biological control strategies. *PLoS ONE* **2014**, *9*, e92072. [CrossRef] [PubMed]
4. Sharma, H.C. Bionomics, host plant resistance, and management of the legume pod borer, *Maruca vitrata*—A review. *Crop Prot.* **1998**, *17*, 373–386. [CrossRef]
5. Chakravarty, S.; Agnihotri, M.; Jagdish, J. Seasonal abundance of predatory bugs, *Eocanthecona furcellata* (Wolff.) and *Rhynocoris fuscipes* (F.) and its olfactory responses towards plant and pest mediated semiochemical cues in pigeonpea ecosystem. *Legum. Res.* **2017**, *40*, 351–357. [CrossRef]
6. Chatterjee, M.; Yadav, J.; Vennila, S.; Shashank, P.R.; Jaiswal, N.; Sreevathsa, R.; Rao, U. Diversity analysis reveals genetic homogeneity among Indian populations of legume pod borer, *Maruca vitrata* (F.). *3 Biotech* **2019**, *9*, 319. [CrossRef]
7. Srinivasan, R.; Tamò, M.; Malini, P. Emergence of *Maruca vitrata* as a major pest of food legumes and evolution of management practices in Asia and Africa. *Annu. Rev. Entomol.* **2021**, *66*, 141–161. [CrossRef]
8. Mahalle, R.; Taggar, G. Insecticides against *Maruca vitrata* (Fabricius) (Lepidoptera: Crambidae) on pigeonpea. *Pestic. Res. J.* **2018**, *30*, 235. [CrossRef]
9. Sambathkumar, S.; Durairaj, C.; Mohankumar, S.; Preetha, B.; Aravintharaj, R.; Ganapathy, N. Spatial genetic vriation in south Indian larval populations of legume pod borer, *Maruca vitrata* Geyer (Lepidoptera: Crambidae). *Indian J. Exp. Biol.* **2019**, *57*, 923–930.
10. Periasamy, M.; Schafleitner, R.; Muthukalingan, K.; Ramasamy, S. Phylogeographical structure in mitochondrial DNA of legume pod borer (*Maruca vitrata*) population in tropical Asia and Sub-Saharan Africa. *PLoS ONE* **2015**, *10*, e0124057. [CrossRef]
11. Malini, P.; Ramasamy, S.; Schafleitner, R.; Muthukalingan, K. Pheromone-binding proteins based phylogenetics and phylogeography of *Maruca* spp. from Asia, Africa, Oceania, and South America. *Ecol. Evol.* **2019**, *9*, 9239–9272. [CrossRef] [PubMed]
12. Chakravarty, S.; Padwal, K.P.; Srivastava, C.P. Molecular Characterization of intraspecific variations in *Helicoverpa armigera* (Hübner) populations across India. *J. Environ. Biol.* **2021**, *42*, 1320–1329. [CrossRef]
13. Karthika, P.; Vadivalagan, C.; Thirumurugan, D.; Murugan, K. Intra-species variation and geographic differentiation among the populations of the quarantine agricultural pest *Leucinoides orbonalis* (Lepidoptera: Crambidae) in the global assemblage—A prospective of DNA barcoding. *Mitochondrial DNA Part A* **2019**, *30*, 682–693. [CrossRef] [PubMed]
14. Hebert, P.D.N.; Cywinska, A.; Ball, S.L.; deWaard, J.R. Biological identifications through DNA barcodes. *Proc. Biol. Sci.* **2003**, *270*, 313–321. [CrossRef]
15. Folmer, O.; Black, M.; Hoeh, W.; Lutz, R.; Vrijenhoek, R. DNA Primers for amplification of mitochondrial *cytochrome c oxidase* subunit I from diverse metazoan invertebrates. *Mol. Mar. Biol. Biotechnol.* **1994**, *3*, 294–299. [PubMed]
16. Hebert, P.D.N.; deWaard, J.R.; Landry, J.F. DNA Barcodes for 1/1000 of the animal kingdom. *Biol. Lett.* **2010**, *6*, 359–362. [CrossRef] [PubMed]
17. Shashank, P.R.; Thomas, A.; Ramamurthy, V.V. DNA Barcoding and Phylogenetic Relationships of *Spodoptera litura* and *S. exigua* (Lepidoptera: Noctuidae). *Fla. Entomol.* **2015**, *98*, 223–228. [CrossRef]
18. Padwal, K.G.; Chakravarty, S.; Srivastava, C.P. Genetic variability and population structure of *Leucinodes orbonalis* (Guenée), a severe insect pest of brinjal in India. *J. Environ. Biol.* **2022**, *43*, 59–65. [CrossRef]
19. Murray, M.G.; Thompson, W.F. Rapid isolation of high molecular weight plant DNA. *Nucleic Acids Res.* **1980**, *8*, 4312–4325. [CrossRef]
20. Tamura, K.; Stecher, G.; Peterson, D.; Filipski, A.; Kumar, S. MEGA6: Molecular evolutionary genetics analysis version 6.0. *Mol. Biol. Evol.* **2013**, *30*, 2725–2729. [CrossRef]
21. Librado, P.; Rozas, J. DnaSP v5: A software for comprehensive analysis of DNA polymorphism data. *Bioinformatics* **2009**, *25*, 1451–1452. [CrossRef] [PubMed]
22. Kimura, M. A simple method for estimating evolutionary rates of base substitutions through comparative studies of nucleotide sequences. *J. Mol. Evol.* **1980**, *16*, 111–120. [CrossRef] [PubMed]
23. Excoffier, L.; Lischer, H.E.L. Arlequin Suite Ver 3.5: A new series of programs to perform population genetics analyses under Linux and Windows. *Mol. Ecol. Resour.* **2010**, *10*, 564–567. [CrossRef] [PubMed]
24. Peakall, R.; Smouse, P.E. GenAlEx 6.5: Genetic analysis in excel. Population genetic software for teaching and research—An update. *Bioinformatics* **2012**, *28*, 2537–2539. [CrossRef] [PubMed]
25. Bandelt, H.J.; Forster, P.; Röhl, A. Median-joining networks for inferring intraspecific phylogenies. *Mol. Biol. Evol.* **1999**, *16*, 37–48. [CrossRef] [PubMed]

26. Sridhar, J.; Chinna Babu Naik, V.; Ghodke, A.; Kranthi, S.; Kranthi, K.R.; Singh, B.P.; Choudhary, J.S.; Krishna, M.S.R. Population genetic structure of cotton pink bollworm, *Pectinophora gossypiella* (Saunders) (Lepidoptera: Gelechiidae) using mitochondrial cytochrome oxidase I (*COI*) gene sequences from India. *Mitochondrial DNA Part A* **2017**, *28*, 941–948. [CrossRef] [PubMed]
27. Arias, M.C.; Sheppard, W.S. Phylogenetic relationships of honey bees (Hymenoptera: Apinae: Apini) inferred from nuclear and mitochondrial DNA sequence data. *Mol. Phylogenet. Evol.* **2005**, *37*, 25–35. [CrossRef]
28. Tajima, F. Statistical method for testing the neutral mutation hypothesis by DNA polymorphism. *Genetics* **1989**, *123*, 585–595. [CrossRef]
29. Fu, Y.X. Statistical tests of neutrality of mutations against population growth, hitchhiking and background selection. *Genetics* **1997**, *147*, 915–925. [CrossRef]
30. Fu, Y.X. Statistical properties of segregating sites. *Theor. Popul. Biol.* **1995**, *48*, 172–197. [CrossRef]
31. Hudson, R.R. Gene genealogies and the coalescent process. In *Oxford Surveys in Evolutionary Biology*; Futuyuma, D., Antonouics, J., Eds.; Oxford University Press: New York, NY, USA, 1990; Volume 7, pp. 1–44.
32. Srivastava, C.P.; Pimbert, M.P.; Jadhav, D.R. Monitoring adult populations of *Maruca testulalis* (Geyer) with light traps at Patancheru and Hisar in India. *Int. Pigeonpea Newsl.* **1992**, *15*, 27–28.
33. Taggar, G.K.; Singh, R.; Cheema, H.K.; Singh, P. Relative abundance, population dynamics and damage potential of spotted pod borer, *Maruca vitrata* (Fabricius) on early pigeonpea in Punjab. *Int. J. Trop. Insect Sci.* **2019**, *39*, 229–234. [CrossRef]
34. Chaitanya, T.; Sreedevi, K.; Navatha, L.; Krishna, T.M.; Prasanti, L. Bionomics and population dynamics of legume pod borer, *Maruca vitrata* (Geyer) in *Cajanus cajan* (L.) Millsp. *Curr. Biot.* **2012**, *5*, 446–453.
35. Sreekanth, M.; Ratnam, M.; Seshamahalakshmi, M.; Rao, Y.K.; Narayana, E. Population build-up and seasonal abundance of spotted pod borer, *Maruca vitrata* (Geyer) on pigeonpea (*Cajanus cajan* (L) Millsp.). *J. App. Biol. Biotechnol.* **2015**, *3*, 43–45. [CrossRef]
36. Sampathkumar, S.; Durairaj, C. Relative abundance of legume pod borer, *Maruca vitrata* Geyer (Lepidoptera: Crambidae) on pigeonpea and its relationship with weather parameters. *Madras Agric. J.* **2015**, *102*, 67–70.
37. Sambathkumar, S.; Durairaj, C.; Mohankumar, S.; Preetha, B.; Aravintharaj, R.; Ganapathy, N.; Surendran, R. Host induced genetic variation in legume pod borer, *Maruca Vitrata*. *J. Environ. Biol.* **2017**, *38*, 1281–1291. [CrossRef]
38. Subramanian, S.; Mohankumar, S. Genetic variability of the bollworm, *Helicoverpa armigera*, occurring on different host plants. *J. Insect Sci.* **2006**, *6*, 26. [CrossRef] [PubMed]
39. Emelianov, I.; Mallet, J.; Baltensweiler, W. Genetic differentiation in *Zeiraphera diniana* (Lepidoptera: Tortricidae, the Larch budmoth): Polymorphism, host races or sibling species? *Heredity* **1995**, *75*, 416–424. [CrossRef]

diversity

MDPI

Perspective

Environmental DNA-Based Methods in Biodiversity Monitoring of Protected Areas: Application Range, Limitations, and Needs

Kathrin Pascher [1,2,*], Vid Švara [1] and Michael Jungmeier [1]

1 UNESCO Chair on Sustainable Management of Conservation Areas, Engineering & IT, Carinthia University of Applied Sciences, Europastraße 4, 9524 Villach, Austria; v.svara@fh-kaernten.at (V.Š.); m.jungmeier@fh-kaernten.at (M.J.)
2 Department of Integrative Biology and Biodiversity Research, Institute of Zoology, University of Natural Resources and Life Sciences, Vienna (BOKU), Gregor-Mendel-Straße 33, 1180 Vienna, Austria
* Correspondence: kathrin.pascher@boku.ac.at; Tel.: +43-1-47654-83328; Fax: +43-1-47654-83309

Abstract: Novel methods for species detection based on collection of environmental DNA (eDNA) are not only important in biodiversity assessment in a scientific context, but are also increasingly being applied in conservation practice. The eDNA-based biodiversity detection methods have significant potential for regular use in biodiversity status assessments and conservation actions in protected areas (PAs) and other effective area-based conservation measures (OECMs) worldwide. Species detection based on DNA from environmental samples, such as water, sediment, soil, air, or organic material, has a broad application scope with precise, comprehensive, and rapid species identification. Here, we provide an overview of the application range of eDNA-based methods for biodiversity monitoring in PAs, evaluate environmental assessments in which this technology has already been implemented for nature conservation, and examine the challenges that can hamper further application in real world practice. Based on the outcomes of two projects, practical experience, and current scientific literature focusing on their application, we conclude that eDNA-based species detection methods provide promising novel approaches that have strong potential as supplement methods, or in some cases even as substitutes for the conventional monitoring methods used for PAs. This advancement is expected to affect decision-making in biodiversity conservation efforts in PAs and OECMs.

Keywords: eDNA; eDNA metabarcoding; biodiversity assessment; nature conservation; protected area management

Citation: Pascher, K.; Švara, V.; Jungmeier, M. Environmental DNA-Based Methods in Biodiversity Monitoring of Protected Areas: Application Range, Limitations, and Needs. *Diversity* **2022**, *14*, 463. https://doi.org/10.3390/d14060463

Academic Editor: Stephan Koblmüller

Received: 21 April 2022
Accepted: 6 June 2022
Published: 9 June 2022

1. Overview of eDNA-Based Methods in the Context of Biodiversity Monitoring

As global ecosystems face increasing pressure from human development and accompanying climate change, biodiversity loss has become the key ecological challenge worldwide [1,2]. Especially during the last decades, a dramatic global decline in species richness and abundance due to degraded habitat quality and diversity has become indisputable in **terrestrial** [3–5] as well as **freshwater** [6–10] and **marine ecosystems** [11]. Protected areas (PAs) and areas with special management status are key for the conservation of local and regional biodiversity, as they harbor a higher species richness than most areas without such status. However, even PAs are not immune to negative trends in biodiversity. This is clearly illustrated by the dramatic decline in flying insect biomass, a loss of over 75%, even in PAs in Germany, over a time span of only 27 years [12]. This reflects not only the major decrease in species diversity and abundance, but also the effects on ecosystem services, such as pollination, pest control, and nutrient supply across the food chain. Habitat destruction and fragmentation, decreasing flower supply, land-use change including agricultural intensification or abandonment of traditional farming, excessive use of pesticides, increased occurrence of pathogens, introduction and spread of invasive species, light pollution, increase in carbon dioxide, climate change, and the interactions of these factors have been recognized as main drivers for the ongoing biodiversity decline [13,14]. However, by applying sufficient measures and practices, such as effective, traceable, and accountable implementation of biodiversity policies, monitoring of biodiversity trends providing evidence-based indicators, and implementing sufficient measures in targeted areas, dramatic species decline can be mitigated [15,16]. Best practice can be promoted by applying recent advances in interdisciplinary biodiversity research.

The methodologies used must overcome several hurdles for effective implementation of different biodiversity monitoring programs. Monitoring on a large scale is costly and time consuming because it must occur at regular time intervals over a long period of time at the same observation sites e.g., [17]. Compounding financial and time constraints, a large number of organism groups would need to be monitored to obtain comprehensive biodiversity data for accurate ecological assessment allowing follow-up conservation action. Hence, biodiversity monitoring programs must typically target taxonomic indicator groups with high informative value [18], and in addition, must cope with a limited number of experts for several taxonomic groups [19]. Moreover, in several surveys, very short phenological timeslots for observation must be considered [20]. In inaccessible or dangerous terrain due to political conflicts, wild animals, or prevailing extreme weather conditions and high elevation, e.g., Bhutan or Himalaya [21], surveys are particularly challenging. On the other hand, biodiversity monitoring methods must meet high methodological standards. Methods must be reliable, reproducible, standardized, applicable to different taxonomic groups, usable across different geographic regions, consider different spatial scales, and be operator-independent, flexible, and applicable to different challenges over the course of a biodiversity monitoring program. Due to the multitude of benefits that a robust biodiversity monitoring program provides to human society, there is significant demand and interest in accommodating these challenges among different users, spanning the scientific community, industry, NGOs, and national, sub-national, and international clients [22]. For targeted implementation of biodiversity surveys, conventional and novel techniques are already available, which, however, have reached different degrees of maturity [17].

One of the most promising approaches that can help overcome challenges of biodiversity monitoring and has the potential to facilitate field surveys and improve conservation measures in PAs is **species detection with DNA-based methods**. These methods enable species- and taxon-specific identification of organisms by aligning genetic sequences (i.e., barcodes) with reference sequences in a database (e.g., IBOL: https://ibol.org/, ABOL: https://abol.ac.at/; accessed on 5 April 2022) [23]. When more than a single species is targeted applying universal primers, the method is referred to as DNA metabarcoding [24]. Taberlet et al. 2012 [25] define DNA metabarcoding as a method for *'the automated identification of multiple species from a single bulk sample—containing entire organisms or from a single environmental sample containing degraded DNA'*. The method can be applied in at least three different target applications: identification of single organisms, characterizing the diversity of a bulk sample—an environmental sample containing organisms from different taxonomic groups [26] being studied, e.g., from a Malaise trap, or species identification from environmental DNA (eDNA). In contrast to organismal DNA, which is extracted directly from collected specimens, eDNA is regarded as a DNA target in the environment. DNA is emitted into the surrounding medium by organisms via skin, hair, gametes, urine, or feces [25,27,28]. Media that can be surveyed for species presence include water, soil, sediment, air, or organic materials, such as stomach content, feces, bird pellets, or even honey, which may contain a variety of plant pollen as well as DNA signatures of visiting pollinators [29]. The released DNA stays in the medium and may persist for periods varying from several days up to months [30] or even longer, as shown for lake sediments or arctic permafrost [31]. It enables the detection of species in a medium without the need for sighting, capturing, or acoustic detection.

Based on eDNA metabarcoding, a **spectrum of taxonomic groups and species** generally found in an environment can be identified from a single environmental sample. This approach provides an overview of the species composition that is principally present in a surveyed habitat [24]. In addition to the detection of broader taxonomic units, eDNA-based methods allow targeted detection of **single species** [32]. Using species-specific primers, the presence or absence of a threatened species (e.g., amphibians in the Atlantic forest [33]), an indicator species (e.g., bioindicators of alpine freshwater environments [34]), an exotic species (e.g., Red-swamp crayfish, *Procambarus clarkii,* native in northern America and invasive in Europe [35]), or a parasite (e.g., the Rana virus *Batrachochytrium dendrobatidis* [36]) in a protected habitat can be verified. In PAs, species-specific eDNA-based methods have been applied for tracing large mammals, e.g., wild cats, lynx, wolves, or bears, based on hair samples [37] or excrement. Activities of protected species, e.g., migration of North American salmonids [38], are also assayed. In addition, eDNA metabarcoding analyses and species-specific assays are applicable not only in biodiversity monitoring but also in several other fields, such as ecology, e.g., in the analysis of stomach and gut contents [39]; paleobiology and palaeontology [40]; archaeoecology [26]; environmental impact assessment; environmental quality; citizen science (rapid test kits, e.g., 'frog in the water drop': https://www.uibk.ac.at/; accessed on 6 April 2022); agriculture and forestry; forensics; controls of food and traditional medical products as well as customs inspections on endangered and protected animals [41]; traceability of food; food safety; seed controls for specific ingredients or allergens; wood industry; and various industrial usages.

It is most likely that these multi-use methodologies will be increasingly applied to different monitoring projects and assessment of biodiversity, not only in scientific research, but also in practice. For example, within the framework of international guidelines and international reporting obligations, repeated biodiversity status analyses of PAs, such as national parks, UNESCO biosphere reserves, UNESCO world heritage sites, and European protected area networks (Natura 2000, Habitats Directive), must be conducted to evaluate the success of applied management activities. In general, any approach in the adaptive management of PAs requires accurate evidence of conservation outcomes [42,43]. Hence, eDNA-based assessments could be used for species detection and identification for this purpose, either as a supplement, or in some cases as a substitute, for conventional approaches as a part of regular monitoring campaigns [44].

In the synopsis of our paper, we provide examples of current and potential future PA monitoring programs that involve assessment of eDNA. For this purpose, we review possible applications, highlight particular cases of the practical implementation of species and biodiversity monitoring in nature conservation areas, identify major challenges, and finally list future goals and needs for effective implementation of eDNA collection in PAs. Our analyses and considerations are based on a literature search (e.g., search engine: Scopus, Google Scholar), lessons learned from the two projects E.DNA (KWF/EFRE UiG 2019/20, KWF No 16048-31819-45776) and BioMONITec (Biodiversity Monitoring Technologies—Transfer of disruptive engineering technologies into conservation practice: COIN FFG 2021-2024, No 884138), as well as long-time experience in national and worldwide conventional nature conservation approaches and applied biomonitoring.

2. eDNA-Based Methodology—Advantages, Disadvantages and Requirements for Use in Protected Areas

eDNA-based methods are particularly advantageous, as they can enable simultaneous assessment of the entire species composition in a comparatively short time and with little effort, making them an ideal tool to support and complement **biodiversity monitoring** of a defined area [45]. The application of eDNA-based methods for species detection from environmental samples has significant potential in comprehensive surveys of various taxonomic groups, from single cell organisms to large mammals. Depending on the respective investigation, eDNA sample collection is generally fast, relatively cheap, and easy [46]. The costs of applying eDNA analyses for biodiversity monitoring are highly dependent on the respective costs per sample offered by the particular laboratory providers, as well as on the total number of samples, since the per-sample cost drastically decreases, once a certain threshold value is met. In addition, the costs for DNA sequencing have been decreasing over recent years [47] as the methods have become widely available [48].

As a tool for **practical nature conservation**, there are several benefits of eDNA metabarcoding compared to conventional, morphology-based identification methods. The main advantages include the possibility of carrying out more comprehensive taxonomic surveys, the ability to cover spatially larger sampling areas, which is particularly crucial for PAs, the possibility of conducting non-invasive sampling of sensitive species in vulnerable ecosystems, and the ability to record traces of protected macroscopic organisms [49]. In addition to taxonomically comprehensive surveys, standardized approaches that enable comparisons of data between PAs should be available for biodiversity monitoring [50,51]. Regular biodiversity monitoring necessary for assessing and managing the status of protected goods—species and habitats—can be performed more economically using eDNA-based detection methods, which is also a key factor for PAs. Typically, taxon specific experts are not needed for sample collection. It can be conducted by non-geneticists, e.g., ecologists without knowledge of genetic analyses, geneticists, or even citizen scientists. This is of special importance for PAs, which often face a shortage of staff, especially of professional ecologists. Nevertheless, sampling training is mandatory for achieving desired outcomes. Thorough sampling is thus the prerequisite for accurate data analyses and species determination, which can then be outsourced. Finally, results of the eDNA metabarcoding analysis can be stored online and are accessible from any part of the world (e.g., IBOL, ABOL [52]). When successfully applied, eDNA-based species detection and identification may in some cases be even more reliable than expert taxonomic work, for example, in identifying larval stage insects, and may be much more effective when dealing with cryptic species e.g., deWaard et al., 2008 [53].

Aside from the listed advantages of eDNA detection methods, several major challenges must be considered when conducting species and community monitoring using eDNA [54]. The reliability of the assessment strongly depends on the sampled medium. Generally, eDNA detection is particularly successful when acquired from aquatic environments, but less reliable when collected from sediments and soil [55]. In this respect, the quality and quantity of the sample also rely on how much DNA is released from each species. Large amounts of DNA are, for instance, discharged from fish and amphibians [56,57]. In general, species can be better traced in particular habitats. The presence of frogs is preferentially and more easily confirmed in aquatic habitats compared to their terrestrial habitats. Species identification also depends on the densities of the organism group present in the investigated medium [49], also taking the spatial and temporal dynamics of eDNA into account [58,59]. Hence, in aquatic environments, assessment of species assembly proved to be more successful in small stagnant freshwater habitats, such as lakes or ponds, than in large running waterways, such as streams and rivers, because of the higher DNA concentrations in the stagnant water bodies [60]. However, challenges such as representative sampling, eDNA capture, and PCR inhibition still hamper complete species diversity detection in aquatic habitats. The success of species identification also varies among taxonomic groups due to the specificity of the primers used and differences in the completeness of the reference database, which in turn also depends on the level of taxonomic knowledge. Further challenges in applying eDNA-based methods include quantifying species abundance, relating species detections to the actual species assemblage of the habitat, and identifying species interaction. For habitat classification, there is also a need to assess the ecological status of key species [61]. In addition, the lack of experts is a major obstacle for data analysis and interpretation of eDNA metabarcoding results. A high risk of bias will result from the collection of samples by non-experts without adequate quality controls.

Apart from the requirement to outsource wet lab and bioinformatics expertise for sample analysis, eDNA sampling can, in general, be conducted by non-experts, if several prerequisites are fulfilled to ensure successful implementation of eDNA sampling. First, sound ecological knowledge, species-specific expertise on the behavior and biology of the sampled organisms, and experience with sampling in the field should be present. To be able to evaluate species lists obtained by eDNA metabarcoding, basic knowledge on laboratory practices e.g., Dully et al., 2021 [62], including DNA extraction, amplification, and sequencing is advantageous. On the other hand, understanding of workflows, basic bioinformatics experience, and knowledge on barcode alignment with reference databases are basic requirements for the expert entrusted with eDNA assessment. Thus, consultation or involvement of highly qualified experts is mandatory in eDNA-based biodiversity assessment. **Ecologists** are needed to identify and implement the sampling strategy, while **technicians** who are trained in the state-of-the-art laboratory work and in using bioinformatics pipelines, and **molecular biologists** who are experienced in interpreting the genetic results, should also be involved. In practice, the majority of eDNA samples collected by non-experts are processed and analysed by external technique providers. eDNA metabarcoding is still relatively cost-intensive, due to the required specialized equipment and expert handling in the context of regular monitoring. However, these methods may still be applied in PAs, as DNA-based methods are becoming increasingly standardized, and often the expertise of samplers is combined with that from companies specialized in performing molecular analyses (e.g., www.aimethods-lab.com; www.naturemetrics.co.uk; www.sinsoma.com; accessed on 7 March 2022).

3. Utilization of eDNA Metabarcoding in Biomonitoring in Protected Areas

Due to its advantages and application possibilities, these novel molecular methods are expected to have an immense implementation potential in future, including biodiversity monitoring practices in natural protection sites worldwide. As of April 2022, there are 251,947 terrestrial areas plus another 478 OECMs (other effective area-based conservation measures) under protection worldwide, covering nearly 17% of the global terrestrial area, including inland waters (https://www.protectedplanet.net/en; accessed on 20 April 2022). Another 17,910 marine PAs and OECMs cover 8% of the area of the world's oceans. Biodiversity includes ecosystems, biotopes and habitats, vegetation units, and ecological interactions, as well as almost all taxa and organismic categories. In the management of PAs, a shift towards evidence-based management and governance can be observed [63], which requires new monitoring capacities. For many categories, such as UNESCO sites or European protected area networks, **monitoring is mandatory**. That means biodiversity assessment must be carried out for evaluation of plant and animal diversity status, including habitat quality. In order to meet these requirements, ecological monitoring must occur regularly. Moreover, to enhance positive development of conservation targets, it is necessary to verify the effectiveness and success of the management measures applied [64]. Consequently, there is a high demand for applicable monitoring practices and related conventional as well as novel survey tools to facilitate these challenges and to achieve the desired conservation results [17].

However, as mentioned previously, PAs in particular have limited financial and staff capacities. Especially in the area of biodiversity assessment, they are largely dependent on external expertise. The number of existing experts is limited, and due to the peripheral location of many PAs, there are usually no experts available on-site. Consequently, eDNA-based methods open up completely new possibilities in this respect. Besides sampling, which can be performed by specially trained non-professional personnel, the required data expertise (taxonomic analysis) can be carried out by external experts at any time and from any location. Thus, for the first time, a basic prerequisite for systematic monitoring of conservation outcomes is being established.

To integrate eDNA sampling into standard conservation practice, a major focus is placed on the development of DNA-based methods applicable across ecosystems. In many cases, methods are established for the optimization of species-specific targets and for the investigation of species communities in different ecosystems [65]. However, there remains a large gap between testing and standard application in PAs, according to the published literature [66].

For the management of PAs, proof of target achievement and thus of management effectiveness cannot be provided without solid evidence of the conservation status. For this purpose, eDNA-based methods have already proven to be applicable in the monitoring of PAs. For example, in the study results on airborne pollen patterns in Natura 2000 sites in the Italian Alps, eDNA metabarcoding was a '*powerful molecular tool to complement traditional biodiversity monitoring*' [67] as it enabled rapid detection of regional plant species. In this study, analyses of pollen DNA with metabarcoding allowed 68 taxa of 32 plant families to be determined, with finer taxonomic resolution than with the use of classical techniques, such as light microscopy. In addition, initial data on plant species composition were obtained. eDNA metabarcoding has also been applied in analyses of *soil* samples, as little has yet been ascertained regarding the composition of soil fauna in general using conventional approaches. For instance, in alpine environments in the Italian Gran Paradiso National Park, the edaphic soil fauna diversity and its composition related to environmental features, such as habitats, vegetation, soil, and topographic features, were surveyed with eDNA metabarcoding [68]. With the application of this method, 18 arthropod families could clearly be distinguished and identified. Key factors for forest soil community composition could be related to parameters such as vegetation and altitude of location, whereas soil pH and slope inclination had the most influential effect on species composition in the prairie soil, revealing the environmental needs of different alpine habitats. Moreover, novel molecular techniques enable comprehensive identification of soil microbial diversity. Fungi, for example, provide key functions in ecosystems in their role as decomposers or plant symbionts. Using classical determination approaches, however, it is challenging to observe and taxonomically identify fungal species. The study of Yan et al. [69] showed the response of soil fungi to ecological restoration in an active restoration site at Mt. Bold in Australia, indicating a shift of fungal communities towards a more natural species composition within only few years. This example shows how eDNA allows for accurate quantification of environmental changes, which makes it a useful monitoring tool in restoration campaigns. eDNA soil analyses can also be applied for confirmation of terrestrial distribution of animals. Two examples are the recording of the endangered sharp-tailed snake (*Contia tenuis*) on Salt Spring Island, British Columbia, Canada [70], and monitoring of the endangered parrot species kākāpō (*Strigops habroptilus*), in New Zealand [71]. However, species identification of mammals, birds, reptiles, and amphibians in their terrestrial habitats remains a major challenge, since the concentration of DNA traces on land is lower and the DNA residues are comparatively more difficult to detect than, for example, in a water medium, because the DNA is bonded to soil particles and immobile, requiring analysis of several soil samples to increase the confidence of species evidence. Moreover, several other environmental factors influence eDNA detection, including abiotic variables that may affect DNA degradation. Consequently, there are relatively few applications of eDNA-based methods in terrestrial environments. eDNA metabarcoding for status assessment was also performed in the Kruger National Park, South Africa, where bacteria communities, including pathogens in waterholes, are monitored to provide a baseline of bacterial diversity, which in future could serve as an indicator to identify ecosystem disturbance [72].

Application of eDNA analysis can be especially promising for monitoring in remote and dangerous terrain. For instance, cave salamanders (*Proteus anguinus*) are challenging to explore because their habitats are dangerous and difficult to access [73]. In PAs where large wild animals occur, for instance in South-African national parks, eDNA metabarcoding of animal traces, such as hair or feces, enables species identification without risking human or animal safety through direct interactions [30]. Even saliva on twigs, e.g., of giraffes, provides information about the presence and variety of browsing animals [49]. Aquatic eDNA samples from waterholes resulted in data about their visitors without requiring visual identification [74]. The assessment of species diversity based on eDNA metabarcoding of aquatic samples is just starting to be explored, and results are compared with conventional animal monitoring methods [60]; (T. Schenekar, pers. comm.).

Methods of eDNA collection have been most successfully applied in PAs in **freshwater ecosystems**, such as ponds, lakes, rivers, and streams, for example, in [75]. For the most accurate and comprehensive assessment possible, eDNA collection in freshwater ecosystems should also be combined with conventional ecological surveys on the ground ('**ground truthing**' [76]). Such combined datasets enable a comprehensive overview about the quality and biodiversity status of the ecosystem under evaluation, also within the framework of environmental impact assessments. Currently, aquatic samples are taken primarily from freshwater systems where invertebrates, fish, and amphibians are the focus of ecological assessment, as reported in [77]. For example, the produced list of aquatic insects present in a sampled medium can provide an introductory overview on the ecological status of the waterbody and may also be used to identify single indicator species or groups. A useful application area is the survey of macrozoobenthos in flowing waters for the assessment of water quality and ecological status. A prominent example of this is the Himalayan state of Bhutan. The massive expansion of hydropower as a renewable energy source has had a significant impact on the country's remarkable river systems. These need to be systematically monitored [78] in order to mitigate ecological damage. This inspection cannot be guaranteed by applying conventional methods and capacities only. If needed, findings on species abundance, community composition, and ecological role can furthermore be investigated in more detail by conventional approaches. Besides freshwater ecosystems, eDNA metabarcoding has already proved to be particularly useful in saltwater ecosystems in connection with marine PAs [79,80]. Gold et al. 2021 referred to molecular methods as '*a promising alternative for marine ecosystem monitoring*' [69] as there exist large data gaps regarding species identification and species communities in marine habitats. In addition, on-site work in marine habitats is particularly challenging, dangerous, costly, and time-consuming, and hence, any facilitation in this regard is welcomed within the framework of the performed monitoring of the marine fauna. In the study of Gold et al., fish communities were investigated using eDNA metabarcoding in comparison to underwater visual census surveys. Out of 25 visually observed species, 19 could be confirmed with eDNA metabarcoding, providing optimism but also addressing further efforts for future applications in marine environments. However, the strengths and limitations of the different approaches still need to be assessed in more detail and for specific monitoring goals in the future.

In PAs, eDNA surveys can serve as a selection tool in biodiversity assessment for particular indicator species and can support efforts to further engage citizens in nature protection. In this regard, eDNA approaches can be very suitable in regional initiatives that attempt to generate data on the presence or absence of species of different taxonomic groups in PAs, and can contribute to regional barcode reference databases. One example of such an initiative is the Austrian Citizen Science campaign called BioBlitz, in which species are collected in the run of the Days of Biodiversity to generate DNA barcodes of species living in the investigated PAs. Citizen scientists contribute their findings, which are verified by taxonomic experts [81]. The information may, in such cases, act as the starting point for further monitoring programs in PAs.

As demonstrated above with examples of applications in PAs, eDNA-based species detection methods have already proven to be a promising novel approach that is expected to have strong potential as a supplement, or, in some cases, even as a replacement, for conventional monitoring methods in conservation. It is anticipated that conventional methods could be eclipsed, especially for complex monitoring, such as soil fauna investigation.

4. Challenges and Limitations of eDNA-Based Methods in Protected Area Monitoring

Despite a promising outlook for the application of eDNA-based methods in conservation, several challenges remain to be confronted. Currently, different protocols exist for the survey of the same medium and taxonomic groups that do not always produce comparable data sets [50,82,83]. Hence, the first step would be to develop standardized methods for eDNA field sample collection and analysis. Some methodological approaches, e.g., for water ecosystems and soil, already exist in a standardized form [82]. However, specific protocols for different media, types of samples and target groups are still missing. As DNA has proven to be present in sediments and soil and is stable for periods of several days or even months [31], the uncertainty of the actual physical presence of the detected species is high [49]. In aquatic environments, the DNA may be displaced over several km and often cannot be assigned to a specific location [84]. Thus, the selection of sampling locations also influences sampled eDNA quality and requires expert knowledge of species-specific occurrence, ecology, and behavior. Another major challenge is that taxon-specific primers must be identified in advance of laboratory analyses [85]. In addition, DNA inhibitors might prevent amplification of the target genomic region by the associated primer [86]. Furthermore, the quality of results is limited by the quality of existing reference databases. Knowledge about soil bacteria species communities, for example, is currently still scarce. For assessment of biodiversity in such cases, however, (molecular) operational taxonomic units ((M)OTUs) can be used [55]. In some cases, sampling of eDNA is not the optimal solution; for freshwater insects, bulk sampling is suggested, as insects do not shed much DNA into their environment [87]. Moreover, for a precise, correct, and complete species list of an investigated habitat, complete taxonomic databases are required. Thus, regional databases should ideally already be established for correct species assignment in a monitoring campaign; however, they should be collected from the same standardized source. The systematic use of eDNA metabarcoding for monitoring in PAs requires decision-makers to be aware of the importance and possibilities of this methodology. Appropriate capacity should be built and trained. Therefore, it is likely to be several years before eDNA-based methods can become established as a standard tool in nature conservation.

In order to use eDNA metabarcoding for biodiversity monitoring programs across PAs, suitable assessment and research questions must be formulated in advance. It is imperative that the applied eDNA-based method fits into the framework describing the goal of the biodiversity monitoring approach in the PA, and that suitable indicators and related questions are defined in advance (Dalton et al. submitted). The first step is to determine whether the monitoring target is already known, or whether it still needs to be identified. In this context, key questions include [27]: Are the desired taxa well-represented in the environmental sample? What type of material should be collected? Are specific sampling protocols available? Which genetic markers and primer sets should be used? Does a comprehensive reference database of DNA barcodes for the surveyed species group exist? Depending on the research assessment and monitoring task in a PA, DNA-based techniques are generally not applicable for every monitoring objective, and in several cases, conventional approaches will ensure more detailed and reliable assessments. Hence, the success of the applied DNA methods depends on the monitoring goal within the PA.

For biodiversity assessment in PAs, ecological information on the species assembly derived from eDNA metabarcoding is restricted and faces many limitations. **Species abundance** is hardly estimable [77]. Studies suggest, abundance should be assessed only when sufficient reference data are available; however, the data should still be interpreted cautiously in this case [88,89]. Consequently, in most cases only presence/absence data are generated. Information about life stage, demographic structure, reproductive success, and fitness of a species is generally lacking. This information is, however, needed to implement suitable management actions in PAs, requiring 'classical' ecological surveys for comprehensive assessment of the status of the indicator group. **Hybrids** can rarely be distinguished, because in most cases maternal mitochondrial DNA is used for eDNA-based approaches [90]. If hybrids need to be determined, specific primers must be developed and applied. Several animal species transfer very little DNA derived from their prey. This can yield false results of the actual occurrence of species in a medium (e.g., predators [91]). Depending on the DNA concentration and applied methodology, e.g., filter extraction method or sample preservation, species detection probability differs. Hence, harmonized optimally performing sampling protocols must be developed or optimized for use in the field, so that they can be applied globally across the widest possible range of PAs [92], simplifying their application by various experts as well as non-experts. In addition, based on the analyzed biodiversity data, better reporting standards would be needed to compare the ecological status of a similar environmental medium and follow-up protection measures in different PAs [93].

In general, eDNA metabarcoding is expected to be cheaper than applied conventional methods. However, if no protocol exists, establishing a novel metabarcoding methodology could be expensive. Beyond that, cost efficiency largely depends on the targeted taxonomic group, the respective applied method, and the number of samples to be processed. PAs often suffer from very limited budgets for biodiversity monitoring and conservation measures. A study by James et al. 1999 [94], based on a World Conservation Monitoring Center survey in 1996 across 600 PAs of 180 countries, with altogether 3.7 million km^2 under protection, investigated the global mean budget and personnel devoted to PAs worldwide, revealing that in the US PAs, the amount spent was US $893 per km^2. The mean amount in developed countries at that time was reported to be $2058 per km^2, while the mean in developing countries was capped at $157 per km^2. However, biomonitoring in developing countries in tropical PAs presents a particular challenge, as these complex ecosystems harbor an exceptionally large diversity of species, the majority of which are still unknown.

Beyond the application of eDNA-based methods in PAs, this novel method will also contribute to practical decision-making applications, such as environmental impact assessments, which could accelerate and compliment environmental legal procedures. In this application area, eDNA metabarcoding shows limitations, for instance, along linear structures, such as railways, or in construction planning, in which soil sample eDNA analyses would only represent a small fraction of the evaluated area, hence providing only point-based information. Thus, expert consultation is needed to survey target areas and assess the occurrence of priority plant and animal species. Despite these challenges, there are strong initiatives in some European countries to use this technique also in environmental impact assessment. In the North Adriatic Sea of Italy, genetic techniques are used in biomonitoring to survey marine diversity around three offshore gas platforms [95]. Finland, as another example, is preparing a plan to regularly implement it in environmental monitoring [96], while Canada has a guideline on the use of eDNA analysis to manage invasive and at-risk aquatic species [51].

5. Future Perspective

eDNA-based methods represent a promising technology in biodiversity monitoring, and are currently expanding into different fields of applied practice. Use of species-specific assays and eDNA metabarcoding in practical nature conservation is expected to fundamentally change assessment opportunities, services, and workflows, and will provide new answers on research, assessment, and management questions.

Potential areas of application of eDNA collection and analysis in the management of PAs include the following:

- **Implementation of effective long-term monitoring** of changes in species composition, especially in the air (e.g., pollen), water (e.g., zoobenthos, diatoms [97]), and soil (microbes, fungi). These investigations may go beyond taxon-specific monitoring and may cover entire species communities.
- **Early detection of biological threats in vulnerable ecosystems**, such as invasive species (e.g., pathogens [67,98]) or farmland and forest (e.g., spotted lanternfly (*Lycorma delicatula*) in northeastern USA; [99]). Robust analytical protocols may contribute to the implementation of an **early warning system**.
- **Systematic detection of rare or cryptic species** that may be of crucial importance for conservation and thus for management of the sites [80].
- Possibilities for **systematic recording of ephemeral natural phenomena and phenological changes** that can be of outstanding importance in the management of a site (e.g., research on shifts in phenology of bryophytes in relation to meteorological factors over time, https://www.lunduniversity.lu.se; accessed on 11 April 2022).

- **Detecting unexpected or unintended trends** in biodiversity in the context of PA management [100].

In order to successfully implement novel eDNA-based species detection methods into PA monitoring programs and increased efficiency of biodiversity monitoring, the following steps should be taken:

(1) Improvement of **reference data libraries**. Several countries with networks of PAs and national parks still have to build up such libraries, and urgently need to sequence more species before they can even consider applying this method (e.g., West Africa: [101]). (2) Acquisition of comprehensive scientific ecological knowledge to support monitoring planning and application of novel genetic methods to different environments. (3) Standardization of lab and field protocols. (4) Harmonization of several guidelines, which should ideally result in a common worldwide-applicable guideline as an initial guidance [102], [Dalton et al. submitted]. (5) Suitable method selection. The most useful, straightforward and cost-efficient methods should be identified and offered to managers and implemented through local, national, and global standards. (6) Development of common workflows as field data collection, data analysis, taxonomic determination, and data management become increasingly decoupled. (7) National training services. eDNA metabarcoding, especially in repeat applications (biodiversity monitoring), places high demands on data management and data handling. Data science and handling of big data require new capacities at the responsible agencies. In this regard, services should be offered nationally to support staff training. (8) Assessment of the method's suitability in each context. In each case, a critical examination must ascertain whether eDNA metabarcoding is able to support ecological field research and assessment at all, and whether the method is able to provide the desired information about the investigated environment or habitat.

In order to further fuel such implementation, several steps would have to be tackled. A significant gap exists between park management practitioners, academic labwork, and data analysis and interpretation, and this must be bridged in future. For a successful 'real world' application and implementation of DNA-based techniques in biodiversity monitoring, mutual understanding from all working perspectives must be worked on. A basic knowledge would have to be acquired by all parties on all steps in the workflow. Furthermore, these workflows must be simplified, and additional administrative and coordination services must be provided in the PAs to ensure a fluent handling process.

To conclude, eDNA analyses are a promising and applicable tool for a variety of monitoring-associated research and management questions in PAs. Different eDNA-based methods have their advantages and limits, so they should be implemented together by a broader group of experts, including molecular biologists, ecologists, and bioinformaticians. The methods have the potential to systematically support biodiversity monitoring and assessment in PA management cycles worldwide. However, the systematic use of eDNA also places high demands on the management of the PA; systematic workflows ranging from data collection to evaluation (big data) and archiving must be developed, tested, and standardized. Ideally, the workflows can be organized based on a labor-sharing approach in collaborations with experts with an ecological background, as not all steps need to be carried out by the PA management body alone. It is expected that the new technologies will be introduced gradually over the next few years, and will bring about a major change in the key processes of PA management.

Author Contributions: Conceptualization, K.P. and M.J.; Data curation, K.P.; Investigation, K.P. and V.Š.; Validation, K.P., V.Š. and M.J.; Writing—original draft preparation, K.P.; Writing—review & editing, K.P., V.Š. and M.J. All authors have read and agreed to the published version of the manuscript.

Funding: The concrete analyses for this publication did not receive external funding. However, reviews on eDNA metabarcoding and application in the different fields of intended use considered in the manuscript were conducted in the run of the two projects: **E.DNA**, funded by KWF/EFRE 2019/20, KWF grant number 16048-31819-45776; and **BioMONITec**, (Biodiversity Monitoring Technologies—

Transfer of disruptive engineering technologies into conservation practice) funded by COIN FFG (Österreichische Forschungsförderungsgesellschaft. Austria) 2021–2024, No. 884138.

Institutional Review Board Statement: Not applicable.

Informed Consent Statement: Not applicable.

Data Availability Statement: Not applicable.

Acknowledgments: We gratefully acknowledge *Daniel Todd Dalton* for polishing the English language, and *Klaus Steinbauer* for linking references. We would like to thank two anonymous reviewers for their valuable comments on the previous manuscript.

Conflicts of Interest: The authors of the article entitled 'Environmental DNA-based Methods in Biodiversity Monitoring of Protected Areas: Application Range, Limitations, and Needs' hereby declare no conflict of interest.

References

1. IPBES. *Summary for Policymakers of the Global Assessment Report on Biodiversity and Ecosystem Services of the Intergovernmental Science-Policy Platform on Biodiversity and Ecosystem Services*; IPBES: Bonn, Germany, 2019.
2. Rockström, J.; Steffen, W.; Noone, K.; Persson, A.; Chapin, F.S.; Lambin, E.F.; Lenton, T.M.; Scheffer, M.; Folke, C.; Schellnhuber, H.J.; et al. A safe operating space for humanity. *Nature* **2009**, *461*, 472–475. [CrossRef] [PubMed]
3. Brooks, D.R.; Bater, J.E.; Clark, S.J.; Monteith, D.T.; Andrews, C.; Corbett, S.J.; Beaumont, D.A.; Chapman, J.W. Large carabid beetle declines in a United Kingdom monitoring network increases evidence for a widespread loss in insect biodiversity. *J. Appl. Ecol.* **2012**, *49*, 1009–1019. [CrossRef]
4. Field, R.H.; Hill, R.K.; Carroll, M.J.; Morris, A.J. Making explicit agricultural ecosystem service trade-offs: A case study of an English lowland arable farm. *Int. J. Agric. Sustain.* **2016**, *14*, 249–268. [CrossRef]
5. Pascher, K.; Moser, D.; Dullinger, S.; Sachslehner, L.; Gros, P.; Sauberer, N.; Traxler, A.; Grabherr, G.; Frank, T. Setup, efforts and practical experiences of a monitoring program for genetically modified plants—An Austrian case study for oilseed rape and maize. *Environ. Sci. Eur.* **2011**, *23*, 12. [CrossRef]
6. Bolpagni, R.; Poikane, S.; Laini, A.; Bagella, S.; Bartoli, M.; Cantonati, M. Ecological and conservation value of small standing-water ecosystems: A systematic review of current knowledge and future challenges. *Water* **2019**, *11*, 402. [CrossRef]
7. Dudgeon, D.; Arthington, A.H.; Gessner, M.O.; Kawabata, Z.-I.; Knowler, D.J.; Lévêque, C.; Naiman, R.J.; Prieur-Richard, A.-H.; Soto, D.; Stiassny, M.L.J.; et al. Freshwater biodiversity: Importance, threats, status and conservation challenges. *Biol. Rev. Camb. Philos. Soc.* **2006**, *81*, 163–182. [CrossRef]
8. Graf, W.; Leitner, P.; Pletterbauer, F. Short overview on the benthic macroinvertebrate fauna of the Danube River. In *The Danube River Basin*; Liška, I., Aggarwal, P.K., Eds.; Springer: Berlin/Heildelberg, Germany, 2015; pp. 287–315. ISBN 978-3-662-47738-0.
9. Maasri, A.; Jähnig, S.C.; Adamescu, M.C.; Adrian, R.; Baigun, C.; Baird, D.J.; Batista-Morales, A.; Bonada, N.; Brown, L.E.; Cai, Q.; et al. A global agenda for advancing freshwater biodiversity research. *Ecol. Lett.* **2022**, *25*, 255–263. [CrossRef]
10. Reid, A.J.; Carlson, A.K.; Creed, I.F.; Eliason, E.J.; Gell, P.A.; Johnson, P.T.J.; Kidd, K.A.; MacCormack, T.J.; Olden, J.D.; Ormerod, S.J.; et al. Emerging threats and persistent conservation challenges for freshwater biodiversity. *Biol. Rev. Camb. Philos. Soc.* **2019**, *94*, 849–873. [CrossRef]
11. Elahi, R.; O'Connor, M.I.; Byrnes, J.E.K.; Dunic, J.; Eriksson, B.K.; Hensel, M.J.S.; Kearns, P.J. Recent trends in local-scale marine biodiversity reflect community structure and human impacts. *Curr. Biol.* **2015**, *25*, 1938–1943. [CrossRef]
12. Hallmann, C.A.; Sorg, M.; Jongejans, E.; Siepel, H.; Hofland, N.; Schwan, H.; Stenmans, W.; Müller, A.; Sumser, H.; Hörren, T.; et al. More than 75 percent decline over 27 years in total flying insect biomass in protected areas. *PLoS ONE* **2017**, *12*, e0185809. [CrossRef]
13. Brühl, C.A.; Zaller, J.G. Biodiversity decline as a consequence of an inappropriate environmental risk assessment of pesticides. *Front. Environ. Sci.* **2019**, *7*, 4. [CrossRef]
14. Ollerton, J.; Erenler, H.; Edwards, M.; Crockett, R. Pollinator declines. extinctions of aculeate pollinators in Britain and the role of large-scale agricultural changes. *Science* **2014**, *346*, 1360–1362. [CrossRef] [PubMed]
15. Kühl, H.S.; Bowler, D.E.; Bösch, L.; Bruelheide, H.; Dauber, J.; Eichenberg, D.; Eisenhauer, N.; Fernández, N.; Guerra, C.A.; Henle, K.; et al. Effective biodiversity monitoring needs a culture of integration. *One Earth* **2020**, *3*, 462–474. [CrossRef]
16. Perino, A.; Pereira, H.M.; Felipe-Lucia, M.; Kim, H.; Kühl, H.S.; Marselle, M.R.; Meya, J.N.; Meyer, C.; Navarro, L.M.; van Klink, R.; et al. Biodiversity post-2020: Closing the gap between global targets and national-level implementation. *Conserv. Lett.* **2021**, *16*, 16. [CrossRef]
17. Dalton, D.T.; Pascher, K.; Berger, V.; Steinbauer, K.; Jungmeier, M. Novel technologies and their application for protected area management: A supporting approach in biodiversity monitoring. In *Protected Area Management—Recent Advances*; Suratman, M.N., Ed.; IntechOpen Publishing: London, UK, 2021; p. 24. [CrossRef]
18. Mihoub, J.-B.; Henle, K.; Titeux, N.; Brotons, L.; Brummitt, N.A.; Schmeller, D.S. Setting temporal baselines for biodiversity: The limits of available monitoring data for capturing the full impact of anthropogenic pressures. *Sci. Rep.* **2017**, *7*, 41591. [CrossRef]

19. Coleman, C.O. Taxonomy in times of the taxonomic impediment—Examples from the community of experts on amphipod crustaceans. *J. Crustacean Biol.* **2015**, *35*, 729–740. [CrossRef]
20. Segrestin, J.; Bernard-Verdier, M.; Violle, C.; Richarte, J.; Navas, M.-L.; Garnier, E. When is the best time to flower and disperse? A comparative analysis of plant reproductive phenology in the Mediterranean. *Funct. Ecol.* **2018**, *32*, 1770–1783. [CrossRef]
21. Wangchuk, S.; Bolch, T.; Zawadzki, J. Towards automated mapping and monitoring of potentially dangerous glacial lakes in Bhutan Himalaya using Sentinel-1 Synthetic Aperture Radar data. *Int. J. Remote Sens.* **2019**, *40*, 4642–4667. [CrossRef]
22. Navarro, L.M.; Fernández, N.; Guerra, C.; Guralnick, R.; Kissling, W.D.; Londoño, M.C.; Muller-Karger, F.; Turak, E.; Balvanera, P.; Costello, M.J.; et al. Monitoring biodiversity change through effective global coordination. *Curr. Opin. Environ. Sustain.* **2017**, *29*, 158–169. [CrossRef]
23. Hebert, P.D.N.; Cywinska, A.; Ball, S.L.; deWaard, J.R. Biological identifications through DNA barcodes. *Proc. Biol. Sci.* **2003**, *270*, 313–321. [CrossRef]
24. Taberlet, P.; Coissac, E.; Hajibabaei, M.; Rieseberg, L.H. Environmental DNA. *Mol. Ecol.* **2012**, *21*, 1789–1793. [CrossRef] [PubMed]
25. Taberlet, P.; Coissac, E.; Pompanon, F.; Brochmann, C.; Willerslev, E. Towards next-generation biodiversity assessment using DNA metabarcoding. *Mol. Ecol.* **2012**, *21*, 2045–2050. [CrossRef] [PubMed]
26. Taberlet, P.; Bonin, A.; Zinger, L.; Coissac, E. *Environmental DNA—For Biodiversity Research and Monitoring*; Oxford University Press: Oxford, UK, 2018; ISBN 9780198767220.
27. Pawlowski, J.; Apothéloz-Perret-Gentil, L.; Mächler, E.; Altermatt, F. *Environmental DNA Applications for Biomonitoring and Bioassessment in Aquatic Ecosystems*; Federal Office for the Environment: Bern, Switzerland, 2020.
28. Ficetola, G.F.; Miaud, C.; Pompanon, F.; Taberlet, P. Species detection using environmental DNA from water samples. *Biol. Lett.* **2008**, *4*, 423–425. [CrossRef] [PubMed]
29. Harper, L.R.; Niemiller, M.L.; Benito, J.B.; Paddock, L.E.; Knittle, E.; Molano-Flores, B.; Davis, M.A. BeeDNA: Microfluidic environmental DNA metabarcoding as a tool for connecting plant and pollinator communities. *bioRxiv* **2021**. [CrossRef]
30. Dejean, T.; Valentini, A.; Duparc, A.; Pellier-Cuit, S.; Pompanon, F.; Taberlet, P.; Miaud, C. Persistence of environmental DNA in freshwater ecosystems. *PLoS ONE* **2011**, *6*, e23398. [CrossRef]
31. Giguet-Covex, C.; Ficetola, G.F.; Walsh, K.; Poulenard, J.; Bajard, M.; Fouinat, L.; Sabatier, P.; Gielly, L.; Messager, E.; Develle, A.L.; et al. New insights on lake sediment DNA from the catchment: Importance of taphonomic and analytical issues on the record quality. *Sci. Rep.* **2019**, *9*, 14676. [CrossRef]
32. Blackman, R.C.; Ling, K.K.S.; Harper, L.R.; Shum, P.; Hänfling, B.; Lawson-Handley, L. Targeted and passive environmental DNA approaches outperform established methods for detection of quagga mussels, Dreissena rostriformis bugensis in flowing water. *Ecol. Evol.* **2020**, *10*, 13248–13259. [CrossRef]
33. Sasso, T.; Lopes, C.M.; Valentini, A.; Dejean, T.; Zamudio, K.R.; Haddad, C.F.; Martins, M. Environmental DNA characterization of amphibian communities in the Brazilian Atlantic forest: Potential application for conservation of a rich and threatened fauna. *Biol. Conserv.* **2017**, *215*, 225–232. [CrossRef]
34. Blattner, L.; Ebner, J.N.; Zopfi, J.; Fumetti, S. von. Targeted non-invasive bioindicator species detection in eDNA water samples to assess and monitor the integrity of vulnerable alpine freshwater environments. *Ecol. Indic.* **2021**, *129*, 107916. [CrossRef]
35. Filipová, L.; Grandjean, F.; Chucholl, C.; Soes, D.M.; Petrusek, A. Identification of exotic North American crayfish in Europe by DNA barcoding. *Knowl. Managt. Aquat. Ecosyst.* **2011**, *401*, 14. [CrossRef]
36. Kamoroff, C.; Goldberg, C.S. Using environmental DNA for early detection of amphibian chytrid fungus Batrachochytrium dendrobatidis prior to a ranid die-off. *Dis. Aquat. Organ.* **2017**, *127*, 75–79. [CrossRef] [PubMed]
37. Steyer, K.; Kraus, R.H.S.; Mölich, T.; Anders, O.; Cocchiararo, B.; Frosch, C.; Geib, A.; Götz, M.; Herrmann, M.; Hupe, K.; et al. Large-scale genetic census of an elusive carnivore, the European wildcat (*Felis s. silvestris*). *Conserv. Genet.* **2016**, *17*, 1183–1199. [CrossRef]
38. Wood, Z.T.; Lacoursière-Roussel, A.; LeBlanc, F.; Trudel, M.; Kinnison, M.T.; Garry McBrine, C.; Pavey, S.A.; Gagné, N. Spatial heterogeneity of eDNA transport improves stream assessment of threatened salmon presence, abundance, and location. *Front. Ecol. Evol.* **2021**, *9*, 16. [CrossRef]
39. Guenay-Greunke, Y.; Bohan, D.A.; Traugott, M.; Wallinger, C. Handling of targeted amplicon sequencing data focusing on index hopping and demultiplexing using a nested metabarcoding approach in ecology. *Sci. Rep.* **2021**, *11*, 1–15. [CrossRef] [PubMed]
40. Thomsen, P.F.; Willerslev, E. Environmental DNA—An emerging tool in conservation for monitoring past and present biodiversity. *Biol. Conserv.* **2015**, *183*, 4–18. [CrossRef]
41. Staats, M.; Arulandhu, A.J.; Gravendeel, B.; Holst-Jensen, A.; Scholtens, I.; Peelen, T.; Prins, T.W.; Kok, E. Advances in DNA metabarcoding for food and wildlife forensic species identification. *Anal. Bioanal. Chem.* **2016**, *408*, 4615–4630. [CrossRef] [PubMed]
42. Gillson, L.; Biggs, H.; Smit, I.P.J.; Virah-Sawmy, M.; Rogers, K. Finding Common Ground between Adaptive Management and Evidence-Based Approaches to Biodiversity Conservation. *Trends Ecol. Evol.* **2019**, *34*, 31–44. [CrossRef]
43. Akçakaya, H.R.; Bennett, E.L.; Brooks, T.M.; Grace, M.K.; Heath, A.; Hedges, S.; Hilton-Taylor, C.; Hoffmann, M.; Keith, D.A.; Long, B.; et al. Quantifying species recovery and conservation success to develop an IUCN Green List of Species. *Conserv. Biol.* **2018**, *32*, 1128–1138. [CrossRef]
44. Bohmann, K.; Evans, A.; Gilbert, M.T.P.; Carvalho, G.R.; Creer, S.; Knapp, M.; Yu, D.W.; de Bruyn, M. Environmental DNA for wildlife biology and biodiversity monitoring. *Trends Ecol. Evol.* **2014**, *29*, 358–367. [CrossRef]

45. Veilleux, H.D.; Misutka, M.D.; Glover, C.N. Environmental DNA and environmental RNA: Current and prospective applications for biological monitoring. *Sci. Total Environ.* **2021**, *782*, 146891. [CrossRef]
46. Biggs, J.; Ewald, N.; Valentini, A.; Gaboriaud, C.; Dejean, T.; Griffiths, R.A.; Foster, J.; Wilkinson, J.W.; Arnell, A.; Brotherton, P.; et al. Using eDNA to develop a national citizen science-based monitoring programme for the great crested newt (*Triturus cristatus*). *Biol. Conserv.* **2015**, *183*, 19–28. [CrossRef]
47. Barba, M.; Czosnek, H.; Hadidi, A. Historical perspective, development and applications of next-generation sequencing in plant virology. *Viruses* **2014**, *6*, 106–136. [CrossRef] [PubMed]
48. Jones, M.R.; Good, J.M. Targeted capture in evolutionary and ecological genomics. *Mol. Ecol.* **2016**, *25*, 185–202. [CrossRef] [PubMed]
49. Herder, J.; Valentini, A.; Bellemain, E.; Dejean, T.; van Delft, J.J.; Thomsen, P.; Taberlet, P. *Environmental DNA—A Review of the Possible Applications for the Detection of (Invasive) Species*; Netherlands Food and Consumer Product Safety Authority: Nijmegen, The Netherlands, 2014.
50. Pawlowski, J.; Apothéloz-Perret-Gentil, L.; Altermatt, F. Environmental DNA: What's behind the term? Clarifying the terminology and recommendations for its future use in biomonitoring. *Mol. Ecol.* **2020**, *29*, 4258–4264. [CrossRef] [PubMed]
51. Abbott, C.; Coulson, M.; Gagné, N.; Lacoursière-Roussel, A.; Parent, G.J.; Bajno, R.; Dietrich, C.; May-McNally, S. *Guidance on the Use of Targeted Environmental DNA (eDNA) Analysis for the Management of Aquatic Invasive Species and Species at Risk*; Canadian Science Advisory Secretariat (CSAS): Ottawa, ON, Canada, 2021; 42p.
52. Fonseca, V.G. Pitfalls in relative abundance estimation using eDNA metabarcoding. *Mol. Ecol. Resour.* **2018**, *18*, 923–926. [CrossRef]
53. deWaard, J.R.; Ivanova, N.V.; Hajibabaei, M.; Hebert, P.D.N. Assembling DNA barcodes. Analytical protocols. *Methods Mol. Biol.* **2008**, *410*, 275–293. [CrossRef]
54. Deiner, K.; Bik, H.M.; Mächler, E.; Seymour, M.; Lacoursière-Roussel, A.; Altermatt, F.; Creer, S.; Bista, I.; Lodge, D.M.; de Vere, N.; et al. Environmental DNA metabarcoding: Transforming how we survey animal and plant communities. *Mol. Ecol.* **2017**, *26*, 5872–5895. [CrossRef]
55. Blaxter, M.; Mann, J.; Chapman, T.; Thomas, F.; Whitton, C.; Floyd, R.; Abebe, E. Defining operational taxonomic units using DNA barcode data. *Philos. Trans. R. Soc. Lond. B Biol. Sci.* **2005**, *360*, 1935–1943. [CrossRef]
56. Wang, S.; Yan, Z.; Hänfling, B.; Zheng, X.; Wang, P.; Fan, J.; Li, J. Methodology of fish eDNA and its applications in ecology and environment. *Sci. Total Environ.* **2021**, *755*, 142622. [CrossRef]
57. Bálint, M.; Nowak, C.; Márton, O.; Pauls, S.U.; Wittwer, C.; Aramayo, J.L.; Schulze, A.; Chambert, T.; Cocchiararo, B.; Jansen, M. Accuracy, limitations and cost efficiency of eDNA-based community survey in tropical frogs. *Mol. Ecol. Resour.* **2018**, *18*, 1415–1426. [CrossRef]
58. Deiner, K.; Altermatt, F. Transport distance of invertebrate environmental DNA in a natural river. *PLoS ONE* **2014**, *9*, e88786. [CrossRef]
59. Jerde, C.L.; Olds, B.P.; Shogren, A.J.; Andruszkiewicz, E.A.; Mahon, A.R.; Bolster, D.; Tank, J.L. Influence of stream bottom substrate on retention and transport of vertebrate environmental DNA. *Environ. Sci. Technol.* **2016**, *50*, 8770–8779. [CrossRef] [PubMed]
60. Harper, L.R.; Lawson Handley, L.; Carpenter, A.I.; Ghazali, M.; Di Muri, C.; Macgregor, C.J.; Logan, T.W.; Law, A.; Breithaupt, T.; Read, D.S.; et al. Environmental DNA (eDNA) metabarcoding of pond water as a tool to survey conservation and management priority mammals. *Biol. Conserv.* **2019**, *238*, 108225. [CrossRef]
61. Beng, K.C.; Corlett, R.T. Applications of environmental DNA (eDNA) in ecology and conservation: Opportunities, challenges and prospects. *Biodivers. Conserv.* **2020**, *29*, 2089–2121. [CrossRef]
62. Dully, V.; Balliet, H.; Frühe, L.; Däumer, M.; Thielen, A.; Gallie, S.; Berrill, I.; Stoeck, T. Robustness, sensitivity and reproducibility of eDNA metabarcoding as an environmental biomonitoring tool in coastal salmon aquaculture—An inter-laboratory study. *Ecol. Indic.* **2021**, *121*, 107049. [CrossRef]
63. Ruoss, E.; Alfare, L.T. Shifting protected area strategies to evidence based governance and management. In Proceedings of the 6th Symposium for Research in Protected Areas, Salzburg, Austria, 2–3 November 2017; pp. 561–564.
64. Stokes, E.J.; Strindberg, S.; Bakabana, P.C.; Elkan, P.W.; Iyenguet, F.C.; Madzoké, B.; Malanda, G.A.F.; Mowawa, B.S.; Moukoumbou, C.; Ouakabadio, F.K.; et al. Monitoring great ape and elephant abundance at large spatial scales: Measuring effectiveness of a conservation landscape. *PLoS ONE* **2010**, *5*, e10294. [CrossRef]
65. Veldhoen, N.; Hobbs, J.; Ikonomou, G.; Hii, M.; Lesperance, M.; Helbing, C.C. Implementation of novel design features for qPCR-based eDNA assessment. *PLoS ONE* **2016**, *11*, e0164907. [CrossRef]
66. Schenekar, T. The current state of eDNA research in freshwater ecosystems: Are we shifting from the developmental phase to standard applicatin in biomonitoring? *Hydrobiologia* **2022**, 20. [CrossRef]
67. Leontidou, K.; Vokou, D.; Sandionigi, A.; Bruno, A.; Lazarina, M.; de Groeve, J.; Li, M.; Varotto, C.; Girardi, M.; Casiraghi, M.; et al. Plant biodiversity assessment through pollen DNA metabarcoding in Natura 2000 habitats (Italian Alps). *Sci. Rep.* **2021**, *11*, 18226. [CrossRef]
68. Rota, N.; Canedoli, C.; Ferrè, C.; Ficetola, G.F.; Guerrieri, A.; Padoa-Schioppa, E. Evaluation of soil biodiversity in alpine habitats through eDNA metabarcoding and relationships with environmental features. *Forests* **2020**, *11*, 738. [CrossRef]

69. Yan, D.; Mills, J.G.; Gellie, N.J.; Bissett, A.; LOWE, A.J.; Breed, M.F. High-throughput eDNA monitoring of fungi to track functional recovery in ecological restoration. *Biol. Conserv.* **2018**, *217*, 113–120. [CrossRef]
70. Matthias, L.; Allison, M.J.; Maslovat, C.Y.; Hobbs, J.; Helbing, C.C. Improving ecological surveys for the detection of cryptic, fossorial snakes using eDNA on and under artificial cover objects. *Ecol. Indic.* **2021**, *131*, 108187. [CrossRef]
71. Urban, L.; Miller, A.; Eason, D.; Vercoe, D.; Shaffer, M.; Wilkinson, S.; Guhlin, J.; Dearden, P.; Jeunen, G.-J.; Gemmell, N.; et al. Genomic monitoring of the critically endangered Kākāpō by real-time targeted nanopore sequencing of environmental DNA. *Curr. Biol.* **2021**, 19. [CrossRef]
72. Farrell, M.J.; Govender, D.; Hajibabaei, M.; van der Bank, M.; Davies, T.J. Bacterial diversity in the waterholes of the Kruger National Park: An eDNA metabarcoding approach 1. *Genome* **2019**, *62*, 229–242. [CrossRef] [PubMed]
73. Gorički, Š.; Stanković, D.; Snoj, A.; Kuntner, M.; Jeffery, W.R.; Trontelj, P.; Pavićević, M.; Grizelj, Z.; Năpăruş-Aljančič, M.; Aljančič, G. Environmental DNA in subterranean biology: Range extension and taxonomic implications for Proteus. *Sci. Rep.* **2017**, *7*, 45054. [CrossRef] [PubMed]
74. Ushio, M.; Fukuda, H.; Inoue, T.; Makoto, K.; Kishida, O.; Sato, K.; Murata, K.; Nikaido, M.; Sado, T.; Sato, Y.; et al. Environmental DNA enables detection of terrestrial mammals from forest pond water. *Mol. Ecol. Resour.* **2017**, *17*, e63–e75. [CrossRef]
75. Fernandez, S.; Sandin, M.M.; Beaulieu, P.G.; Clusa, L.; Martinez, J.L.; Ardura, A.; García-Vázquez, E. Environmental DNA for freshwater fish monitoring: Insights for conservation within a protected area. *PeerJ* **2018**, *6*, e4486. [CrossRef]
76. Li, J.; Hatton-Ellis, T.W.; Lawson Handley, L.-J.; Kimbell, H.S.; Benucci, M.; Peirson, G.; Hänfling, B. Ground-truthing of a fish-based environmental DNA metabarcoding method for assessing the quality of lakes. *J. Appl. Ecol.* **2019**, *56*, 1232–1244. [CrossRef]
77. Pilliod, D.S.; Goldberg, C.S.; Arkle, R.S.; Waits, L.P. Estimating occupancy and abundance of stream amphibians using environmental DNA from filtered water samples. *Can. J. Fish. Aquat. Sci.* **2013**, *70*, 1123–1130. [CrossRef]
78. Jorde, K.; Jungmeier, M.; Schneider, M.; Peter, A.; Watzal, M.; Dorji, C.; Haas, C. *Guideline to Determine Minimum Environmental Flow Regulations for Dewatered Reaches of Hydropower Projects in Bhutan*; National Environment Commission: Thimphu, Bhutan, 2018; p. 128.
79. Liu, Q.; Zhang, Y.; Wu, H.; Liu, F.; Peng, W.; Zhang, X.; Chang, F.; Xie, P.; Zhang, H. A review and perspective of eDNA application to eutrophication and HAB Control in Freshwater and Marine Ecosystems. *Microorganisms* **2020**, *8*, 417. [CrossRef]
80. Gold, Z.; Sprague, J.; Kushner, D.J.; Zerecero Marin, E.; Barber, P.H. eDNA metabarcoding as a biomonitoring tool for marine protected areas. *PLoS ONE* **2021**, *16*, e0238557. [CrossRef]
81. Michaela, S.; Sabine, S.; Oliver, M.; Christoph, L.; Christian, B.; Elisabeth, H.; Stefan, D.; Andreas, E.; Rupert, F.; Elisabeth, G.; et al. Beitrag der ABOL-BioBlitze zur österreichischen Biodiversitäts-Erfassung: DNA-Barcodes aus 2019 und 2020. *Acta Zoo Bot. Austria* **2022**, *158*, 81–95.
82. Bruce, K.; Blackman, R.; Bourlat, S.J. *A Practical Guide to DNA-Based Methods for Biodiversity Assessment*; Pensoft Publishing: Sofia, Bulgaria, 2021; ISBN 978-619-248-053-0.
83. Minamoto, T.; Miya, M.; Sado, T.; Seino, S.; Doi, H.; Kondoh, M.; Nakamura, K.; Takahara, T.; Yamamoto, S.; Yamanaka, H.; et al. An illustrated manual for environmental DNA research: Water sampling guidelines and experimental protocols. *Environ. DNA* **2021**, *3*, 8–13. [CrossRef]
84. Shogren, A.J.; Tank, J.L.; Andruszkiewicz, E.; Olds, B.; Mahon, A.R.; Jerde, C.L.; Bolster, D. Controls on eDNA movement in streams: Transport, retention, and resuspension. *Sci. Rep.* **2017**, *7*, 5065. [CrossRef] [PubMed]
85. Zhang, S.; Zhao, J.; Yao, M. A comprehensive and comparative evaluation of primers for metabarcoding eDNA from fish. *Methods Ecol. Evol.* **2020**, *11*, 1609–1625. [CrossRef]
86. Schenekar, T.; Schletterer, M.; Lecaudey, L.A.; Weiss, S.J. Reference databases, primer choice, and assay sensitivity for environmental metabarcoding: Lessons learnt from a re-evaluation of an eDNA fish assessment in the Volga headwaters. *River Res. Appl.* **2020**, *36*, 1004–1013. [CrossRef]
87. Blackman, R.; Mächler, E.; Altermatt, F.; Arnold, A.; Beja, P.; Boets, P.; Egeter, B.; Elbrecht, V.; Filipe, A.F.; Jones, J.; et al. Advancing the use of molecular methods for routine freshwater macroinvertebrate biomonitoring—The need for calibration experiments. *Metabarcoding Metagenomics* **2019**, *3*, 49–57. [CrossRef]
88. Lacoursière-Roussel, A.; Côté, G.; Leclerc, V.; Bernatchez, L. Quantifying relative fish abundance with eDNA: A promising tool for fisheries management. *J. Appl. Ecol.* **2016**, *53*, 1148–1157. [CrossRef]
89. Yates, M.C.; Fraser, D.J.; Derry, A.M. Meta-analysis supports further refinement of eDNA for monitoring aquatic species-specific abundance in nature. *Environ. DNA* **2019**, *1*, 5–13. [CrossRef]
90. Clusa, L.; Ardura, A.; Fernández, S.; Roca, A.A.; García-Vázquez, E. An extremely sensitive nested PCR-RFLP mitochondrial marker for detection and identification of salmonids in eDNA from water samples. *PeerJ* **2017**, *5*, e3045. [CrossRef]
91. Nichols, R.V.; Königsson, H.; Danell, K.; Spong, G. Browsed twig environmental DNA: Diagnostic PCR to identify ungulate species. *Mol. Ecol. Resour.* **2012**, *12*, 983–989. [CrossRef]
92. Lock, M.; van Duren, I.; Skidmore, A.K.; Saintilan, N. Harmonizing forest conservation policies with essential biodiversity variables Incorporating Remote Sensing and Environmental DNA Technologies. *Forests* **2022**, *13*, 445. [CrossRef]
93. Fediajevaite, J.; Priestley, V.; Arnold, R.; Savolainen, V. Meta-analysis shows that environmental DNA outperforms traditional surveys, but warrants better reporting standards. *Ecol. Evol.* **2021**, *11*, 4803–4815. [CrossRef] [PubMed]

94. James, A.N.; Green, M.J.B.; Paine, J. *A Global Review of Protected Area Budgets and Staff*; WCMC—World Conservation Press: Cambridge, UK, 1999; p. 46.
95. Cordier, T.; Frontalini, F.; Cermakova, K.; Apothéloz-Perret-Gentil, L.; Treglia, M.; Scantamburlo, E.; Bonamin, V.; Pawlowski, J. Multi-marker eDNA metabarcoding survey to assess the environmental impact of three offshore gas platforms in the North Adriatic Sea (Italy). *Mar. Environ. Res.* **2019**, *146*, 24–34. [CrossRef] [PubMed]
96. Norros, V.; Laamanen, T.; Meissner, K.; Lehtinen, S.; Lohtander-Buckbee, K.; Nygård, H.; Ruohonen-Lehto, M.; Sirkiä, P.; Suikkanen, S.; Tolkkinen, M.; et al. *Roadmap for Implementing Environmental DNA (eDNA) and Other Molecular Monitoring Methods in Finland: Vision and Action Plan for 2022–2025*; Draft of the Reports of the Finnish Environment Institute XX/2022; Finnish Environment Institute: Helsinki, Finland, 2022; p. 50.
97. Apothéloz-Perret-Gentil, L.; Bouchez, A.; Cordier, T.; Cordonier, A.; Guéguen, J.; Rimet, F.; Vasselon, V.; Pawlowski, J. Monitoring the ecological status of rivers with diatom eDNA metabarcoding: A comparison of taxonomic markers and analytical approaches for the inference of a molecular diatom index. *Mol. Ecol.* **2021**, *30*, 2959–2968. [CrossRef]
98. Suarez-Menendez, M.; Planes, S.; Garcia-Vazquez, E.; Ardura, A. Early alert of biological risk in a coastal lagoon through eDNA metabarcoding. *Front. Ecol. Evol.* **2020**, *8*, 10. [CrossRef]
99. Valentin, R.E.; Fonseca, D.M.; Gable, S.; Kyle, K.E.; Hamilton, G.C.; Nielsen, A.L.; Lockwood, J.L. Moving eDNA surveys onto land: Strategies for active eDNA aggregation to detect invasive forest insects. *Mol. Ecol. Resour.* **2020**, *20*, 746–755. [CrossRef]
100. Boulanger, E.; Loiseau, N.; Valentini, A.; Arnal, V.; Boissery, P.; Dejean, T.; Deter, J.; Guellati, N.; Holon, F.; Juhel, J.-B.; et al. Environmental DNA metabarcoding reveals and unpacks a biodiversity conservation paradox in Mediterranean marine reserves. *Proc. Biol. Sci.* **2021**, *288*, 20210112. [CrossRef]
101. Echi, C.P.; Suresh, U.K.; George, S.; Ratheesh, V.R.; Vinitha, R.M.; Ejere, C.V.; Iyaji, O.F.; Nnamonu, I.E. Contribution towards the development of a DNA barcode reference library for West African mammals. *Afr. J. Biotechnol.* **2013**, *12*, 6704–6708. [CrossRef]
102. Jungmeier, M.; Arpa, Y.N.; Pechacek, P. *The Guidelines for Biodiversity Monitoring: Conservation and Sustainable Management of Turkey's Steppe Ecosystems Project–GCP/TUR/061/GFF*; FAO; MAF: Ankara, Turkey, 2022; p. 76.

diversity

MDPI

Article

High-Throughput DNA Metabarcoding as an Approach for Ichthyoplankton Survey in Oujiang River Estuary, China

Rijin Jiang [1], James Leonard Lusana [2,3,*] and Yongjiu Chen [2,*]

1 Zhejiang Marine Fisheries Research Institute, Zhoushan 316021, China
2 College of Marine Science and Technology, Zhejiang Ocean University, Zhoushan 316022, China
3 School of Aquatic Sciences and Fisheries Technology, University of Dar es Salaam, Dar es Salaam P.O. Box 60091, Tanzania
* Correspondence: lusanaj@gmail.com or lusana.james@udsm.ac.tz (J.L.L.); yongjiuchen@zjou.edu.cn or yongjiu.chen@gmail.com (Y.C.)

Abstract: High-throughput DNA metabarcoding of mitochondrial 12S rRNA and Cyt b gene sequences was coupled with a morphology-based identification tool to assess ichthyoplankton community structure in Oujiang River Estuary, China. The performances of 12S and Cyt b barcoding markers were compared in terms of taxonomic resolution, detection and coverage, and their suitability was established for use as a quick and powerful ichthyoplankton assessment tool. A total of 30,138 ichthyoplankton (2462 eggs and 27,676 larvae) samples were collected from April to August 2015 and identified to 145 taxa belonging to 57 families and 105 genera. June and July were the main spawning months. Ichthyoplankton were more abundant around Lingkun and Qidu Islands and the upper parts of Oujiang River Estuary. The 12S gene marker presented higher species coverage and detection rate than Cyt b. DNA metabarcoding exhibited more representative species identification power than morphology. The findings reported in this study provided a key attempt towards the development of time-efficient and cost-effective ichthyoplankton identification and assessment tool.

Keywords: ichthyoplankton; Oujiang River Estuary; metabarcoding; morphology; 12S; Cyt b

Citation: Jiang, R.; Lusana, J.L.; Chen, Y. High-Throughput DNA Metabarcoding as an Approach for Ichthyoplankton Survey in Oujiang River Estuary, China. *Diversity* **2022**, *14*, 1111. https://doi.org/10.3390/d14121111

Academic Editor: Stephan Koblmüller

Received: 13 September 2022
Accepted: 20 November 2022
Published: 14 December 2022

1. Introduction

Ichthyoplankton are the early life stages (eggs and larvae) of marine fishes found in the sunlight zone of the water column usually less than 200 m deep [1,2]. Ichthyoplankton research is important because it provides information about both juvenile and adult fishes, such as spawning seasons and locations, recruitment strength, migration history, and spatial and temporal structures [3,4]. This information is essential for effective fish stock management and conservation particularly in light of anthropogenic disturbances and rapid climate changes [2]. Correct and accurate identification of fish eggs and larvae is a crucial step for fish ecological studies and conservation planning. Misinterpretation of fish biology and ecology derived from inaccurate ichthyoplankton identification could lead to biased fish stock evaluations and subsequently, poor conservation and management policies [5,6].

The rapid advance of next-generation DNA sequencing (NGS) analysis [4,7,8] has revolutionized genetic approaches for biodiversity research by providing an alternative tool for fish identification and assessment across all life stages. NGS is often more cost-effective, rapid and accurate than traditional methods [7,9–12]. This technology is rapidly transforming aquatic research to the genomic level, and combatting various challenges in the marine environment, from food security and biodiversity loss to climate change [13].

DNA metabarcoding using NGS has recently emerged as a potentially powerful method for assessing and monitoring the community structure of fishes, including eggs and larvae [4,14]. In order to achieve higher and more accurate species resolution and detection, DNA metabarcoding requires a heedful selection of barcode markers and primers.

Conserved barcode markers targeting certain regions of the mitochondrial genes, e.g., 12S rRNA [11,15–17] and Cyt b [18] provide a broad taxonomic resolution and coverage for fishes, even when DNA is degraded or present at a very low concentration [19]. In particular, 12S has been recommended for animal metabarcoding due to the presence of highly conserved regions that flank variable regions, permitting the design of primers with high taxon resolution power for the aimed taxonomic groups, and allowing concurrent identification of massive sets of existing organisms in a single sample of pooled DNA [12,20].

The purpose of this study was to assess the spatial and temporal distribution, species identity and composition of ichthyoplankton community structure in Oujiang River Estuary using 12S and Cyt b DNA metabarcoding coupled with a morphology-based identification approach. The performances of 12S and Cyt b metabarcoding markers were also compared in terms of taxonomic resolution, detection and coverage, and their suitability was established for use as a quick and powerful ichthyoplankton assessment tool.

2. Materials and Methods

2.1. River Estuary Survey and Sample Collection

A total of five surveys were carried out across Oujiang River Estuary from April to August, 2015 using a commercial fishing boat. One survey was conducted every month for the period of four consecutive days during monthly tide flood. Fish eggs and larvae were collected using the shallow horizontal planktonic net or oblique drag sampling survey with a digital flow meter with the density index (Ind./100 m^3). Eleven sampling locations (F1–11) were established for ichthyoplankton surveys. Plankton samples were sieved through smaller sized meshes and washed with sea water, from which fish eggs and larvae were picked out, placed into separate jars according to sampling sites and months, and preserved in 100% ethanol prior to morphological and molecular analyses.

2.2. Morphological Assessment

Ichthyoplankton morphological identification followed [21] and used dissecting microscope attached with a camera (Nikon SMZ800- Tokyo, Japan). All fish eggs and larvae from all sampling sites and months were identified to the lowest possible taxonomic rank using morphological features.

2.3. Metabarcoding Assessment

2.3.1. DNA Extraction

A total of 22 tubes (samples) containing various eggs and larvae collected from six stations (F2, F3, F4, F5, F7, and F9) in all sampling months were sequenced for 12S and Cyt b DNA metabarcoding analysis. Total DNA was extracted using DNeasy® Blood and Tissue Kit (Qiagen, Valencia, CA, USA) from fish eggs and larvae in a batch of 15 samples. The resulting DNA samples were then pooled together for each specific site and month. Extracted DNA was visualized by agarose gel electrophoresis using 1% agarose gel in 1X TAE buffer stained with DNA Green fluorescent dye for band characterization through Gel Imaging System.

2.3.2. PCR Amplification and DNA Sequencing

A 12S gene fragment (<100 bp) was amplified by PCR using the primer set of teleo_F_L1848 and teleo_R_H1913 [12]. A second marker of Cyt b gene fragment (<460 bp) was amplified using the primer pair of L14841 and H15149 [18]. All PCRs were conducted in a Thermo Cycler with a 25 μL reaction volume containing 8.5 μL sterile nuclease-free water in analytical grade, 12.5 μL GoTaq® Green Master Mix (Promega Inc., Madison, WI, USA), 1 μL each of the primer set, and 2 μL template DNA. The thermal profile included a preliminary denaturation for 2 min at 95 °C followed by 35 cycles of denaturation at 95 °C for 30 s, annealing at 50 °C for 30 s, extension at 72 °C for 60 s and finally a single extra extension at 72 °C for 10 min. PCR products were confirmed via gel electrophoresis using

1.5% agarose gel in 1X TAE buffer stained with DNA Green fluorescent dye for band characterization through Gel Imaging System. PCR products for all genes were sent for NGS analysis at LC Science (Hangzhou, China) following standard protocols for PE300 library construction and sequencing on an Illumina MiSeq platform.

2.4. Data Analysis

2.4.1. Morphological Data

The spatial and temporal distribution patterns of eggs and larvae were visualized in 2D graphs produced in Surfer®. For quantitative analysis, the abundance of eggs and larvae was estimated by density, D (using D = N/V) in the number of individuals per cubic meter where N is the number of eggs/larvae per catch, and V is the filtration volume. Species percentage was used to measure the level of species dominance. The species with the highest percentage of total catch was considered the dominant species in each sampling period and area.

2.4.2. DNA Metabarcoding Data

All raw data from the MiSeq sequencing platform were received in FASTQ format and preprocessed by trimming the barcodes and the adapter sequences. Extended reads were produced by merging the paired ends of the sequences using FLASH software [22]. Trimmomatic [23] was performed on the merged data for quality filtering by discarding all tags that have an "N" base percentage higher than 5%, a low-quality base percentage $\geq$ 20% or a short sequence length. A quality control check was performed by visually analyzing a QC report generated in FastQC [24]. The resulting reads were then imported into the QIIME pipeline [25] using MacQIIME version 1.9.1 (http://www.wernerlab.org/software/macqiime (accessed on 26 October 2017). Low-quality reads and short sequences were removed; then clean reads were assigned to samples, or demultiplexed, based on their nucleotide barcode using the split_libraries_fastq.py script. The demultiplexed sequences were clustered into OTUs with CD-HIT at $\geq$97% similarities using the pick_otus.py script.

For both markers, a representative set of sequences were selected from each OTU using the pick_rep_set.py script. The 12S and Cyt b QIIME compatible databases were created in MacQIIME following the standard method by Baker [26] after downloading all available fish 12S and Cyt b sequences in the GenBank database. The representative sequences of each gene marker were then blasted against the created QIIME compatible reference database and assigned taxonomic names using the assign_taxonomy.py script (minimum percent identity = 95%, maximum e-value = 0.001). Finally, OTU tables were built (make_otu_table.py) and singletons were removed from OTUs (filter_otus_from_otu_table.py). The summarize_taxa_through_plots.py script was used to summarize species taxonomy.

A phylogenetic tree was constructed after aligning and filtering the representative set of sequences in MacQIIME. The community structure of fish eggs and larvae was determined by calculating within-community diversity (alpha diversity) and between-community diversity (beta diversity). The level of alpha diversity was determined by calculating Shannon (also known as Shannon-Weiner), Simpson and Chao1 indices (alpha_diversity.py). Beta diversity among sampling sites and months was compared for each metabarcoding gene using Bray-Curtis distance and visualized using principal coordinate analysis plots generated by beta_diversity_through_plots.py and make_2d_plots.py scripts.

3. Results

3.1. Morphology

Species Identification and Composition

A total of 30,138 (2462 eggs and 27,676 larvae) ichthyoplankton samples were collected and identified. The highest number of eggs and larvae were recorded in June (Table 1). Ichthyoplankton samples were morphologically classified into 38 fish groups, including eight groups of eggs and 30 groups of larvae. Twenty-eight groups were identified to species level, four groups to genus level and six groups to family level. *Coilia mystus*

(79.73%), *C. nasus* (11.86%), and Cyprinidae (7.55%) were the dominant fish egg species, and *C. nasus* (57.67%), *C. mystus* (33.30%), and Gobiidae (7.27%) were the dominant larva species in Oujiang River Estuary (Figure 1).

Table 1. Species number, quantity and average density of fish eggs and larvae in Oujiang River Estuary from April to August 2015.

	Fish Eggs			Fish Larvae		
	Number of Species	Number of Eggs	Average Density Eggs/100 m^3	Number of Species	Number Larvae	Average Density Larvae/100 m^3
April	0	0	0	9	71	1.82
May	7	184	5.27	15	1997	46.65
June	4	1585	24.33	13	18,882	299.89
July	3	332	4.26	17	5054	66
August	3	361	5.19	12	1672	21.32
Total	9	2462	7.84	36	27,676	87.1

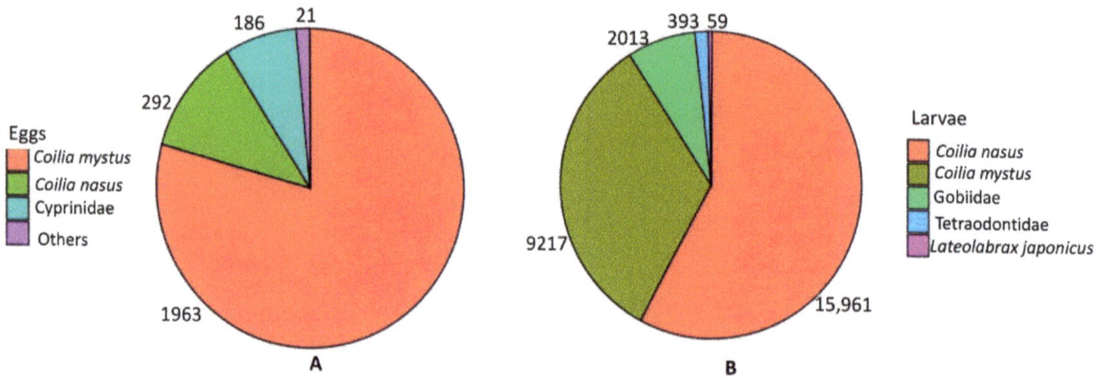

Figure 1. Number of dominant fish groups in Oujiang River Estuary. (**A**) Fish eggs and (**B**) fish larvae.

3.2. Ichthyoplankton Spatial and Temporal Distribution

Ichthyoplankton density varied among sampling points and seasons. In April, the average density of fish larvae was 1.82 ind./100 m^3. Sampling station F8 (4.73 ind./100 m^3), F7 (2.33 ind./100 m^3) and F3 (2.27 ind./100 m^3) had the highest fish larva densities. Other sampling stations had densities less than 2 ind./100 m^3 (Figure 2). April was dominated by Mugilidae 3.59 ind./100 m^3, *Lateolabrax japonicus* 3.56 ind./100 m^3 and Engraulidae 2.63 ind./100 m^3. No eggs were collected in April (Figure 3).

In May, the average density of fish larvae was 46.65 ind./100 m^3. Sampling stations in the upper parts of Qidu Island, i.e., F10 (145.56 ind./100 m^3), F1 (124.22 ind./100 m^3) and F2 (110.78 ind./100 m^3) had the greatest larva densities. Other sampling stations had larva densities less than 60 ind./100 m^3. The dominant fish larvae in May were Gobiidae (254.24 ind./100 m^3), *C. mystus* (90.22 ind./100 m^3) and *C. nasus* (53.76 ind./100 m^3 (Figure 2). The average egg density was 5.27 ind./100 m^3 dominated by *C. mystus* (24.71 ind./100 m^3), *C. nasus* (16.31 ind./100 m^3) and *Cyprinus carpio* (8.05 ind./100 m^3). The highest egg density was observed in the upper part of the river at sampling points of F1 (19.94 ind./100 m^3), F4 (17.35 ind./100 m^3) and F3 (14.12 ind./100 m^3). Other sampling locations had egg densities less than 1 ind./100 m^3 (Figure 3).

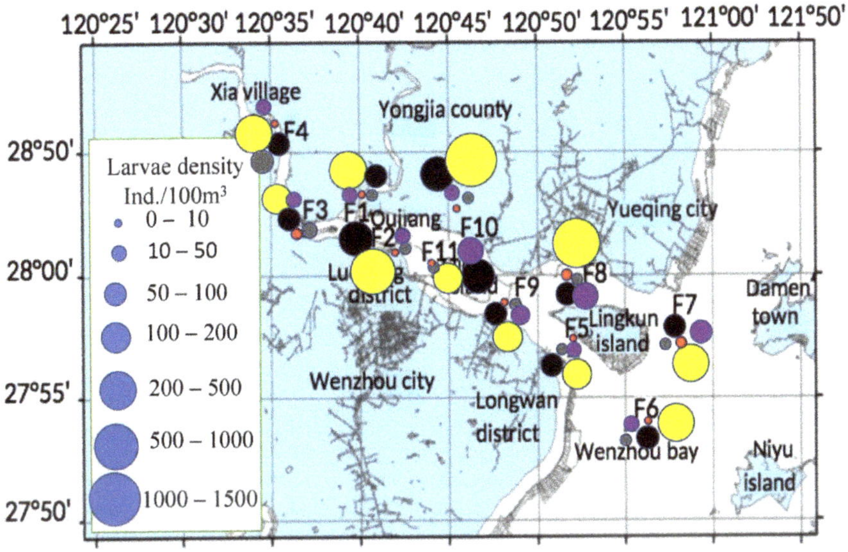

Figure 2. Spatial-temporal distribution of fish larvae in Oujiang River Estuary: each sampling month is indicated with a specific color: red = April, black = May, yellow = June, grey = July and purple = August. The size of circle reflects the density of larvae in a specific site.

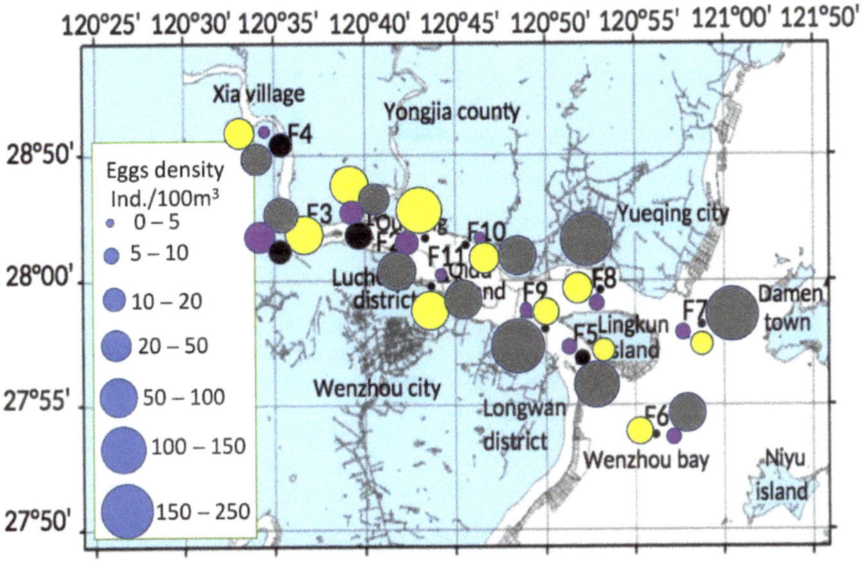

Figure 3. Spatial-temporal distribution of fish eggs in Oujiang River Estuary: Each sampling month is indicated with a specific color: black = May, yellow = June, grey = July and purple = August. The size of circle reflects the density of eggs in a specific site.

In June, the average density of 299.89 ind./100 m^3 for fish larvae was recorded, dominated by *C. nasus* (2241.27 ind./100 m^3) and *C. mystus* (743.28 ind./100 m^3). The highest larva density was observed in areas around Qidu Island, i.e., F10 (1008.79 ind./100 m^3), F11 (822.79 ind./100 m^3) and F2 (755.18 ind./100 m^3). Other sampling locations had larva densities less than 160 ind./100 m^3 (Figure 2). The average egg density was 24.33 ind./100 m^3

dominated by *C. nasus* (203.14 ind./100 m^3), Cyprinidae (25.61 ind./100 m^3) and *C. mystus* (13.76 ind./100 m^3). The highest egg density was observed in the upper part of the River at F2 (110.60 ind./100 m^3), F3 (53.16 ind./100 m^3) and F1 (48.13 ind./100 m^3). Other sampling locations had egg densities less than 30 ind./100 m^3 (Figure 3).

In July, the average density of 66 ind./100 m^3 for fish larvae was collected. Areas around Lingkun and Qidu Islands at F4 (37.01 ind./100 m^3) had the greatest larva densities. Other sampling stations had larva densities less than 30 ind./100 m^3. The dominant larvae were *C. mystus* (457.32 ind./100 m^3) and *C. nasus* (162.55 ind./100 m^3; Figure 2). The average egg density was 4.26 ind./100 m^3 dominated by *C. nasus* (46.7 ind./100 m^3) and *Stolephorus chinensis* (0.21 ind./100 m^3). The highest egg density was observed around the Lingkun Island at F8 (226.19 ind./100 m^3), F7 (175.88 ind./100 m^3) and F9 (164.24 ind./100 m^3; Figure 3).

In August, the average density of 21.32 ind./100 m^3 for fish larvae was collected. Sampling stations around Qidu Island to Lingkun Island, i.e., F8 (58.93 ind./100 m^3), F11 (44.54 ind./100 m^3), F7 (37.20 ind./100 m^3) and F9 (32.64 ind./100 m^3) had the greatest larva densities. The densities of other sites were less than 20 ind./100 m^3. The dominant fish larvae were *C. mystus* 172.90 ind./100 m^3 and *C. nasus* 42.62 ind./100 m^3 (Figure 2). The average eggs density was 5.9 ind./100 m^3 dominated by *C. mystus* 29.25 ind./100 m^3 and *C. nasus* 27.82 ind./100 m^3. Higher egg density stations were in upper parts of the River Estuary at F3 (24.12 ind./100 m^3), F2 (17.01 ind./100 m^3 and F1 (12.25 ind./100 m^3) (Figure 3).

3.3. *DNA Metabarcoding*

3.3.1. Sequencing and Reads Quality

The raw data generated libraries for 12S and Cyt b. Of 12S, 788,906 tags resulted from 79.81 Mb sequences. The quality control yielded 784,064 clean tags in 78.01 Mb, with an average GC content of 44.83% and the sequence length distribution of <200 bases. A second library of Cyt b generated 487,697 tags in 293.37 Mb. The quality control subsequently produced 433,420 tags in 148.48 Mb with an average 43.15% GC content that subsequently contributed to <315 bp sequence length distribution.

3.3.2. Species Identification and Composition

The DNA metabarcoding provided results about species identity, diversity, abundance, distribution, and composition of ichthyoplankton in Oujiang River Estuary. After taxonomic assignment of OTUs, about 0.01% of 12S sequences and 61.8% of Cyt b sequences had no BLAST hits. The 12S sequence dataset was assigned to 82 taxon groups from 661 OTUs with taxonomic coverage of 23 orders, 33 families, and 68 genera, of which 77 were identified to species level and five groups to genus level. The Cyt b metabarcoding recovered 412 OTUs, of which 46 taxa were identified to 22 orders, 34 families, and 51 genera. Within the 46 taxa, 45 were identified to species level and one taxon to genus level. The 12S metabarcoding analysis indicated that *C. nanus* (32.4%) and *C. mystus* (14.6%) were the dominant fish species, while the Cyt b metabarcoding revealed that *Moringua microchir* (20.6%) and *C. nanus* (10.7%) were the dominant fish species in Oujiang River Estuary, although 61% of the sequences had no BLAST hits and were not identified (Figure 4).

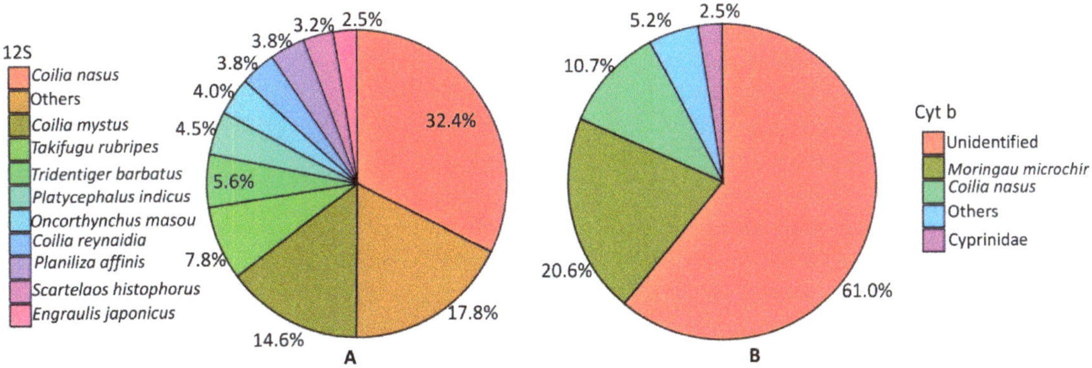

Figure 4. The species percentage composition of fish eggs and larvae in Oujiang River Estuary identified by molecular analysis: (**A**) 12S DNA metabarcoding, (**B**) Cyt b DNA metabarcoding.

3.3.3. Ichthyoplankton Community Structure and Diversity Patterns

In Oujiang River Estuary, the analysis of 12S alpha diversity reflected by Shannon and Simpson indices indicated that the highest species abundance and diversity were found in August, followed by July, June and April. F5 (July) located around Lingkun Island had the most diverse composition, followed by F5 (April), F3 (August), F9 (August) and F9 (June), while F4 (July) was the least diverse (Table 2). The Cyt b analysis showed that May was the most diverse month followed by June. F2 (May) was the most diverse followed by F3 (June), while F7 (August) was the least diverse (Table 2). Generally, the composition and number of species varied among the sampling seasons and sites. The highest number of species was found in July followed by June. The maximum numbers of species were detected in areas around Lingkun Island and Qidu Island.

3.3.4. Comparison of Assessment Tools and Markers

A total of 22 samples of 12S and 18 samples of Cyt b were successful amplified. The total number of sequences read counts passed quality control per library was 784,064 (99.38%) for 12S and 433,420 (88.87%) for Cyt b. After demultiplexing, 757,888 (12S) and 353,581(Cyt b) sequences resulted for taxonomic analysis. A large proportion of Cyt b sequences (61.8%) had no species identity because they had no BLAST hits mostly due to lack of reference sequences available on the GenBank database, while few of 12S (0.01%) had no BLAST hits. The obtained results indicated that 12S marker was more efficient in identifying fish species than Cyt b.

Based on molecular and morphology analysis, a total of 145 species were identified in Oujiang River Estuary, belonging to 57 families and 105 genera. In total, 128 taxa were identified to species level, 11 to genus level and 6 to family level. Based on morphological criteria, 38 taxa were observed from all the samples collected in all these months (55 subsamples) representing 16 families and 27 genera. The 12S metabarcoding dataset identified 82 taxa from 22 subsamples belonging to 46 families and 83 genera. The Cyt b metabarcoding dataset detected 46 taxa from 18 subsamples representing 34 families and 31 genera. The 12S detected 67, Cyt b detected 36, and morphology identified 24 unique species. The number of species in common revealed by the 12S and morphology, by the 12S and Cyt b and by the Cyt b and morphology were eight, four and three, respectively. The three tools identified three species in common (Figure 5 and Supplementary Table S1).

Table 2. 12S/Cyt b OTU statistics and alpha diversity indices for ichthyoplankton in Oujiang River Estuary. The value in parenthesis indicates sampling month (4–8 represents April–August). 12S—non-bold values above the slash and Cyt b—bold values below the slash.

	OTU Statistics			OTU Diversity and Abundance		
Sample	**Number of Clean Reads**	**Number of OTUs**	**Identified OTUs (%)**	**Chao1**	**Simpson**	**Shannon**
F5 (4)	57,291/**39,856**	201/**116**	99.94/**2.40**	236.04/**136.31**	0.79/**0.11**	2.97/**0.61**
F3 (5)	35,468/**37,845**	126/**137**	99.99/**83.10**	198.06/**168.32**	0.19/**0.46**	0.79/**1.54**
F5 (7)	19,155	165	99.94	201.12	0.83	3.32
F5 (8)	32,023/**10,867**	124/**98**	99.95/**5.90**	148.23/**205.63**	0.59/**0.25**	1.94/**1.2**
F3 (4)	3559/**21,629**	128/**73**	99.99/**97.7**	179.75/**143.2**	0.57/**0.17**	1.75/**0.8**
F3 (6)	15,119/**11,770**	69/**119**	99.99/**35.70**	128.5/**150.71**	0.14/**0.54**	0.58/**1.86**
F2 (5)	49,057/**5850**	175/**105**	99.98/**55.00**	246.32/**150.56**	0.68/**0.74**	2.15/**2.85**
F2 (6)	24,758/**8542**	118/**88**	99.98/**21.00**	151.79/**113**	0.55/**0.41**	1.74/**1.65**
F2 (7)	92,620/**13,868**	184/**54**	99.99/**9.30**	203.12/**67.2**	0.66/**0.27**	2.24/**1.11**
F3 (7)	67,985	214	99.99	254.53	0.74	2.42
F3 (8)	21,767/**7532**	191/**62**	99.97/**94.60**	262.36/**83.11**	0.81/**0.27**	2.92/**1.19**
F4 (5)	56,326/**9371**	174/**77**	100/**97.90**	244/**100.21**	0.69/**0.63**	2.29/**2**
F4 (6)	17,017	117	100	247.71	0.42	1.67
F4 (7)	49,670/**10,904**	96/**31**	100/**98.00**	137.35/**41.5**	0.02/**0.12**	0.15/**0.57**
F4 (8)	10,946	53	100	80.08	0.04	0.23
F9 (5)	31,835	135	100	227.81	0.32	1.14
F9 (6)	17,341/**11,816**	154/**86**	99.98/**12.90**	197.56/**123.19**	0.7/**0.32**	2.82/**1.25**
F9 (7)	38,062/**90,137**	137/**213**	99.99/**34.30**	168.95/**228.4**	0.71/**0.56**	2.32/**1.9**
F7 (4)	14,804/**15,338**	98/**47**	100/**18.10**	152.38/**58.25**	0.77/**0.34**	2.41/**1.33**
F7 (8)	11,196/**21,467**	111/**34**	99.95/**3.20**	144/**45**	0.61/**0.08**	2.35/**0.39**
F7 (6)	45,304/**14,296**	131/**61**	100/**18.20**	136.83/**70**	0.68/**0.34**	2.18/**1.25**
F9 (8)	14,553/**9977**	167/**94**	99.60/**9.90**	238.5/**172**	0.71/**0.26**	2.88/**1.2**
F5 (5)	**12,316**	**98**	**5.30**	**112.29**	**0.24**	**1.1**

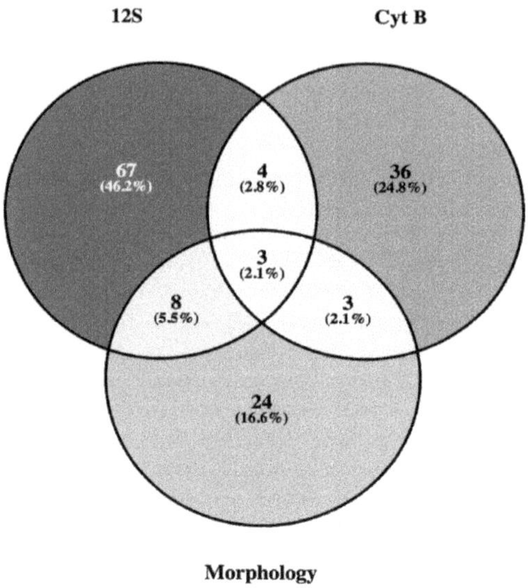

Figure 5. Venn diagram showing the overlap numbers of species identified by morphology and molecular tools in Oujiang River Estuary.

4. Discussion

4.1. Species Identification and Composition

Fish can be identified using distinguishable morphometric and meristic characteristics; however, the latter are typically used for quick identification [27]. Morphological features commonly used to identify adult fish species are absent at the early development stages, making ichthyoplankton identification more tedious and difficult [28,29]. The species spawn in the area were freshwater, coastal and estuarine related, including Mugilidae, Cyprinidae, Sciaenidae, Lateolabracidae, Gobiidae and pelagic species such as Engraulidae indicated by their high presence.

DNA metabarcoding analysis of ichthyoplankton from Oujiang River Estuary was successful in discerning several fish species and provided biodiversity and abundance data within and between communities. The study identified fish species that commonly and rarely inhabit Oujiang River Estuary. The detected fish species matched previous observations [30–34]. The highest abundance and composition of dominant and common egg and larva species reflected the spawning localities and seasons of adult fish stocks in Oujiang River Estuary. The observed number of unidentified OTUs due to lack of BLAST hit can be explained by sequence data gaps in GenBank database [35]. These findings demonstrated that DNA metabarcoding is a suitable tool for analyzing and monitoring a large scale of pooled samples, due to its ability to produce and detect millions of DNA reads that allow concurrent species identification and analysis [9,36].

4.2. Ichthyoplankton Community Structure and Diversity Patterns

The results that most of fish eggs and larvae in Oujiang River Estuary were caught consistently during every sampling month, with the highest catch of eggs and larvae in June and July, indicating that fish species reproduce throughout all five of the months. Fish egg and larva catch increased from April to June and decreased from July, indicating that June is the spawning peak in the river. The findings are consistent with the reports that many fishes in Oujiang River Estuary spawn in June, July and August [31,32]. The community composition, diversity, spatial and temporal distribution of ichthyoplankton varied among sampling sites and months, as revealed by 12S, Cyt b, and morphology. The variation could be the result of changes in oceanographic conditions, specifically a rise in water temperature that favored spawning activities for many fish species [37].

4.3. Comparison of Assessment Tools

There was a difference in the PCR amplification success between 12S (<100 bp) and Cyt b (<460 bp) markers. The 12S was more successfully amplified than the Cyt b gene. This could be an effect of the size for the targeted barcode markers [18]. Rees et al. [19] urged that DNA degradation and mismatch of PCR primers in the DNA binding sites affected DNA amplification process that subsequently affected DNA sequencing success. Our results that suggest the 12S marker is more efficient in detecting fish species than Cyt b are in consistence with Hänfling et al. [18]. In this study, 61% of Cyt b and 0.06% of 12S OTUs were unidentified. The variability in species detection could be due to the difference in reference sequences available on GenBank database, and fragment size and persistence. The complete set of fish references obtained for this study included a total of 30,719 (12S) and 4211 (Cyt b) sequences available in GenBank database, thus Cyt b references lacked for many species. Cyt b DNA metabarcoding was therefore unable to detect some common and dominant fish species detected by 12S and morphology in common. A lack of suitable GenBank databases and barcode misidentification accounted for a large proportion of unidentified OTUs. A small proportion of disparity were probably derived from PCR and sequencing errors that could be avoided by improving read preprocessing and quality filtering [12,38,39].

The findings demonstrated that NGS-based metabarcoding is a suitable approach for assessing and analyzing a pooled sample of ichthyoplankton communities. Despite that morphological and molecular species identification often disagree each other but display a certain common degree of taxonomic overlapping [7], each approach can miss the taxa identified by the other [40]. The correct ichthyoplankton identification to species level is possible under molecular identification tools [28]. With suitable primer selection [12], the power of DNA metabarcoding to detect fish species is superior to all conventional fish assessment and monitoring methods. The results pointed out the potentials and bottlenecks of DNA metabarcoding in identifying fish eggs and larvae and emphasized on the importance of combining molecular and morphological tools in assessing ichthyoplankton community structures.

5. Conclusions

The study addressed key issues associated with fisheries management and conservation by providing data regarding fish spawning localities and seasons. Despite a relatively small-scale assessment survey, confidence can be gained in generalizability of spawning seasons is between May and August based on the spatial and temporal analysis. Apparently, DNA metabarcoding is a promising approach for ichthyoplankton ecological and biological survey that expands our current knowledge of fisheries resources. As the 12S genetic marker presented higher species coverage and detection than the Cyt b, the study highlighted the importance of having a complete and accurate reference database for better and more accurate species detection. Generally, the findings reported here provide another key attempt towards the development of powerful and cost-effective ichthyoplankton identification tool, as well as opportunities to overcome the high cost and time consumption in morphological identification.

Supplementary Materials: The following supporting information can be downloaded at: https://www.mdpi.com/article/10.3390/d14121111/s1, Table S1. List of identified fish species in Oujiang River Estuary; taxonomy description; identification method; sampling month.

Author Contributions: All authors contributed to the study conception and design. Material preparation, data collection and morphological analysis were performed by R.J. DNA extraction, PCR and molecular analysis were performed by J.L.L., Y.C. and R.J. supervised and acquired fundings for the project. The first draft of the manuscript was written by J.L.L. and all the authors commented on previous versions of the manuscript. All authors have read and agreed to the published version of the manuscript.

Funding: This study was supported by the National Key R&D Program of China (Grant/Award Numbers: 2018YFD0900903 and 2018YFD0900904) and Zhejiang Ocean University Independent Voyage for Sophisticated Ocean Front and Fisheries Investigation (SOPHI).

Institutional Review Board Statement: Not applicable.

Data Availability Statement: Cyt b gene: https://www.ncbi.nlm.nih.gov/bioproject/PRJNA911478/; 12S gene: https://www.ncbi.nlm.nih.gov/bioproject/PRJNA911479/.

Acknowledgments: We are grateful to Rui Yin and Jing-Xiang Liang of Zhejiang Ocean University for their assistance in sampling and data collection. We thank the anonymous reviewers for their critical and insightful comments on the manuscript.

Conflicts of Interest: The authors declare that they have no conflict of interest to declare.

References

1. Lewis, L.A.; Richardson, D.E.; Zakharov, E.V.; Hanner, R. Integrating DNA barcoding of fish eggs into ichthyoplankton monitoring programs. *Fish. Bull.* **2016**, *114*, 153–165. [CrossRef]
2. Rodríguez, J.M.; Alemany, F.; García, A. *A Guide to the Eggs and Larvae of 100 Common Western Mediterranean Sea Bony Fish Species*; FAO: Rome, Italy, 2017; Available online: http://www.fao.org/3/a-i7708e.pdf (accessed on 15 April 2016).
3. Ellis, J.R.; Milligan, S.P.; Readdy, L.; Taylor, N.; Brown, M.J. Spawning and nursery grounds of selected fish species in UK waters. *Sci. Ser. Tech. Rep.* **2012**, *147*, 56.
4. Gleason, L.U.; Burton, R. High-throughput molecular identification of fish eggs using multiplex suspension bead arrays. *Mol. Ecol. Resour.* **2012**, *12*, 57–66. [CrossRef] [PubMed]
5. Armstrong, M.J.; Connolly, P.; Nash, R.D.M.; Pawson, M.G.; Alesworth, E.; Coulahan, P.J.; Dickey-Collas, M.; Milligan, S.P.; O'Neill, M.F.; Witthames, P.R.; et al. An application of the annual egg production method to estimate the spawning biomass of cod (*Gadus morhua* L.), plaice (*Pleuronectes platessa* L.) and sole (*Solea solea* L.) in the Irish Sea. *ICES J. Mar. Sci.* **2001**, *58*, 183–203. [CrossRef]

6. Fox, C.J.; Taylor, M.I.; Pereyra, R.; Villasana, M.I.; Rico, C. TaqMan DNA technology confirms likely overestimation of cod (*Gadus morhua* L.) egg abundance in the Irish Sea: Implications for the assessment of the cod stock and mapping of spawning areas using egg-based methods. *Mol. Ecol.* **2005**, *14*, 879–884. [CrossRef] [PubMed]
7. Harvey, J.B.; Johnson, S.B.; Fisher, J.L.; Peterson, W.T.; Vrijenhoek, R.C. Comparison of morphological and next generation DNA sequencing methods for assessing zooplankton assemblages. *J. Exp. Mar. Biol. Ecol.* **2017**, *487*, 113–126. [CrossRef]
8. Kumar, G.; Kocour, M. Applications of next-generation sequencing in fisheries research: A review. *Fish. Res.* **2017**, *186*, 11–22. [CrossRef]
9. Bucklin, A.; Lindeque, P.K.; Rodriguez-Ezpeleta, N.; Albaina, A.; Lehtiniemi, M. Metabarcoding of marine zooplankton: Prospects, progress and pitfalls. *J. Plankton Res.* **2016**, *38*, 393–400. [CrossRef]
10. Thomsen, P.F.; Willerslev, E. Environmental DNA—An emerging tool in conservation for monitoring past and present biodiversity. *Biol. Conserv.* **2015**, *183*, 4–18. [CrossRef]
11. Valdez-Moreno, M.; Vásquez-Yeomans, L.; Elias-Gutierrez, M.; Ivanova, N.V.; Hebert, P. Using DNA barcodes to connect adults and early life stages of marine fishes from the Yucatan Peninsula, Mexico: Potential in fisheries management. *Mar. Freshw. Res.* **2010**, *61*, 655–671. [CrossRef]
12. Valentini, A.; Taberlet, P.; Miaud, C.; Civade, R.; Herder, J.; Thomsen, P.F.; Bellemain, E.; Besnard, A.; Coissac, E.; Boyer, F.; et al. Next-generation monitoring of aquatic biodiversity using environmental DNA metabarcoding. *Mol. Ecol.* **2016**, *25*, 929–942. [CrossRef] [PubMed]
13. Ashburner, M.; Ball, C.A.; Blake, J.A.; Botstein, D.; Butler, H.; Cherry, J.M.; Davis, A.P.; Dolinski, K.; Dwight, S.S.; Eppig, J.T.; et al. Gene ontology: Tool for the unification of biology. *Nat. Genet.* **2014**, *25*, 25–29. [CrossRef]
14. Beng, K.C.; Tomlinson, K.W.; Shen, X.H.; Surget-Groba, Y.; Hughes, A.C.; Corlett, R.; Slik, J.W.F. The utility of DNA metabarcoding for studying the response of arthropod diversity and composition to land-use change in the tropics. *Sci. Rep.* **2016**, *6*, 24965. [CrossRef]
15. Kelly, R.P.; Port, J.A.; Yamahara, K.M.; Crowder, L.B. Using Environmental DNA to Census Marine Fishes in a Large Mesocosm. *PLoS ONE* **2014**, *9*, e86175. [CrossRef]
16. Miya, M.; Sato, Y.; Fukunaga, T.; Sado, T.; Poulsen, J.Y.; Sato, K.; Minamoto, T.; Yamamoto, S.; Yamanaka, H.; Araki, H.; et al. MiFish, a set of universal PCR primers for metabarcoding environmental DNA from fishes: Detection of more than 230 subtropical marine species. *R. Soc. Open Sci.* **2015**, *2*, 150088. [CrossRef]
17. Thomsen, P.F.; Kielgast, J.; Iversen, L.L.; Møller, P.R.; Rasmussen, M.; Willerslev, E. Detection of a Diverse Marine Fish Fauna Using Environmental DNA from Seawater Samples. *PLoS ONE* **2012**, *7*, e41732. [CrossRef] [PubMed]
18. Hänfling, B.; Handley, L.L.; Read, D.S.; Hahn, C.; Li, J.; Nichols, P.; Blackman, R.C.; Oliver, A.; Winfield, I.J. Environmental DNA metabarcoding of lake fish communities reflects long-term data from established survey methods. *Mol. Ecol.* **2016**, *25*, 3101–3119. [CrossRef] [PubMed]
19. Rees, H.C.; Maddison, B.C.; Middleditch, D.J.; Patmore, J.R.; Gough, K. REVIEW: The detection of aquatic animal species using environmental DNA—A review of eDNA as a survey tool in ecology. *J. Appl. Ecol.* **2014**, *51*, 1450–1459. [CrossRef]
20. Deagle, B.E.; Jarman, S.N.; Coissac, E.; Pompanon, F.; Taberlet, P. DNA metabarcoding and the cytochrome c oxidase subunit I marker: Not a perfect match. *Biol. Lett.* **2014**, *10*, 20140562. [CrossRef]
21. Zukan, N. *An Atlas of the Early-Stage Fishes in Japan*; Okiyama, M., Ed.; University of Tokyo Press: Tokyo, Japan, 1989.
22. Magoč, T.; Salzberg, S.L. FLASH: Fast length adjustment of short reads to improve genome assemblies. *Bioinformatics* **2011**, *27*, 2957–2963. [CrossRef]
23. Bolger, A.M.; Lohse, M.; Usadel, B. Trimmomatic: A flexible trimmer for Illumina sequence data. *Bioinformatics* **2014**, *30*, 2114–2120. [CrossRef] [PubMed]
24. Andrews, S. FastQC: A Quality Control Tool for High throughput Sequence Data *Babraham Bioinformatics*. 2010. Available online: https://doi.org/citeulike-article-id:11583827 (accessed on 20 October 2017).
25. Caporaso, J.G.; Kuczynski, J.; Stombaugh, J.; Bittinger, K.; Bushman, F.D.; Costello, E.K.; Fierer, N.; Peña, A.G.; Goodrich, J.K.; Gordon, J.I.; et al. QIIME allows analysis of high- throughput community sequencing data Intensity normal-ization improves color calling in SOLiD sequencing. *Nat. Methods* **2010**, *7*, 335–336. [CrossRef] [PubMed]
26. Baker, C.C. Entrez Qiime: A Utility for Generating QIIME Input Files from the NCBI Databases. 2017. Available online: https://zenodo.org/record/159607#.Y3mh5HZBw2w (accessed on 26 October 2017).
27. Froese, R.; Papasissi, C. The use of modern relational databases for identification of fish larvae. *J. Appl. Ichthyol.* **1990**, *6*, 37–45. [CrossRef]
28. Ko, H.-L.; Wang, Y.-T.; Chiu, T.-S.; Lee, M.-A.; Leu, M.-Y.; Chang, K.-Z.; Chen, W.-Y.; Shao, K.-T. Evaluating the Accuracy of Morphological Identification of Larval Fishes by Applying DNA Barcoding. *PLoS ONE* **2013**, *8*, e53451. [CrossRef] [PubMed]
29. Strauss, R.E.; Bond, C.E. *Methods for Fish Biology, Taxonomic Methods: Morphology*; Moyle, P., Schreck, C., Eds.; American Fisheries Society: Bethesda, MD, USA, 1971; pp. 109–140.
30. Lin, H.-Y.; Chiu, M.-Y.; Shih, Y.-M.; Chen, I.-S.; Lee, M.-A.; Shao, K.-T. Species composition and assemblages of ichthyoplankton during summer in the East China Sea. *Cont. Shelf Res.* **2016**, *126*, 64–78. [CrossRef]
31. Zhou, M.; Lin, Y.; Yang, S.; Cao, W.; Zheng, L. Ecological distribution of ichthyoplankton around Oujiang estuary in spring and autumn in 2015. *J. Mar. Sci.* **2015**, *30*, 94–105.

32. Zhou, Y.D.; Jin, H.W.; Zhang, H.L.; Jiang, R.J.; Pan, G.L. The category composition and abundance distributions of ichthyoplankton along the north central coast of Zhejiang province in spring and summer. *J. Fish. China* **2011**, *35*, 880–889. [CrossRef]
33. Wan, R.J.; Zeng, D.Y.; Bian, X.D.; Ni, X.B. Species composition and abundance distribution pattern of ichthyoplankton and their relationship with environmental factors in the East China Sea ecosystem. *J. Fish. China* **2015**, *38*, 1375–1398. [CrossRef]
34. Zhou, M.; Lin, Y.; Yang, S.; Cao, W.; Zheng, L. Composition and ecological distribution of ichthyoplankton in eastern Beibu Gulf. *Acta Oceanol. Sin.* **2011**, *30*, 94–105. [CrossRef]
35. Andruszkiewicz, E.A.; Starks, H.A.; Chavez, F.P.; Sassoubre, L.M.; Block, B.A.; Boehm, A.B. Biomonitoring of marine vertebrates in Monterey Bay using eDNA metabarcoding. *PLoS ONE* **2017**, *12*, e0176343. [CrossRef]
36. Cowart, D.A.; Pinheiro, M.; Mouchel, O.; Maguer, M.; Grall, J.; Miné, J.; Arnaud-Haond, S. Metabarcoding Is Powerful yet Still Blind: A Comparative Analysis of Morphological and Molecular Surveys of Seagrass Communities. *PLoS ONE* **2015**, *10*, e0117562. [CrossRef] [PubMed]
37. Okazaki, Y.; Nakata, H. Effect of the mesoscale hydrographic features on larval fish distribution across the shelf break of East China Sea. *Cont. Shelf Res.* **2007**, *27*, 1616–1628. [CrossRef]
38. Leray, M.; Knowlton, N. DNA barcoding and metabarcoding of standardized samples reveal patterns of marine benthic diversity. *Proc. Natl. Acad. Sci. USA* **2015**, *112*, 2076–2081. [CrossRef]
39. Lindeque, P.; Parry, H.E.; Harmer, R.A.; Somerfield, P.; Atkinson, A. Next Generation Sequencing Reveals the Hidden Diversity of Zooplankton Assemblages. *PLoS ONE* **2013**, *8*, e81327. [CrossRef] [PubMed]
40. Shaw, J.L.; Clarke, L.J.; Wedderburn, S.D.; Barnes, T.C.; Weyrich, L.S.; Cooper, A. Comparison of environmental DNA metabarcoding and conventional fish survey methods in a river system. *Biol. Conserv.* **2016**, *197*, 131–138. [CrossRef]

diversity

MDPI

Article

DNA Barcoding of *Trichobilharzia* (Trematoda: Schistosomatidae) Species and Their Detection in eDNA Water Samples

Nikolaus Helmer [1,2], Christoph Hörweg [3], Helmut Sattmann [3], Susanne Reier [4,5], Nikolaus U. Szucsich [1], Jana Bulantová [6] and Elisabeth Haring [1,5,*]

1 Central Research Laboratories, Natural History Museum Vienna, Burgring 7, 1010 Vienna, Austria
2 Artificial Life Lab, Institute of Biology, University of Graz, Universitätsplatz 2, 8010 Graz, Austria
3 3rd Zoological Department, Natural History Museum Vienna, Burgring 7, 1010 Vienna, Austria
4 1st Zoological Department, Natural History Museum Vienna, Burgring 7, 1010 Vienna, Austria
5 Department of Evolutionary Biology, University of Vienna, Djerassiplatz 1, 1030 Vienna, Austria
6 Department of Parasitology, Charles University, Vinicna 7, 12844 Prague 2, Czech Republic
* Correspondence: elisabeth.haring@nhm-wien.ac.at

Abstract: We designed and tested species-specific PCR primers to detect *Trichobilharzia* species via environmental DNA (eDNA) barcoding in selected Austrian water bodies. Tests were performed with eDNA samples from the field as well as with artificial samples from the lab, where snails releasing cercariae were kept in aquariums. From two localities, *Trichobilharzia* was documented based on the release of cercariae from snails, enabling morphological species identification. In both cases, the corresponding species were detected via eDNA: *Trichobilharzia szidati* and *Trichobilharzia physellae*. Nonetheless, the stochasticity was high in the replicates. PCR tests with aquarium water into which the cercariae had been released allowed eDNA detection even after 44 days. As in the PCRs with eDNA samples from the field, positive results of these experiments were not obtained for all samples and replicates. PCR sensitivity tests with dilution series of *T. szidati* genomic DNA as well as of PCR amplification products yielded successful amplification down to concentrations of 0.83 pg/µL and 0.008 pg/µL, respectively. Our results indicate that the presumed species specificity of PCR primers may not be guaranteed, even if primers were designed for specific species. This entails misidentification risks, particularly in areas with incomplete species inventories.

Keywords: trematodes; Europe; environmental DNA barcoding; cercariae; swimmer's itch; *cytochrome c oxidase subunit 1*; *CO1*

Citation: Helmer, N.; Hörweg, C.; Sattmann, H.; Reier, S.; Szucsich, N.U.; Bulantová, J.; Haring, E. DNA Barcoding of *Trichobilharzia* (Trematoda: Schistosomatidae) Species and Their Detection in eDNA Water Samples. *Diversity* **2023**, *15*, 104. https://doi.org/10.3390/d15010104

Academic Editors: Luc Legal and Manuel Elias-Gutierrez

Received: 24 November 2022
Revised: 1 January 2023
Accepted: 7 January 2023
Published: 12 January 2023

1. Introduction

Living organisms naturally release DNA into their environment. This may occur via cells separated from the body surface or by organs communicating with the external environment (e.g., mucus cells and intestinal cells). DNA release to the environment may take place on a large scale even after the death of the organism. Such DNA is termed environmental DNA or "eDNA" and can be detected and analysed using molecular genetic methods such as DNA barcoding [1]. In addition to this intracellular eDNA, free (extracellular) DNA is present in the environment, in the substrate and in water bodies, mostly bound to particles. Extracellular eDNA is generally more short-lived than intracellular eDNA and the speed of degradation differs between free-flowing vs. particle-bound eDNA [2].

Analyses of eDNA using environmental samples as the basis for species detection have gained increasing importance and have been discussed as alternatives or complementary approaches to classical species monitoring, which requires direct observation and/or collection of organisms or their traces. Samples of eDNA from water bodies offer methodological benefits compared to other environmental samples but are also associated with certain difficulties. Unlike in terrestrial systems, where DNA remains predominantly detectable locally, DNA or cells released into water may spread and thus are potentially detectable in wider parts of the water body at places where the organisms themselves do not occur. While in flowing waters the transport of eDNA plays a major role, this is of lesser importance in stagnant waters. Even in the latter, however, factors such as wind direction, water currents, and stratification may influence DNA distribution [3]. A second difference of eDNA in aquatic compared to terrestrial systems is its faster degradation in the former. The degradation speed can vary greatly and depends on factors including water temperature, pH, UV exposure, dissolved ions, as well as the organic activity of macro- and microorganisms [4–7].

The present study focused on eDNA detection by means of DNA barcode sequencing of avian schistosomes (Trematoda, Digenea, Schistosomatidae), obligate parasites of animals that use birds as final hosts. Eggs of the avian schistosomes are released into the water with the faeces of the final host. Swimming, free-living larvae—miracidia—hatch from the eggs and swim in search of the first intermediate host to infect, namely aquatic gastropods. Miracidia penetrate the gastropod host, then transform into mother sporocysts that produce several daughter sporocysts asexually, which finally release multiple infective larvae (cercariae). Once these cercariae are released from the snails, they actively penetrate the final host.

Among avian schistosomes, several species are of medical relevance because humans (and other mammals) may serve as accidental hosts. After penetration, the cercariae (schistosomula) show only limited development in humans and cannot mature [8–11], but they may cause dermatitis symptoms ("Swimmer's itch", "cercarial dermatitis") with medical consequences such as intensively itching maculo–papulovesicular eruptions with possible secondary infections and local oedema. In serious cases, generalized signs such as fever, cough, diarrhoea, and local lymph node swelling have been observed [11,12]. The occurrence of cercariae and cases of cercarial dermatitis concern bathers as well as certain occupation groups such as those working in fishery and aquaculture, biologists working in water habitats, or lifeguards in natural swimming areas. This makes cercariae relevant for public health and the local economy (especially when tourists are concerned) because cercarial dermatitis may lead to the closure of near-natural bathing facilities [11,13]. The causative agents of cercarial dermatitis in Europe are mostly species of the genus *Trichobilharzia* with approximately 35 species worldwide [14]. Six of these species are known in Europe [15,16], namely *Trichobilharzia szidati* Neuhaus 1952 [17], *Trichobilharzia regenti* Horák, Kolářová, & Dvořák 1998 [18], *Trichobilharzia franki* Müller & Kimmig 1994 [19], *Trichobilharzia salmanticensis* Simon-Vicente & Simon-Martin 1999 [20], *Trichobilharzia anseri* Jouet et al. 2015 [15], and *Trichobilharzia mergi* Kolářová, Skírnisson, Ferté, & Jouet 2013 [21]. Recently, an additional species, *Trichobilharzia physellae* (Talbot, 1936) [22], originating from North America, was recorded in a lake in Austria [23]. Besides *Trichobilharzia*, five other genera of avian schistosomes have been reported to infect aquatic birds in Europe: *Allobilharzia*, *Bilharziella*, *Dendritobilharzia*, *Gigantobilharzia*, and *Ornithobilharzia* [13].

Detecting cercariae by direct sampling with nets is often hampered because of their unpredictable distribution in the water body combined with their small size and fragility. The traditional approach is to collect snails of potential host species, put them into separate glass jars filled with water, and incubate them under artificial or natural light to stimulate the shedding of the cercariae; these are subsequently further investigated (see Section 2.2). Due to the low prevalence of parasites in intermediate mollusc hosts in the patent period, however, the detection by such cercarial release experiments usually requires sampling large numbers of gastropods (which might be under nature protection). Such interventions in snail populations are therefore not desirable and may pose legal hurdles as well.

Monitoring by sampling and analysing environmental water would avoid the time-consuming processes of sampling gastropods, conducting releasing experiments, and mounting cercariae for species identification. Thus, eDNA analyses have several advantages for the detection of dermatitis-causing cercariae: they are less labour intensive, have only a minor impact on the environment (compared with collecting hundreds of snails), and are highly informative considering that cercariae of many species are morphologically not distinguishable [24]. Several studies based on eDNA combined with a qPCR approach successfully detected cercariae down to the species level [25–31]. Compared to the qPCR approach, where the presence of a species is indicated by PCR success, the barcoding approach includes sequencing of the obtained PCR product and thus has the potential to reveal false positives.

Several factors must be considered in evaluating the potential of detecting cercarial DNA in freshwater samples. The quantity and duration of cercarial shedding as well as the length of cercarial survival under natural conditions may influence the sensitivity of the analyses as well as the process of DNA degradation after cercarial death. Concerning the quantity and circadian rhythm of shedding, studies on *Trichobilharzia szidati* yielded high numbers (many thousands per day) and a long duration of shedding (several weeks), with maximum shedding depending of the time of day, temperature, and light [32]. Field studies with *T. szidati* revealed maximum shedding in late summer [33]. In laboratory experiments, those authors determined that cercarial life spans are inversely correlated to temperature and that cercaria remained active for up to 60 h at 20 °C. Accordingly, the success of DNA water analyses may depend on different parameters such as cercarial production, temperature, and light conditions, along with other external influences such as water current conditions, water physics/chemistry, and ecological interferences, which are poorly studied [33].

The initial aim of the present pilot study was (1) to design and test species-specific PCR primers for eDNA detection of *Trichobilharzia* species occurring in Austria. (2) We then compared the eDNA results with the results from cercarial release experiments (as described above and in Section 2.2). (3) We also report observations on the release of cercariae in the laboratory.

2. Materials and Methods

2.1. Field Work and eDNA Sampling

During field excursions in the framework of an overarching project in 2019 and 2020, freshwater snails as potential hosts were collected in six water bodies in Upper and Lower Austria, at altogether seven sampling localities (Table 1). The present study focused on species of the family Lymnaeidae because the *Trichobilharzia* species in Europe are almost exclusively found in lymnaeid snail hosts [34]. The water bodies in Table 1 include two tributaries of river systems (locality 1, 2), three lakes (locality 3, 5, 9) and three ponds (locality 4, 7, 8). They were selected based on the occurrence of Lymnaeidae as well as waterfowl. Moreover, swimmer's itch had been recorded in some of them in the past (locality 4, 5, 8, 9).

Table 1. List of locations where host snails and eDNA samples were collected for the present study.

Locality Number	Locality	Coordinates N/E
1	AT, OÖ, Unterer Inn, Reichersberger Au	48°20′20.71″/13°21′35.47″
2	AT, OÖ, Unterer Inn, Hagenauer Bucht	48°16′31.96″/13°05′33.58″
3	AT, OÖ, Eggelsberg, Seeleithensee	48°03′33.19″/12°58′01.56″
4	AT, NÖ, Rekawinkel, pond	48°11′02.90″/16°01′56.50″
5	AT, OÖ, Linz, Lake Pichlinger-See	48°14′22.62″/14°23′00.05″
6	CZ, Prague, Charles University	lab breeding
7	CZ, Jetřichovice, natural swimming pond	50°50′38.50″/14°23′43.00″
8	AT, NÖ, Niederfellabrunn, Löschteich	48°27′40.70″/16°18′42.50″
9	AT, OÖ, Linz, Lake Pleschinger-See	48°19′10.00″/14°19′56.80″

Snails were examined for cercariae using conventional illumination methods (cercarial release experiments; see Section 2.2). In addition, eDNA water samples were taken at these water bodies (1, 2, 3, 4, 5, 8, 9). The aim was to compare the results of the release experiments (Section 2.3) with those from the eDNA analyses. The eDNA samples were taken near aquatic vegetation in shallow water (down to 0.5 m depth, not stirring up the sediment) as close as possible to where the host snails were collected. Sampling was conducted, with one exception (eDNA-9, locality 4), close to the shoreline. At every location visited in 2020, at least two samples were taken. For eDNA sampling, water was squeezed by hand with sterile, DNA-free plastic syringes through Sterivex™ filters with a pore size of 0.45 µm (Merck Millipore, Darmstadt, Germany). Sampling was conducted with fresh sterile gloves for each sampling, taking care that no sampling equipment came into contact with collected snails or transfer bottles. Filters were immediately stored in cool containers and subsequently transported to the laboratory and stored at −80 °C. The initial aim to filter a standardized volume (2 L) of water could not be accomplished. The maximum water volume that could be squeezed through the filters varied considerably among water bodies depending on the number of suspended solids: in rather turbid water, e.g., at locality 4, only 100 mL could be used compared to a maximum of 1200 mL at locality 8. Overall, 18 eDNA samples of water taken from seven localities were analysed: 13 of these were collected directly in the field, four were obtained in the laboratory from water samples (without snails) brought from the field in bottles (eDNA samples taken in the lab). One eDNA sample was taken from transport water containing a living *Lymnaea stagnalis* (which later released *T. szidati* cercariae). In addition, 15 eDNA samples were taken from aquariums in which snails were kept for some time (up to 24 h), three of them when the snails were still present and 12 of them after having removed the snails.

2.2. Release of Cercariae in the Laboratory

Collected snails were placed individually in jars and positioned on the windowsill (exposed to indirect sunlight during day) for up to three days, as light stimulates the release of cercariae. Released cercariae were fixed in 80–96% ethanol. For morphological identification, the cercariae isolated during the hatching experiments were transferred to a glycerol-ethanol (80%) mixture (50:50) including 1 mL borax-carmine solution. After evaporation of the ethanol in a thermo-incubator (approx. after 48 h at 40 °C), cercariae were placed in a drop of glycerol on a micro slide, covered with a cover glass, and sealed. The glycerol increases the lucency, whereas the slight staining with borax-carmine renders the anatomy more visible. Mounted cercariae were morphologically examined under a microscope (Nikon Eclipse Ni-U, Nikon Instruments Inc., New York, NY, USA) and microphotographs were made using the mounted Nikon DSRi2 microscope camera unit and the corresponding application NIS-Elements BR v.5.02.00. Subsequently, the microphotographs were processed with Gimp 2.10.24 (https://www.gimp.org, accessed on 28 May 2021). The specimens were stored in the NHMW collection Evertebrata Varia (inventory numbers for the permanent mounts NHMW-ZOO-EV-Micro 5860-5862 and for the fixed cercariae NHMW-ZOO-EV 21447-21449).

An additional experiment was performed to investigate how long cercariae are shed by an adult *Lymnaea stagnalis* (from Rekawinkel, Lower Austria) infected with *T. szidati*. This specimen had been collected and brought to the laboratory on 20 August 2020 and was kept in a glass jar with water from the pond. On 23 August 2020, the snail was placed into an aquarium for the experiments. On 24 August 2020, the snail was placed into another aquarium and eDNA longevity tests were conducted with the water from the first aquarium (see Section 2.9). All aquariums (25 L) were filled before the experiments with tap water and left to stand for two days before the snail was introduced in order not to stress the snail further. On three days, 2 and 16 September as well as 6 October 2020, the animal was incubated in a jar in 250 mL of tap water at 22 °C for 12 h in daylight to check for shedding of cercariae. The snail from the aquarium was transferred into the jar to observe and record cercarial release and to count numbers of cercariae, which was not possible in the aquarium.

2.3. Cercariae Analysed Genetically

Cercariae of *Trichobilharzia* species released from snails collected in the field were fixated. Some were mounted for morphological examination, and some were genetically analysed (see Section 2.4 following pages.) to obtain DNA barcode sequences of the mitochondrial *cytochrome c oxidase subunit 1* gene (*CO1*). Their *CO1* sequences were used for the procedures of primer design, which also included sequences of *T. franki* and *T. physellae*, which we published earlier [23,35], and of other species (from GenBank). Moreover, for primer design, the Department of Parasitology of Charles University (Prague, Czech Republic) provided further cercariae (*T. franki, T. szidati, T. regenti*, Table 1). This addition was important to include another species probably also present in Austria (*T. regenti*) and to cover a wider range of genetic variability. The five specimens analysed for the primer design in the present study were: *T. franki* (2 specimens; Ra2-4-014, Ra13-16-002), *T. szidati* (2 specimens; Ls17-15-003, Ls36-13-002), and *T. regenti* (1 specimen; Rlag1-15-002) (Table 2). All DNA barcodes of cercariae generated in the present study complemented the collection of reference sequences at the NHMW database (ABOL, Austrian Barcode of Life Initiative) and likewise were uploaded to the NCBI (National Center for Biotechnology Information) gene database GenBank, and BOLD (Barcode of Life Database; see Table 2).

Table 2. List of cercariae analysed genetically and their snail host species. The locality numbers correspond to those in Table 1.

Species	Accession Number/ BOLD-ID	Ind.ID	Host	Life Stage	Locality Code
Trichobilharzia franki	OP347092/NHBP008-21	Ra2-4-014	*Radix auricularia*	cercaria	1
Trichobilharzia franki	OP347091/NHBP016-21	Ra13-16-002	*Radix auricularia*	cercaria	7
Trichobilharzia regenti	OP347089/NHBP010-21	Rlag1-15-002	*Radix lagotis*	cercaria	6
Trichobilharzia szidati	OP347090/NHBP014-21	Ls17-15-003	*Lymnaea stagnalis*	cercaria	6
Trichobilharzia szidati	OP347093/NHBP018-21	Ls36-13-002	*Lymnaea stagnalis*	cercaria	4

2.4. DNA Extraction

DNA was extracted in the clean room of the NHM DNA laboratory, following standard routines against contaminations, e.g., regular overnight UV irradiation of the clean room, treatment of work surfaces with DNA Exitus (AppliChem, Darmstadt, Germany) and/or 10% sodium hypochlorite, UV irradiation of reaction tubes. All post-PCR work was performed in a separate laboratory. All DNA extractions included control extractions without samples to screen for contaminated reagents. Likewise, all PCRs included negative control reactions without template DNA. PCRs with good-quality DNA (e.g., from cercariae) were made independently from those with eDNA. For the DNA analysis of cercariae, we used the QIAamp DNA Micro Kit (Qiagen, Hilden, Germany), which is optimised for small tissue samples. As described in Helmer et al. [23], the cercariae were isolated individually with stainless insect needles under a stereomicroscope (Wild-Leica Heerbrugg M420 macroscope, Leica, Wetzlar, Germany), dried on the needle for about 10 s to remove the ethanol, and then transferred to the lysis buffer. The extraction was performed according to the manufacturer's protocol. In the last step, the DNA was eluted with 25 µL AE buffer.

For eDNA analysis, the collected filter samples were extracted using the dNeasy PowerWater Sterivex Kit (Qiagen, Hilden, Germany). The extraction was performed according to the manufacturer's protocol except omitting the incubation of the Sterivex filter at 90 °C (step 7 of the protocol) because less DNA shearing was desired. In the last step, the DNA was eluted with 50 µL of EB solution.

2.5. PCR Amplification of Marker Sequences

Partial sequences of the following genes were studied: (1) the mitochondrial *cytochrome c oxidase subunit 1* gene (*CO1*), (2) the nuclear-encoded *18S* rRNA gene (*18S*). The latter were used to test eDNA samples for any potential trematode DNA (the *18S* primers are universal primers in contrast to the *CO1* primers, which were aimed to bind *Trichobilharzia* spp.). *CO1* sequences were analysed in single cercariae as well as in eDNA samples (see Section 2.6). First, a partial sequence of the *CO1* gene was amplified by PCR from five cercariae (Table 2) with the primer pair Cox1_schist_5_trich/Tricho_rev_20, which yields a fragment of 1216 bp in length. This section of the *CO1* gene is considerably longer than the usually amplified DNA barcoding region (the 634 bp long, so-called "Folmer region"; [36]). Since sequencing of some sections of this fragment was difficult due to DNA homopolymers as well as due to limitations of Sanger sequencing, additional overlapping shorter fragments were amplified to achieve overall high-quality sequences and to achieve barcode quality according to the requirements of the BOLD database. For this task, the following primer pairs were used for PCR as well as sequencing: Tricho_tRNA_fw/Tricho_tRNA_rv_2 (amplicon length of 396 bp), Cox1_schist_5_trich/CO1560R_modif (amplicon length of 612 bp), Tricho_Fw2/Tricho_Rv2_2 (amplicon length of 491 bp), and ZDOE-COI-fw/Tricho_rev_20 (amplicon length of 486 bp). All primers are listed in Table 3. For an overview of the primer positions see Figure 1.

Table 3. List of *CO1* and *18S* PCR primers used. Primer Tricho_tRNA_fw binds in the adjacent tRNA *Ser* gene. Fwd = forward primer, Rev = reverse primer. Asterisks: primers designed for particular species.

Gene	Primer Name	Orientation	Sequence 5′-3′	Source
CO1	Tricho_tRNA_fw	Fwd	GGTTGTCGCTGCTAACGA	[23]
CO1	Cox1_schist_5_trich	Fwd	GTTRGTTTCTTTGGATCATAAGCG	[23]
CO1	Tricho_tRNA_rv_2	Rev	CCATATAAAACATTGAAGGAACC	[23]
CO1	Tricho_Fw2	Fwd	GGTTCTGTAAAATTTATAACTAC	[23]
CO1	CO1560R_modif	Rev	GCAGTACCAAATTTTCGATC	[23]
CO1	Tricho_Rv2_2	Rev	CCTAACATATACAACCAAG	[23]
CO1	Tricho_rev_20	Rev	GCATTCCTAAATAATGCATAGG	[23]
CO1	ZDOE-CO1-fw	Fwd	TAGTTTGTGCTATGGGTTCTATAGT	[23]
CO1	ZDOE-COI-fw_szidati *	Fwd	TAGTTTGTGCTATGGGTTCTATAGTG	present study
CO1	ZDOE-COI-fw_physellae *	Fwd	TGGTTTGTGCTATGGGTTCTATAGTT	present study
CO1	eDNA-franki-rv *	Rev	CCCCACGCAAATACCTTGTG	present study
CO1	eDNA-regenti-rv *	Rev	CTCCACGTAAATAACTAGTA	present study
CO1	eDNA-szidati-rv *	Rev	CACCCCGCAAGTAGCTAGTC	present study
CO1	eDNA-physellae-rv *	Rev	CTCCTCGCAAATAACTAGTT	present study
18S	Trem-18S-f	Fwd	GGTTCCTTAGATCGTACATGC	[37]
18S	Trem-18S-r	Rev	GTACTCATTCGAATTACGGAGC	[37]

Polymerase chain reaction (PCR) was carried out in a 25 µL reaction volume with TopTaq or (later, when TopTaq was no longer available) Taq DNA polymerase (both from Qiagen, Hilden, Germany) as described in Helmer et al. [23]. For some experiments, i.e., for the tests of primer specificity (see Sections 3.1 and 3.2), another polymerase was used (Qiagen Multiplex PCR Kit). PCR ingredients were, besides reaction buffer provided by the manufacturer, 0.25 units DNA polymerase, 0.5 µM of each primer, and 0.2 mM of each dNTP (Invitrogen, Carlsbad, CA, USA). For the amplifications with the Multiplex PCR Kit, 12.5 µL Multiplex PCR Master Mix was used. For amplification of the *CO1* fragments from cercariae, 1.5 µL of DNA template was used, for eDNA (*CO1* and *18S*) 2 µL of DNA template was used. Thermal cycling conditions: 94 °C for 3 min; 45 cycles of 94 °C for 30 s, annealing temperature for 30 s, and 72 °C for 30 s; final extension at 72 °C for 10 min. Annealing temperatures of primers are listed in Table 4.

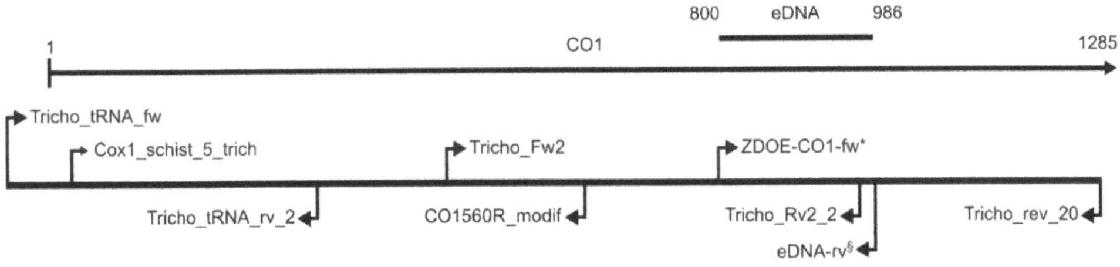

Figure 1. Positions of primers used relative to *CO1* of *Trichobilharzia szidati* (mt genome, GenBank accession number MG570047). The orientations of primers are marked by arrows. The long line on top shows part of the *CO1* gene from the gene start (position 1) until the position of the 3′-end of primer Tricho_rev_20 (position 1285). The short line on top illustrates the position of the 186 bp-amplicon used in the eDNA analyses. The asterisk (*) indicates that this is the position of ZDOE-COI-fw_szidati as well as ZDOE-COI-fw_physellae; eDNA-rv§ indicates the position of the four primers eDNA-franki-rv, eDNA-regenti-rv, eDNA-szidati-rv, and eDNA-physellae-rv.

Table 4. Conditions of PCR reactions as well as amplicon lengths for the different primer pairs. T_{ann} = annealing temperature.

Primer Combination	Amplicon Length	T_{ann}/Elongation Time
Cox1_schist_5_trich/Tricho_rev_20	1216 bp	53 °C/90 s
Tricho_tRNA_fw/Tricho_tRNA_rv_2	396 bp	54 °C/60 s
Cox1_schist_5_trich/CO1560R_modif	612 bp	52 °C/6 0 s
Tricho_Fw2/Tricho_Rv2_2	491 bp	49 °C/60 s
ZDOE-COI-fw/Tricho_rev_20	486 bp	53 °C/60 s
ZDOE-CO1-fw/eDNA-franki-rv	186 bp	54 °C/30 s
ZDOE-CO1-fw/eDNA-regenti-rv	186 bp	50 °C/30 s
ZDOE-CO1-fw/eDNA-szidati-rv	186 bp	57 °C/30 s
ZDOE-CO1-fw/eDNA-physellae-rv	186 bp	51 °C/30 s
ZDOE-COI-fw_szidati/eDNA-szidati-rv	186 bp	57 °C/30 s
ZDOE-COI-fw_physellae/eDNA-physellae-rv	186 bp	51 °C/30 s
Trem-18S-f/Trem-18S-r	413–428 bp	57 °C/60 s

PCR products were extracted from agarose gels with the QIAquick Gel Extraction Kit (Qiagen, Inc.) and sequenced (both directions) at Microsynth Austria (Vienna, Austria) using the PCR primers.

2.6. Establishment of the eDNA PCR Primers

For eDNA analyses, primers for short fragments were designed. A partial sequence of the *18S* gene was amplified with the primer pair Trem-18S-f/Trem-18S-r (Table 3; amplicon length 413–428 bp). Based on available mitochondrial genomes of *Trichobilharzia* species downloaded from GenBank and *CO1* sequences generated in our own lab, specific PCR primers were developed for the species that could be plausibly expected for Upper Austria—*T. franki*, *T. szidati*, and *T. regenti*—allowing to amplify short fragments from these species to detect even highly degraded DNA in eDNA water samples. For each species, the primer pair consisted of ZDOE-COI-fw as a forward primer and a species-specific reverse-oriented eDNA primer (eDNA-franki-rv, eDNA-szidati-rv, eDNA-regenti-rv) (Table 3). These primer pairs yield a PCR product of 141 bp in length that lies outside the established *CO1* barcoding region. This region of the *CO1* gene was selected after preliminary tests of other primers (data not shown) because its high variation enabled differentiation and construction of species-specific primers, each amplifying the section from only one of the *Trichobilharzia* species. The primers were each tested with the extracted cercariae specimens of *T. franki* (Ra2-4-014), *T. regenti* (Rlag1-15-002), and *T. szidati* (Ls36-13-002) to ensure that they yield a PCR product only for the target species and not for the two remaining species. Later during the present study, after the finding of *Trichobilharzia physellae* in Upper Austria [23], a further species-specific reverse primer was designed for *T. physellae*. Subsequently, this primer was tested with the other three *Trichobilharzia* species, and the other species-specific primers were tested in *T. physellae*. After initial PCRs showed that the *physellae* primer also delivered results with *T. szidati*, we designed additional forward primers specific for *T. physellae* and for *T. szidati* (see Results). Furthermore, the eDNA primers were tested not only in PCRs with each of the available *Trichobilharzia* species but also with other trematode species, yielding no PCR products (data not shown).

2.7. Sequence Analysis

PCR products were purified using the QIAquick PCR Purification Kit (Qiagen, Hilden, Germany) according to the manufacturer's protocol and sequenced in both directions at Microsynth Austria using the PCR primers plus additional sequencing primers (Table 3). The obtained electropherograms were checked with FinchTV 1.4.0 (Geospiza Inc.) and edited using GeneDoc 2.7.0 [38]. The edited sequences were subjected to a plausibility assessment by comparison with already published sequences using the BLAST search in the international database of the barcoding initiative BOLD (https://v4.boldsystems.org/ accessed on 6 February 2022) and the NCBI gene database GenBank (https://www.ncbi.nlm.nih.gov/genbank/ accessed on 6 February 2022).

Sequences generated from cercariae are registered in the BOLD database under the accession numbers NHBP008-21, NHBP010-21, NHBP014-21, NHBP016-21, and NHBP018-21, and in NCBI GenBank under the accession numbers OP347089–OP347093 (Table 2).

2.8. Detection of Trematode DNA in eDNA Samples from the Field

The DNA of each water sample was analysed three times by PCR with the species-specific primer pairs for *T. franki*, *T. regenti*, and *T. szidati* as well as once with the species-specific primers for *T. physellae*. The 3-fold repetition was conducted because false-negative results can occur in PCRs of eDNA samples if the amount of target DNA present is very small. If all three runs were negative, the eDNA extracts (from water samples taken in the field or lab) were also tested with the *18S* primers to find possible evidence of other trematode species present. For all positive eDNA samples from the field, PCR products were sequenced (Table 5).

Table 5. List of eDNA samples analysed. Ind. ID = laboratory number of the sample or individual; Locality numbers in parentheses, for eDNA samples from aquariums, indicate the water body from which the snails originated. For all aquarium samples, the column "sample date" gives the day of eDNA sampling (in the lab). "Volume" indicates the amount of water sampled. The column "*CO1* Test *T. szidati*" indicates the number of times the species-specific eDNA primer produced a positive PCR result for the species of concern. Genetic identification based on NCBI-BLAST results (all above 96% identity).

Ind. ID	Locality Number	Sample Date	Sample Type, Volume	Genbank BLAST Hit	*CO1* Test *T. szidati*
1-A-4	1	04.06.19	aquarium sample, 600 mL	*Echinostoma revolutum* (*18S*)	0/3
2-W-13	4	09.07.19	water sample from field, 360 mL		0/3
3-W-14	5	19.07.19	water sample from field, 600 mL	*Lissorchis kritskyi* or *Auridistomum chelydrae* (*18S*)	0/3

Table 5. *Cont.*

Ind. ID	Locality Number	Sample Date	Sample Type, Volume	Genbank BLAST Hit	*CO1* Test*T. szidati*
4-W-8	2	19.08.19	water sample from field, 180 mL	*Echinostoma revolutum (18S)*	0/3
eDNA-1	2	07.07.20	water sample from field, 200 mL	*Sminthurides aquaticus (18S)*	0/3
eDNA-2	2	07.07.20	water sample from field, 300 mL	*Sminthurides aquaticus (18S)*	0/3
eDNA-3	8	17.07.20	water sample from field, 1200 mL	*Opisthioglyphe ranae (18S)*	0/3
eDNA-4	8	17.07.20	water sample from field 350 mL	*Opisthioglyphe ranae (18S)*	0/3
eDNA-5	(4)	24.08.20	aquarium sample(with *L. stagnalis*), 170 mL	*Trichobilharzia szidati (CO1)*	3/3
eDNA-6	(4)	24.08.20	aquarium sample(with *L. stagnalis*), 200 mL	*Trichobilharzia szidati (CO1)*	3/3
eDNA-7	4	24.08.20	water sample from field, 120 mL	*Trichobilharzia szidati (CO1), Parastrigea robusta (18S)*	1/3
eDNA-8	4	24.08.20	water sample from field, 200 mL	*Sminthurides aquaticus (18S)*	0/6
eDNA-9	4	24.08.20	water sample from field, 170 mL	*Trichobilharzia szidati (CO1)*	2/3
eDNA-10	4	24.08.20	water sample from field, 100 mL		0/6
eDNA-11	4	24.08.20	transport water sample(with *L. stagnalis*), 130 mL	*Trichobilharzia szidati (CO1), Trichobilharzia* spp.various identical (*18S*)	3/3
eDNA-12	(4)	25.08.20	aquarium test sample 1, 450 mL	*Trichobilharzia szidati (CO1)*	3/3
eDNA-13	(4)	26.08.20	aquarium test sample 2, 450 mL	*Trichobilharzia szidati (CO1)*	3/3
eDNA-14	(4)	27.08.20	aquarium test sample 3, 350 mL	*Trichobilharzia szidati (CO1)*	3/3
eDNA-15	(4)	28.08.20	aquarium test sample 4, 350 mL	*Trichobilharzia szidati (CO1)*	3/3
eDNA-16	(4)	29.08.20	aquarium test sample 5, 300 mL	*Trichobilharzia szidati (CO1)*	3/3
eDNA-17	(4)	30.08.20	aquarium test sample 6, 260 mL	*Trichobilharzia szidati (CO1)*	3/3
eDNA-18	(4)	31.08.20	aquarium test sample 7, 250 mL	*Trichobilharzia szidati (CO1)*	3/3
eDNA-20	(4)	09.09.20	aquarium test sample 8, 150 mL		0/6
eDNA-21	(4)	10.09.20	aquarium test sample 9, 200 mL		0/6
eDNA-22	(4)	15.09.20	aquarium test sample 10, 60 mL		0/6
eDNA-23	(4)	15.09.20	aquarium test sample 11, 70 mL		0/6
eDNA-24	(4)	07.10.20	aquarium test sample 12, 140 mL	*Trichobilharzia szidati (CO1)*	1/3
eDNA-25	9	24.09.20	water sample from field, 800 mL	*Diplostomum pseudospathaceum (18S)*	0/3
eDNA-26	9	16.10.20	water sample lab, 850 mL		0/3
eDNA-27	9	16.10.20	water sample lab, 800 mL	*Trichobilharzia physellae (CO1)*	0/3
eDNA-28	9	16.10.20	water sample lab, 850 mL		0/3
eDNA-29	9	16.10.20	water sample lab, 950 mL		0/3
eDNA-31	3	01.08.19	water sample from field; 780 mL		0/3

2.9. eDNA from Aquarium Water

To answer the question how long cercariae are detectable via eDNA after their release from snails (and after the death of cercariae), water samples were taken with 0.45 µm Sterivex™ filter and syringes (as described above) from the middle of an aquarium (25 L) in which an individual of *Lymnaea stagnalis* had released *T. szidati* cercariae (not quantified) for one day (for setup of the aquariums see Section 2.2). Sampling of aquarium water (at room temperature, on average 23 °C) was repeated 14 times between 24 August 2020 (day 0) and 7 October 2020 (day 44). Two of these eDNA samples were taken before removal of the host snail from the aquarium on the 24 August (eDNA-5, eDNA-6; "aquarium sample" in Table 5) and thus served as positive controls in this experiment.

As with the eDNA samples taken from the field, several (3–6) PCR runs were conducted with the aquarium samples to detect false-negative results. PCR reactions with the primer pair specific for *T. szidati* (ZDOE-COI-fw/eDNA-szidati-rv) were performed with all the extracts, as this was the species released from the host snail. Additionally, each extract was tested once with the species-specific eDNA primers of *T. franki* and *T. regenti* to further test the specificity of the primers.

2.10. Sensitivity Tests

In another experiment performed to determine the smallest detectable amount of DNA in a water sample, dilution series of DNA were prepared. The concentration was measured by a Qubit fluorometer (Invitrogen). For this purpose, a PCR product (ZDOE-CO1-Fw/eDNA-szidati-rv and ZDOE-CO1-Fw/eDNA-franki-rv, respectively; 141 bp length) was prepared from two individuals: *T. szidati* Ls17-15-003 (concentration of PCR product: 7.98 ng/µL) and *T. franki* Ra13-16-002 (8.93 ng/µL) (Table 2). Furthermore, genomic DNA from these two individuals was used (Ls17-15-003, 0.0850 ng/µL; Ra13-16-002, 0.0830 ng/µL). For each template (both experiments: PCR products as well as genomic DNA), the dilution series consisted of six dilution steps (1/10–1/1,000,000), yielding seven DNA solutions each. For PCR, 1 µL of these DNA solutions was used as template. Species-specific primers were employed: ZDOE-CO1-Fw/eDNA-szidati-rv, ZDOE-CO1-Fw/eDNA-franki-rv). PCR with the dilution series of the PCR products was performed with 30 PCR cycles. PCR with genomic DNA (7 reactions each) was performed 1× with 30 cycles and 1× with 40 cycles.

3. Results

At seven localities in Lower and Upper Austria, a total of 13 eDNA samples were taken in 2019 and 2020 (Table 5). Our aim to compare the eDNA results with records of *Trichobilharzia* spp. in the field was hampered by the fact that only a few reports were available; this also corresponded with the lack of official reports on cercarial dermatitis in these two years.

The following four cases list snails collected in 2019 and 2020 that released *Trichobilharzia* cercariae (Table 6). (1) Cercariae tentatively assigned to *T. szidati* were released by *Lymnaea stagnalis* collected at locality 2 (Hagenauer Bucht 19 August 2019). Due to insufficient fixation, the cercariae were in poor condition. (2) Cercaria identified as *T. franki* were released by *Radix auricularia* collected at locality 1 (Reichersberger Au; 27 May 2019). (3) Cercariae identified as *T. szidati* were released by two *Lymnaea stagnalis* snails collected at locality 4 (Rekawinkel; 20 and 24 August 2020). (4) Cercariae identified as *T. physellae* were released by *Physella acuta* collected at locality 9 (Lake Pleschinger-See; 16 September 2020 and 11 November 2020). For details of the corresponding eDNA results, see Section 3.2.1.

Table 6. List of *Trichobilharzia* species released from snails collected at four different sampling sites in the field during 2019 and 2020. The locality numbers correspond to the sampling sites in Table 1. * *Trichobilharzia szidati* cercaria that were only tentatively assigned due to insufficient fixation.

Locality Number	Cercaria Species	Host Snail	Collection Date (dd.mm.yyyy)	Prevalence
2	*Trichobilharzia szidati* *	*Lymnaea stagnalis*	19.08.2019	9/28 (32.1%)
1	*Trichobilharzia franki*	*Radix auricularia*	27.05.2019	1/10 (10.0%)
4	*Trichobilharzia szidati*	*Lymnaea stagnalis*	20.08.2020 24.08.2020	3/9 (33.3%)1/12 (8.3%)
9	*Trichobilharzia physellae*	*Physella acuta*	16.09.2020 11.11.2020	1/6 (16.7%)1/30 (3.3%)

3.1. Establishment of the eDNA-PCR Primers

CO1 sequences (length of 1216 or 1263 bp) of five cercariae specimens were generated in the present study (Table 2). They formed the basis for eDNA primer design together with published sequences of seven species that were available at the time of the study: *Trichobilharzia anseri, T. franki, T. ocellata* (currently considered a synonym of *T. szidati*), *T. physellae, T. regenti, T. stagnicolae, T. szidati*).

Specific eDNA primers developed in the first phase of the present study for amplifying a 186 bp fragment of the mt *CO1* gene (Table 3) were tested with DNA extracted from *T. franki, T. szidati,* and *T. regenti*. The cross tests performed with these primers showed clear species specificity. No PCR products could be obtained in the other *Trichobilharzia* species tested. Note that after the detection of *T. physellae* in Upper Austria, when test series were performed with eDNA primers designed for that species (primer pair ZDOE-COI-fw/eDNA-physellae-rv), certain problems arose. This indicates that a mismatch at the 3′–end of a primer is not sufficient to exclude species from amplification. Repeated tests with the other three species resulted in PCR products also in *T. szidati* despite a mismatch in the 3′-position of the primer eDNA-physellae-rv as well as additional internal mismatches (in combination with forward primer ZDOE-COI-fw). Thus, this primer pair could not prevent amplification in the non-target species. Subsequently, new forward primers with additional mismatches at the 3′–end were designed to further make the primers more specific (ZDOE-COI-fw_szidati, ZDOE-COI-fw_physellae). Tests with the new forward primers also yielded cross-species results (albeit with lower PCR success compared to the target species): ZDOE-COI-fw_physellae + eDNA-physellae-rv showed results also in *T. szidati* and *T. regenti*, ZDOE-COI-fw_szidati + eDNA-szidati-rv showed results also in *T. physellae*. Sequencing of the PCR products confirmed the authenticity of the sequence case (*T. szidati* or *T. regenti*, respectively). Moreover, it also showed that the non-matching 3′–base was present in both the forward as well as reverse primers while the sequence between the two primers was identical with the template tested in each case (*T. szidati*, *T. physellae*, or *T. regenti*). Thus, the amplification was not due to shorter primers (e.g., failure in primer production). This demonstrated that the *T. physellae* primers indeed amplified in those two other species as well as the *T. szidati* primer amplified in *T. physellae*. Interestingly, when we repeated these experiments with another DNA polymerase (Multiplex PCR Kit, Qiagen) there was no unspecific amplification. To summarize, there was a reproducible non-intentional amplification in some non-target species, which depended on the DNA polymerase used. It occurred with some (but not all) primer combinations, despite considerable mismatches (including the 3′–position) with the primer binding sites.

3.2. eDNA Experiments

Eighteen eDNA samples of water were taken from seven localities (Table 5). Thirteen eDNA samples were collected directly in the field ("water samples from field" in Table 5), four were obtained in the laboratory from water samples (without snails) brought from the field in bottles in which eDNA samples were taken in the lab ("water sample lab" in Table 5). One eDNA sample was taken from transport water ("transport water sample") containing a living *Lymnaea stagnalis* (which later released *T. szidati* cercariae). Furthermore, 15 eDNA samples were taken from aquariums in which snails were kept for some time (up to 24 hours), three of them when the snails were still present ("aquarium sample") and 12 samples after having removed the snails ("aquarium test samples").

3.2.1. eDNA Samples from the Field

Among all water samples from the field, four (eDNA-5, e-DNA-6, eDNA-7, eDNA-9) were positive for *T. szidati* using the species-specific *CO1* primers (the maximum similarity values with BLAST search on 6 February 2022 were 100% for all four samples). They all were from sample sites of one water body (locality 4, Rekawinkel, Lower Austria). Moreover, the transport water sample from this locality delivered a positive result for *T. szidati*. Besides *T. szidati*, a second species—*T. physellae*—was demonstrated with eDNA analyses (similarity values 100%). In September 2020, after completion of the initially planned analyses, *Trichobilharzia physellae* was detected by hatching experiments in a snail collected at locality 9 (Lake Pleschinger-See). Since this was the first record of the species in Europe [23], it had not been covered by our experimental design. The subsequent construction of eDNA primers specific for *T. physellae* yielded a positive result from the eDNA sample taken from Lake Pleschinger-See, yet not from all samples. The eDNA samples from the field taken at the first day of record of *T. physellae* proved negative. From the four water samples transported to the lab on 16 October 2021, only one gave a positive result (eDNA-27).

The detection of an unexpected new species resulted in problems and reconsideration of our experimental design described above. Moreover, testing the *physellae*-specific reverse primer (Primer-Pair: ZDOE-COI-fw/eDNA-physellae-rv) with the eDNA samples yielded a PCR product in one sample that had been positive for *T. szidati* (sample eDNA-5). This confirmed the problems with species specificity mentioned above. Sequencing of the PCR products showed that the *physellae*-specific reverse primer had amplified a sequence of *T. szidati* (the species observed in that water body).

The PCR tests with the primers specific for *T. regenti* and *T. franki* were negative in all eDNA samples from the field. Importantly, even the eDNA sample from 2019 (4-W-8) from the bay Hagenauer Bucht at the lower Inn River (locality code 2) did not yield a PCR product with the *Trichobilharzia*-specific *CO1* primers. At this locality, cercariae, presumably *Trichobilharzia szidati*, had been detected in 2019 by hatching experiments. Unfortunately, they could not be identified unambiguously due to insufficient fixation of the material. Nonetheless, despite of the evidence for the presence of *Trichobilharzia* sp. at that locality at that time, the corresponding eDNA sample did not provide evidence of any of the four species.

In summary (with the exception of sample 4-W-8 from 2019), the results of these eDNA analyses were mostly consistent with the rare detection of *Trichobilharzia* cercariae in the various water bodies.

Besides using *CO1* as a marker sequence, we also tested universal *18S* primers. They indicated the presence of trematodes in 8 of 16 water samples. The *18S* sequences thus obtained were blasted in GenBank and BOLD. Some sequences yielded a single species as best hit based on *18S*, i.e., *Diplostomum pseudospathaceum* (100% similarity), *Echinostoma revolutum* (99.72%), *Opisthioglyphe ranae* (98.68%), and *Parastrigea robusta* (100%). In sample eDNA-11, the identification to *T. szidati* based on *CO1* was confirmed at genus level by *18S* results. The BLAST similarity scores of 100% for various species of *Trichobilharzia* indicated for this genus a lack of taxonomic resolution at species level with the *18S* marker sequence. The *18S* sequences of sample 3-W-14 gave similarity values of up to 96% for species of the superorder Plagiorchiida. In addition, three samples yielded a closest hit with *Sminthurides aquaticus* (98.92–100%), a springtail species (Table 5; all similarity values from GenBank on 6 February 2022).

3.2.2. eDNA from Aquarium Water

Three samples were considered as positive controls because the host snail *L. stagnalis* was still present in the aquarium when the eDNA sample was taken (eDNA-5, eDNA-6, eDNA-11). All PCR reactions with the primers for *T. szidati* were positive in these samples, whereas the primers for *T. franki* and *T. regenti* yielded no PCR products, confirming that these primers were species-specific (these experiments were performed before the first record of *T. physellae* in Austria).

To test how long cercariae are detectable in aquarium water after removal of the host snail (from which the cercariae were released), 12 eDNA samples were taken at different times (Table 5). The first seven samples (first week after snail removal) tested positively for *T. szidati* in all three replicates. Of the samples taken thereafter, only the last one (eDNA-24, from 7 October 2020; day 44) was positive in one of the three PCR tests. The four negative ones (eDNA-20–eDNA-23) were additionally tested three times more with the *CO1* primers for *T. szidati*. None of these subsequent PCR reactions resulted in PCR products. All controls with the eDNA primers for *T. franki* and *T. regenti* were performed with all water samples from the field and proved negative.

3.3. Sensitivity Tests

Dilution series of genomic DNA of two individuals (*T. szidati* Ls17-15-003; *T. franki* Ra13-16-002) were tested in a PCR with 30 thermal cycles. Amplification of the targeted PCR product was accomplished up to a dilution of 10^{-2}, corresponding to a minimum concentration of 0.85 pg/µL for Ls17-15-003 and 0.83 pg/µL for Ra13-16-002. PCR with 40 cycles enabled amplification in dilutions up to 10^{-3} (corresponding to 0.085 pg/µL for Ls17-15-003 and 0.083 pg/µL for Ra13-16-002).

The second kind of sensitivity test was performed using the PCR fragments generated from the two individuals as template, i.e., it was a re-amplification (ZDOE-CO1-Fw/eDNA-szidati-rv and ZDOE-CO1-Fw/eDNA-franki-rv, respectively; 141 bp length). PCR with 40 cycles enabled amplification in the entire dilution series, i.e., gave positive results up to a dilution of 10^{-6} (corresponding to a minimum detectable concentration of 0.008 pg/µL for Ls17-15-003 and of 0.009 pg/µL for Ra13-16-002).

3.4. Cercariae Release Experiments

Cercarial release experiments with a *Lymnaea stagnalis* (from Rekawinkel, Lower Austria) infected with *T. szidati* showed that the snail released cercariae for six weeks. The snail expelled cercariae each time, which were fixed in 80% EtOH. A rough estimate of the released number was made by counting cercariae in small volumes and extrapolating. The estimates of released cercariae within 12 h were 14,000 (2 September 2020), 5000 (16 September 2020), and 900 (6 October 2020). By 16 October 2020, the host snail had died.

4. Discussion

4.1. Detection of Trichobilharzia Species—Correspondence between Observations and eDNA Tests

The results of the PCR detection of *Trichobilharzia* species in eDNA samples from the seven water bodies investigated yielded a positive result in 4 out of 18 samples. *Trichobilharzia szidati* was detected in water of a small pond (Rekawinkel, Lower Austria), whereas *T. physellae* was found in a water sample from Lake Pleschinger-See in Upper Austria (Table 5). This result should be considered in the context of the presumably scarce occurrence of *Trichobilharzia* species, as shown by the lack of reports on cercarial dermatitis in the years 2019–2020. Assuming similar transmission cycles of *Trichobilharzia* across Europe, previous studies on trematode diversity seem to confirm a low prevalence [39–43]. Another case occurred in the bay Hagenauer Bucht of the lower Inn River. *Trichobilharzia* cercariae had been detected there by hatching experiments with *L. stagnalis* collected in locality 2 in 2019, but could not be identified (morphologically) due to insufficient fixation. The analysis of the corresponding eDNA sample (from 2019; conducted in 2020) provided no evidence for the presence of *Trichobilharzia*. The presumably limited stability of eDNA samples after long-term storage as a possible explanation [44–46], however, is contradicted by a positive PCR product obtained with *18S* primers in four samples from 2019. This indicates the presence of DNA, but not (or not enough) DNA from one of the *Trichobilharzia* target species. Another explanation might be that the corresponding eDNA sampling point in the water body (in the riparian area) was several meters away from where the host snail was collected. Two more eDNA samples taken in 2020 at the same position also failed to give a result for trematode DNA (eDNA-1 and eDNA-2). This implies that, generally, a considerable number of sampling sites would be needed to find eDNA of the target species with a high probability. For example, an eDNA study on carp showed the necessity of a high sampling effort if only little environmental DNA is present [47]. The mostly negative results for the eDNA samples from Lake Pleschinger-See, the water body where *T. physellae* could be detected via hatching tests, further confirmed the interpretation of a certain stochasticity of eDNA detection [23]. Only two of the samples proved positive, of which one contained *T. physellae* and the other *Diplostomum pseudospathaceum*.

In summary, eDNA assays based on *CO1* were always negative in localities where no cercariae of *Trichobilharzia* species were found. However, using *18S* primers, several of the samples delivered eDNA detection of other trematode taxa even though the *18S* fragment was considerably longer than the *CO1* fragments. Identification to species level is unreliable based on the *18S* marker sequence due both to insufficient reference data and lack of taxonomic resolution.

The finding that the PCRs with species-specific eDNA primers designed for *T. franki* and *T. regenti* were negative was interpreted as confirmation of their specificity. Nonetheless, the design and testing of species-specific primers for *T. physellae*, after the species had been discovered in Austria, should be treated with caution. Although we still cannot explain why the primers for *T. physellae* amplified (despite several mismatches) in other species, too, we conclude that presumed species-specificity of primers may not be guaranteed. A detection system based on species-specific primers developed on a presumably known species inventory may be compromised when new (or formerly not detected) species appear. We therefore recommend checking PCR results by sequencing PCR products, or including other control procedures such as nested PCR, ensuring that species identification is reliable. Future experiments (e.g., mixing two species into one PCR reaction) should test in more detail how the properties of various DNA polymerases effect primer specificity.

4.2. Stochasticity of eDNA Detection

In general, our results underline the potential of the eDNA approach. However, we observed considerable stochasticity in detection success. This is in accordance with reports by Jothikumar et al. 2015 [30] about eDNA samples taken from lakes with suspected cases of swimmer's itch. That analysis yielded only 32 % positive results for the presence of avian schistosomes. Similarly, in the study by Sato et al. 2018 [25] on *Schistosoma mansoni*, only one out of 14 environmental water samples was positive. In another study, however, *S. mansoni* was detected in water samples from four of five sites in central Kenya with known ongoing transmission [28]. Wind may be an important factor for the distribution of cercariae in a water body [48]. Those authors found that wind direction, wind speed, and time of day were the best predictors for the risk of contracting swimmer's itch. Thus, even sampling at the same site over time might yield varying results. Accordingly, the location (and perhaps the circumstances) of sampling may be relevant, even if the water body is relatively small as in the case of site 11 (pond in Rekawinkel, about 300 m^2). Of the four sampling sites in this water body, only two were positive (and here only 1 and 2 of each of the three PCR replicates, respectively). Interestingly, one (eDNA-7) of the positive samples was collected from the shore, the other one (eDNA-9) from the middle of the pond. In contrast, all PCR tests (six replicates each) from the other two sample collection sites were negative (one, eDNA-8, also from the shore, the other, eDNA-10, in a marshy area of the pond). The sample in the marshy area was possibly affected by the large amount of suspended sediment (only 100 mL water filtered). This was not the case, however, for the other negative sampling site (200 mL filtered water, compared to 120 and 170 mL for the two positive samples). This demonstrates, albeit with a small sample, that the sampling regime (position in the lake), the turbidity of the water, as well as the number of PCR repetitions can affect the results.

The fact that—among all the PCR reactions carried out with the five samples from Lake Pleschinger-See (where *T. physellae* cercariae were found, see [23])—only one yielded a positive PCR result for *T. physellae* demonstrates once again that PCR detection has a high stochasticity. One way to improve detection success would be to increase the amount of target eDNA by sampling larger volumes of water by pooling multiple eDNA filters [49]. However, besides a possibly low concentration of *Trichobilharzia* in the lake, our results reflect recent experiences of other studies reporting a heterogeneous distribution of eDNA in natural waters [6,50]. This somewhat relativizes a major expected advantage of eDNA detection compared to conventional screening of host snails by releasing experiments. Without knowledge of the specific zones that the host snails inhabit, their detection might be missed in still waters.

4.3. DNA Degradation and Sensitivity of PCR Detection

Our laboratory experiments with aquarium water showed that cercariae could still be detected with high reliability after one week. It remains open whether whole cercariae, their remains, or suspended DNA were sucked into the filters during sampling. Note, however, that this issue also arises with every eDNA sampling from waters in the field. The results were consistent because the samples of the first week were positive and the samples taken at a later time were, with one exception, negative. This result is consistent with a previous study that detected cercariae that had been shed by snails 8 days earlier [28]. Nonetheless, it is somewhat surprising that the very last sample provided a positive result (1 out of 3 PCRs). Here, a contamination cannot be ruled out completely. Another possible explanation is that the water had cleared up in the last week of sampling, enabling more water to be sampled through the filter. The negative results of the aquarium samples after 2 weeks and the single positive result of the last sample taken (after 6 weeks) suggest that a PCR response at very low concentrations is subject to a strong random factor. This result is interesting with regard to the reported degradation times for eDNA from different species, which can vary between 72 h and 25 days [6,51–53]. Since a positive result was obtained at extremely low concentrations in the PCR experiments with the dilution series, similar to those found by Kane et al. 2013 [29], a major factor potentially negatively influencing the results of eDNA analyses might be inhibitory substances and DNA from other organisms (e.g., algae, bacteria). These could be co-extracted from the eDNA filter (in the present case from the eDNA from aquarium water) and exert a competitive or inhibitory effect in the PCR. In this sense, the PCR experiments with the dilution series (which represent an ideal system) cannot be transferred directly to natural aquatic samples. In recent years, many factors have been shown to influence the rate of eDNA degradation in both limnic and marine habitats. For example, evidence is available that high water temperature boosts the activity of the microbiome and thus increases the degradation rate of both mitochondrial (e.g., *CO1*) and nuclear DNA (e.g., *18S*) [4,54,55]. Accordingly, some of our negative results might be explained by the relatively warm water temperature of around 23 °C in our experiment. There also appears to be a relationship between degradation rate and the overall biomass of the water body studied. The more biomass a water body contains, the higher the degradation rate [4,54]. This could also have influenced our samples because the pond near Rekawinkel and the bay Hagenauer Bucht in particular contained abundant suspended matter, which also meant that less water was pressed through the filter while sampling. The pH as well as UV-B radiation also have an effect because degradation is faster in acidic environments and at high UV-B levels [5,56].

4.4. eDNA Barcoding for the Detection of Trichobilharzia in Practice

In recent years, studies in a variety of animal groups demonstrated the effective use of eDNA to detect animals/species groups that are otherwise difficult to detect or to monitor [57–59]. Our results confirmed the possibility to detect cercariae in water samples as well as the stability of DNA in aquarium water over several days, but also demonstrated the uncertainty of detection regarding the location of sampling. This could also be true for waters in the field, which raises certain caveats regarding the applicability for monitoring and assessment of current situations. The single observation of a laboratory-kept *L. stagnalis* individual shedding cercariae in large quantities over six weeks suggests that even when access to a water body is approved after a negative eDNA test, high levels of cercariae can reappear very quickly. For example, warm weather conditions could once again boost cercariae output. Therefore—even if our conclusions are currently based on few samples and data—both a warning system as well as an all-clear system based on eDNA do not seem reasonable (at least without an immense sampling effort). This, however, does not mean that eDNA tools cannot be usefully applied or expanded. Environmental DNA as a tool for identifying the causative agent in water bodies (where cercarial dermatitis was reported) is no doubt extremely helpful instead of collecting and examining (possibly protected) snails. In future experiments, we aim to optimize our species-specific primers and test them with the qPCR method.

5. Conclusions

We demonstrated that it is possible to detect trematodes in eDNA water samples, and in combination with matching primers, a barcoding approach is also plausible. Nonetheless, uncritical detection with supposedly species-specific primers based solely on a PCR product of the desired size is short-sighted and risky. This is largely because species specificity may not be guaranteed, especially if the species inventory of an area is incompletely known. This makes sequencing of every positive result still advisable. The collection of eDNA samples is additionally hampered in the case of potentially locally restricted tiny animals such as cercariae, as this creates stochasticity in the results. At the same time, however, the relatively long detectability of at least one week would enable recording the species population of trematodes of a water body through regular but not excessive sample collection. This calls for further studies, especially for improving the localisation of sampling.

Author Contributions: Conceptualisation: N.H., J.B., E.H., C.H., S.R., H.S. and N.U.S.; Field work: N.H., E.H., C.H., S.R., H.S. and N.U.S.; Morphological investigations: N.H., J.B., C.H. and H.S.; Laboratory procedures: N.H. and S.R.; Data analysis, N.H., C.H., H.S. and E.H.; contributed intellectually to the interpretation and discussion of results: N.H., J.B., E.H, C.H., S.R., H.S. and N.U.S.; Writing—Original draft preparation: N.H. and E.H.; Writing—Review & Editing, N.H., J.B., E.H, C.H., S.R., H.S. and N.U.S.; Project administration, E.H. and N.U.S.; Funding acquisition, E.H., C.H. and N.U.S. All authors have read and agreed to the published version of the manuscript.

Funding: This research received partial funding by the Institutional Grant Charles University Cooperation Biology 2022–2026 and by the Grant of Ministry of Education, Youth and Sports of the Czech Republic (8J21AT011). Furthermore, the study was supported by WTZ Programme (Scientific & Technological Cooperation) (CZ18/2021), by the Upper Austrian Provincial Government (Reference number: WW-583711-2019/KR), and the "Freunde des Naturhistorischen Museums Wien".

Institutional Review Board Statement: Not applicable.

Data Availability Statement: Raw and processed data will be shared on personal request directly from the corresponding author.

Acknowledgments: We thank Wolfgang Heinisch and Hubert Blatterer (Department of Water Management, Upper Austria Provincial Government) for initiating and promoting the Upper Austrian cercaria project in the course of which the first field trips took place and the eDNA study was initiated. We are grateful to the Upper Austrian Provincial Government and the "Freunde des Naturhistorischen Museums Wien" for financial support. Thanks are also due to Julia Schindelar, Marcia Sittenthaler, and Alexandra Wanka for technical assistance in the lab. Finally, we thank Michael Stachowitsch for scientific English copyediting.

Conflicts of Interest: The authors declare no conflict of interest.

References

1. Pawlowski, J.; Apothéloz-Perret-Gentil, L.; Mächler, E.; Altermatt, F. *Anwendung von eDNA-Methoden in Biologischen Untersuchungen und bei der Biologischen Bewertung von Aquatischen Ökosystemen: Richtlinien*; Bundesamt für Umwelt: Bern, Switzerland, 2020.
2. Taberlet, P.; Bonin, A.; Zinger, L.; Coissac, E. *Environmental DNA: For Biodiversity Research and Monitoring*; Online Edi.; Oxford University Press: Oxford, UK, 2018; ISBN 9780198767220.
3. Harrison, J.B.; Sunday, J.M.; Rogers, S.M. Predicting the fate of eDNA in the environment and implications for studying biodiversity. *Proc. R. Soc. B Biol. Sci.* **2019**, *286*, 20191409. [CrossRef] [PubMed]
4. Jo, T.; Arimoto, M.; Murakami, H.; Masuda, R.; Minamoto, T. Estimating shedding and decay rates of environmental nuclear DNA with relation to water temperature and biomass. *Environ. DNA* **2020**, *2*, 140–151. [CrossRef]
5. Strickler, K.M.; Fremier, A.K.; Goldberg, C.S. Quantifying effects of UV-B, temperature, and pH on eDNA degradation in aquatic microcosms. *Biol. Conserv.* **2015**, *183*, 85–92. [CrossRef]
6. Troth, C.R.; Sweet, M.J.; Nightingale, J.; Burian, A. Seasonality, DNA degradation and spatial heterogeneity as drivers of eDNA detection dynamics. *Sci. Total Environ.* **2021**, *768*, 144466. [CrossRef]
7. Bochove, K.; Bakker, F.T.; Beentjes, K.K.; Hemerik, L.; Vos, R.A.; Gravendeel, B. Organic matter reduces the amount of detectable environmental DNA in freshwater. *Ecol. Evol.* **2020**, *10*, 3647–3654. [CrossRef]
8. Loker, E.S.; DeJong, R.J.; Brant, S.V. Scratching the Itch: Updated Perspectives on the Schistosomes Responsible for Swimmer's Itch around the World. *Pathogens* **2022**, *11*, 587. [CrossRef]
9. Auer, H.; Aspöck, H. "Vogelbilharzien" als Erreger einer Hautkrankheit: Die Zerkarien-Dermatitis. *Denisia* **2002**, *6*, 321–331.
10. Sattmann, H.; Hörweg, C.; Konecny, R. Zerkariendermatitis in Österreich—Rückblick und Perspektiven. *Denisia* **2004**, *13*, 457–461.
11. Horák, P.; Mikeš, L.; Lichtenbergová, L.; Skála, V.; Soldánová, M.; Brant, S.V. Avian Schistosomes and Outbreaks of Cercarial Dermatitis. *Clin. Microbiol. Rev.* **2015**, *28*, 165–190. [CrossRef]
12. Kolářová, L.; Horák, P.; Skírnisson, K.; Marečková, H.; Doenhoff, M. Cercarial Dermatitis, a Neglected Allergic Disease. *Clin. Rev. Allergy Immunol.* **2013**, *45*, 63–74. [CrossRef]
13. Soldánová, M.; Selbach, C.; Kalbe, M.; Kostadinova, A.; Sures, B. Swimmer's itch: Etiology, impact, and risk factors in Europe. *Trends Parasitol.* **2013**, *29*, 65–74. [CrossRef]

14. Brant, S.V.; Loker, E.S. Discovery-based studies of schistosome diversity stimulate new hypotheses about parasite biology. *Trends Parasitol.* **2013**, *29*, 449–459. [CrossRef]
15. Jouet, D.; Kolářová, L.; Patrelle, C.; Ferté, H.; Skírnisson, K. *Trichobilharzia anseri* n. sp. (Schistosomatidae: Digenea), a new visceral species of avian schistosomes isolated from greylag goose (*Anser anser* L.) in Iceland and France. *Infect. Genet. Evol.* **2015**, *34*, 298–306. [CrossRef]
16. Christiansen, A.Ø.; Olsen, A.; Buchmann, K.; Kania, P.W.; Nejsum, P.; Vennervald, B.J. Molecular diversity of avian schistosomes in Danish freshwater snails. *Parasitol. Res.* **2016**, *115*, 1027–1037. [CrossRef]
17. Neuhaus, W. Biologie und Entwicklung von *Trichobilharzia szidati* n. sp. (Trematoda, Schistosomatidae), einem Erreger von Dermatitis beim Menschen. *Z. Parasitenkd.* **1952**, *15*, 203–266. [CrossRef]
18. Horák, P.; Kolářová, L.; Dvořák, J. *Trichobilharzia regenti* n. sp. (Schistosomatidae, Bilharziellinae), a new nasal schistosome from Europe. *Parasite* **1998**, *5*, 349–357. [CrossRef]
19. Müller, V.; Kimmig, P. *Trichobilharzia franki* n. sp.—Die Ursache für Badedermatitiden in südwestdeutschen Baggerseen [*Trichobilharzia franki* n. sp.—A causative agent of swimmer's itch in south-western Germany]. *Appl. Parasitol.* **1994**, *35*, 12–31.
20. Simon-Martin, F.; Simon-Vicente, F. The life cycle of *Trichobilharzia salmanticensis* n. sp. (Digenea: Schistosomatidae), related to cases of human dermatitis. *Res. Rev. Parasitol.* **1999**, *59*, 13–18.
21. Kolářová, L.; Skírnisson, K.; Ferté, H.; Jouet, D. *Trichobilharzia mergi* sp. nov. (Trematoda: Digenea: Schistosomatidae), a visceral schistosome of *Mergus serrator* (L.) (Aves: Anatidae). *Parasitol. Int.* **2013**, *62*, 300–308. [CrossRef]
22. Talbot, S.B. Studies on schistosome dermatitis: II. Morphological and life history studies on three dermatitis-producing schistosome cercariae, *C. elvae* Miller, 1923, *C. stagnicolae* n. sp., and C. physellae n. sp. *Am. J. Epidemiol.* **1936**, *23*, 372–384. [CrossRef]
23. Helmer, N.; Blatterer, H.; Hörweg, C.; Reier, S.; Sattmann, H.; Schindelar, J.; Szucsich, N.U.; Haring, E. First Record of *Trichobilharzia physellae* (Talbot, 1936) in Europe, a Possible Causative Agent of Cercarial Dermatitis. *Pathogens* **2021**, *10*, 1473. [CrossRef] [PubMed]
24. McPhail, B.A.; Froelich, K.; Reimink, R.L.; Hanington, P.C. Simplifying Schistosome Surveillance: Using Molecular Cercariometry to Detect and Quantify Cercariae in Water. *Pathogens* **2022**, *11*, 565. [CrossRef]
25. Sato, M.O.; Rafalimanantsoa, A.; Ramarokoto, C.; Rahetilahy, A.M.; Ravoniarimbinina, P.; Kawai, S.; Minamoto, T.; Sato, M.; Kirinoki, M.; Rasolofo, V.; et al. Usefulness of environmental DNA for detecting Schistosoma mansoni occurrence sites in Madagascar. *Int. J. Infect. Dis.* **2018**, *76*, 130–136. [CrossRef] [PubMed]
26. Rudko, S.P.; Turnbull, A.; Reimink, R.L.; Froelich, K.; Hanington, P.C. Species-specific qPCR assays allow for high-resolution population assessment of four species avian schistosome that cause swimmer's itch in recreational lakes. *Int. J. Parasitol. Parasites Wildl.* **2019**, *9*, 122–129. [CrossRef]
27. Alzaylaee, H.; Collins, R.A.; Shechonge, A.; Ngatunga, B.P.; Morgan, E.R.; Genner, M.J. Environmental DNA-based xenomonitoring for determining *Schistosoma* presence in tropical freshwaters. *Parasit. Vectors* **2020**, *13*, 63. [CrossRef]
28. Sengupta, M.E.; Hellström, M.; Kariuki, H.C.; Olsen, A.; Thomsen, P.F.; Mejer, H.; Willerslev, E.; Mwanje, M.T.; Madsen, H.; Kristensen, T.K.; et al. Environmental DNA for improved detection and environmental surveillance of schistosomiasis. *Proc. Natl. Acad. Sci. USA* **2019**, *116*, 8931–8940. [CrossRef] [PubMed]
29. Kane, R.A.; Stothard, J.R.; Rollinson, D.; Leclipteux, T.; Evraerts, J.; Standley, C.J.; Allan, F.; Betson, M.; Kaba, R.; Mertens, P.; et al. Detection and quantification of schistosome DNA in freshwater snails using either fluorescent probes in real-time PCR or oligochromatographic dipstick assays targeting the ribosomal intergenic spacer. *Acta Trop.* **2013**, *128*, 241–249. [CrossRef] [PubMed]
30. Jothikumar, N.; Mull, B.J.; Brant, S.V.; Loker, E.S.; Collinson, J.; Secor, W.E.; Hill, V.R. Real-Time PCR and Sequencing Assays for Rapid Detection and Identification of Avian Schistosomes in Environmental Samples. *Appl. Environ. Microbiol.* **2015**, *81*, 4207–4215. [CrossRef] [PubMed]
31. Rudko, S.P.; Reimink, R.L.; Froelich, K.; Gordy, M.A.; Blankespoor, C.L.; Hanington, P.C. Use of qPCR-Based Cercariometry to Assess Swimmer's Itch in Recreational Lakes. *Ecohealth* **2018**, *15*, 827–839. [CrossRef]
32. Soldánová, M.; Selbach, C.; Sures, B. The Early Worm Catches the Bird? Productivity and Patterns of *Trichobilharzia szidati* Cercarial Emission from *Lymnaea stagnalis*. *PLoS ONE* **2016**, *11*, e0149678. [CrossRef]
33. Al-Jubury, A.; Kania, P.W.; Bygum, A.; Buchmann, K. Temperature and light effects on *Trichobilharzia szidati* cercariae with implications for a risk analysis. *Acta Vet. Scand.* **2020**, *62*, 54. [CrossRef] [PubMed]
34. Horák, P.; Kolářová, L.; Adema, C.M. Biology of the schistosome genus *Trichobilharzia*. *Adv. Parasitol.* **2002**, *52*, 155–233. [CrossRef] [PubMed]
35. Reier, S.; Haring, E.; Billinger, F.; Blatterer, H.; Duda, M.; Gorofsky, C.; Grasser, H.-P.; Heinisch, W.; Hörweg, C.; Kruckenhauser, L.; et al. First confirmed record of *Trichobilharzia franki* Müller & Kimmig, 1994, from *Radix auricularia* (Linnaeus, 1758) for Austria. *Parasitol. Res.* **2020**, *119*, 4135–4141. [CrossRef] [PubMed]
36. Folmer, O.; Black, M.B.; Hoeh, W.R.; Lutz, R.A.; Vrijenhoek, R.C. DNA primers for amplification of mitochondrial *cytochrome c oxidase subunit I* from diverse metazoan invertebrates. *Mol. Mar. Biol. Biotechnol.* **1994**, *3*, 294–299.
37. Hohensee, N. *Digenean Trematodes in Freshwater Snails in the Surroundings of Vienna with a Focus on Species Involved in Human Infections*; University of Vienna: Vienna, Austria, 2016.

38. Nicholas, K.B.; Nicholas, H.B.J. GeneDoc: A tool for editing and annotating multiple sequence alignments. *Embnet. News* **1997**, *4*, 1–4.
39. Al-Jubury, A.; Duan, Y.; Kania, P.W.; Tracz, E.S.; Bygum, A.; von Jørgensen, L.G.; Horák, P.; Buchmann, K. Avian schistosome species in Danish freshwater lakes: Relation to biotic and abiotic factors. *J. Helminthol.* **2021**, *95*, e22. [CrossRef]
40. Gaub, L.; Sattmann, H.; Hörweg, C.; Walochnik, J. Molecular validation of species determination of larval trematodes from freshwater snail hosts in Austria, with special emphasis on the genus *Trichobilharzia* Skrjabin & Zakharow, 1920. *Arianta* **2020**, *8*, 13–19.
41. Duan, Y.; Al-Jubury, A.; Kania, P.W.; Buchmann, K. Trematode diversity reflecting the community structure of Danish freshwater systems: Molecular clues. *Parasit. Vectors* **2021**, *14*, 43. [CrossRef]
42. Gordy, M.A.; Hanington, P.C. A fine-scale phylogenetic assessment of digenean trematodes in central Alberta reveals we have yet to uncover their total diversity. *Ecol. Evol.* **2019**, *9*, 3153–3238. [CrossRef]
43. Selbach, C.; Soldánová, M.; Feld, C.K.; Kostadinova, A.; Sures, B. Hidden parasite diversity in a European freshwater system. *Sci. Rep.* **2020**, *10*, 2694. [CrossRef]
44. Mauvisseau, Q.; Halfmaerten, D.; Neyrinck, S.; Burian, A.; Brys, R. Effects of preservation strategies on environmental DNA detection and quantification using ddPCR. *Environ. DNA* **2021**, *3*, 815–822. [CrossRef]
45. Allison, M.J.; Round, J.M.; Bergman, L.C.; Mirabzadeh, A.; Allen, H.; Weir, A.; Helbing, C.C. The effect of silica desiccation under different storage conditions on filter-immobilized environmental DNA. *BMC Res. Notes* **2021**, *14*, 106. [CrossRef] [PubMed]
46. Spens, J.; Evans, A.R.; Halfmaerten, D.; Knudsen, S.W.; Sengupta, M.E.; Mak, S.S.T.; Sigsgaard, E.E.; Hellström, M. Comparison of capture and storage methods for aqueous macrobial eDNA using an optimized extraction protocol: Advantage of enclosed filter. *Methods Ecol. Evol.* **2017**, *8*, 635–645. [CrossRef]
47. Furlan, E.M.; Gleeson, D.; Wisniewski, C.; Yick, J.; Duncan, R.P. eDNA surveys to detect species at very low densities: A case study of European carp eradication in Tasmania, Australia. *J. Appl. Ecol.* **2019**, *56*, 2505–2517. [CrossRef]
48. Sckrabulis, J.P.; Flory, A.R.; Raffel, T.R. Direct onshore wind predicts daily swimmer's itch (avian schistosome) incidence at a Michigan beach. *Parasitology* **2020**, *147*, 431–440. [CrossRef] [PubMed]
49. Hunter, M.E.; Ferrante, J.A.; Meigs-Friend, G.; Ulmer, A. Improving eDNA yield and inhibitor reduction through increased water volumes and multi-filter isolation techniques. *Sci. Rep.* **2019**, *9*, 5259. [CrossRef]
50. Beentjes, K.K.; Speksnijder, A.G.C.L.; Schilthuizen, M.; Hoogeveen, M.; van der Hoorn, B.B. The effects of spatial and temporal replicate sampling on eDNA metabarcoding. *PeerJ* **2019**, *7*, e7335. [CrossRef]
51. Barnes, M.A.; Turner, C.R.; Jerde, C.L.; Renshaw, M.A.; Chadderton, W.L.; Lodge, D.M. Environmental Conditions Influence eDNA Persistence in Aquatic Systems. *Environ. Sci. Technol.* **2014**, *48*, 1819–1827. [CrossRef]
52. Dejean, T.; Valentini, A.; Duparc, A.; Pellier-Cuit, S.; Pompanon, F.; Taberlet, P.; Miaud, C. Persistence of Environmental DNA in Freshwater Ecosystems. *PLoS ONE* **2011**, *6*, e23398. [CrossRef]
53. Pilliod, D.S.; Goldberg, C.S.; Arkle, R.S.; Waits, L.P. Factors influencing detection of eDNA from a stream-dwelling amphibian. *Mol. Ecol. Resour.* **2014**, *14*, 109–116. [CrossRef]
54. Jo, T.; Murakami, H.; Yamamoto, S.; Masuda, R.; Minamoto, T. Effect of water temperature and fish biomass on environmental DNA shedding, degradation, and size distribution. *Ecol. Evol.* **2019**, *9*, 1135–1146. [CrossRef]
55. Goldberg, C.S.; Strickler, K.M.; Fremier, A.K. Degradation and dispersion limit environmental DNA detection of rare amphibians in wetlands: Increasing efficacy of sampling designs. *Sci. Total Environ.* **2018**, *633*, 695–703. [CrossRef]
56. Seymour, M.; Durance, I.; Cosby, B.J.; Ransom-Jones, E.; Deiner, K.; Ormerod, S.J.; Colbourne, J.K.; Wilgar, G.; Carvalho, G.R.; de Bruyn, M.; et al. Acidity promotes degradation of multi-species environmental DNA in lotic mesocosms. *Commun. Biol.* **2018**, *1*, 4. [CrossRef] [PubMed]
57. Balasingham, K.D.; Walter, R.P.; Mandrak, N.E.; Heath, D.D. Environmental DNA detection of rare and invasive fish species in two Great Lakes tributaries. *Mol. Ecol.* **2018**, *27*, 112–127. [CrossRef]
58. Dougherty, M.M.; Larson, E.R.; Renshaw, M.A.; Gantz, C.A.; Egan, S.P.; Erickson, D.M.; Lodge, D.M. Environmental DNA (eDNA) detects the invasive rusty crayfish *Orconectes rusticus* at low abundances. *J. Appl. Ecol.* **2016**, *53*, 722–732. [CrossRef] [PubMed]
59. Ushio, M.; Murata, K.; Sado, T.; Nishiumi, I.; Takeshita, M.; Iwasaki, W.; Miya, M. Demonstration of the potential of environmental DNA as a tool for the detection of avian species. *Sci. Rep.* **2018**, *8*, 4493. [CrossRef] [PubMed]

diversity

MDPI

Opinion

The Future of DNA Barcoding: Reflections from Early Career Researchers

Danielle M. Grant [1,2,3], Ole Bjørn Brodnicke [4], Ann M. Evankow [5], André O. Ferreira [6,7], João T. Fontes [6,7], Aslak Kappel Hansen [8,9,10], Mads Reinholdt Jensen [8], Tuğba Ergül Kalaycı [11], Alexandra Leeper [12,13], Shalaka Kiran Patil [3], Sebastian Prati [14,15], Anna Reunamo [16], Aradhana J. Roberts [17], Rajesh Shigdel [18], Valentina Tyukosova [19], Mika Bendiksby [5,19], Rakel Blaalid [20], Filipe O. Costa [6,7], Peter M. Hollingsworth [21], Elisabeth Stur [19] and Torbjørn Ekrem [19,*]

1 NORCE Norwegian Research Centre, NO-5012 Bergen, Norway; dani@norceresearch.no
2 Bjerknes Centre for Climate Research, NO-5007 Bergen, Norway
3 Department of Biological Sciences, University of Bergen, NO-5020 Bergen, Norway; shalaka.patil@uib.no
4 Department of Biology, University of Copenhagen, DK-2200 Copenhagen, Denmark; ole.brodnicke@gmail.com
5 Natural History Museum, University of Oslo, NO-0318 Oslo, Norway; ann.evankow@nhm.uio.no (A.M.E.); mika.bendiksby@nhm.uio.no (M.B.)
6 CBMA—Centre of Molecular and Environmental Biology, University of Minho, 4710-057 Braga, Portugal; alof446@gmail.com (A.O.F.); jtadeusfontes@gmail.com (J.T.F.); fcosta@bio.uminho.pt (F.O.C.)
7 Institute of Science and Innovation for Bio-Sustainability (IB-S), University of Minho, 4710-057 Braga, Portugal
8 Department of Biology, Aarhus University, DK-8000 Aarhus, Denmark; akhansen@snm.ku.dk (A.K.H.); mrj@bio.au.dk (M.R.J.)
9 Natural History Museum of Denmark, DK-1350 Copenhagen, Denmark
10 Natural History Museum Aarhus, DK-8000 Aarhus, Denmark
11 Department of Biology, University of RTE, 53100 Rize, Turkey; tugba.ergul@erdogan.edu.tr
12 Aquaculture Department, Matís ohf. Food and Biotechnology, 113 Reykjavík, Iceland; alexandral@matis.is
13 Department of Animal and Aquaculture Sciences, Faculty of Biosciences, Norwegian University of Life Sciences, NO-1432 Aas, Norway
14 Department of Arctic and Marine Biology, University of Tromsø, NO-9019 Tromsø, Norway; sebastian.prati@uni-due.de
15 Department of Aquatic Ecology, University of Duisburg-Essen, D-45141 Essen, Germany
16 Finnish Environment Institute SYKE, 00790 Helsinki, Finland; anna.reunamo@syke.fi
17 Department of Physical Geography and Ecosystem Science, Lund University, S-223 62 Lund, Sweden; aradhana.roberts@nateko.lu.se
18 Department of Clinical Science, University of Bergen, NO-5020 Bergen, Norway; rajesh.shigdel@uib.no
19 Department of Natural History, NTNU University Museum, Norwegian University of Science and Technology, NO-7491 Trondheim, Norway; v.tyukosova@zoznam.sk (V.T.); elisabeth.stur@ntnu.no (E.S.)
20 Department of Natural History, University Museum of Bergen, NO-5020 Bergen, Norway; Rakel.Blaalid@uib.no
21 Royal Botanic Garden Edinburgh, Edinburgh EH3 5LR, UK; PHollingsworth@rbge.org.uk
* Correspondence: torbjorn.ekrem@ntnu.no; Tel.: +47-73597812

Citation: Grant, D.M.; Brodnicke, O.B.; Evankow, A.M.; Ferreira, A.O.; Fontes, J.T.; Hansen, A.K.; Jensen, M.R.; Kalaycı, T.E.; Leeper, A.; Patil, S.K.; et al. The Future of DNA Barcoding: Reflections from Early Career Researchers. *Diversity* **2021**, *13*, 313. https://doi.org/10.3390/d13070313

Academic Editor: Stephan Koblmüller

Received: 14 June 2021
Accepted: 5 July 2021
Published: 9 July 2021

Abstract: Over the last two decades, the use of DNA barcodes has transformed our ability to identify and assess life on our planet. Both strengths and weaknesses of the method have been exemplified through thousands of peer-reviewed scientific articles. Given the novel sequencing approaches, currently capable of generating millions of reads at low cost, we reflect on the questions: What will the future bring for DNA barcoding? Will identification of species using short, standardized fragments of DNA stand the test of time? We present reflected opinions of early career biodiversity researchers in the form of a SWOT analysis and discuss answers to these questions.

Keywords: biodiversity; cryptic species; metabarcoding; reference libraries; high-throughput sequencing; biomonitoring

1. Introduction

The use of short, standardized DNA sequences to identify species (i.e., DNA barcoding [1]) has considerably changed how we assess, analyze, and monitor biodiversity within all ecosystems (e.g., [2,3]). Since its initiation, DNA barcoding has significantly contributed to our understanding of species boundaries and the composition of biological communities across the world [4]. In addition, it has paved the way for national and international biodiversity research programs. Notable examples include biodiversity biomonitoring [5], food industry surveillance [6], and detecting substitution in the herbal medicine industry [7].

The foundation of species identification by DNA barcoding is a curated barcode reference library, enabling comparisons of DNA sequences from unidentified organisms to sequences from previously identified taxa. The largest database for this purpose is the Barcode of Life Data Systems (BOLD [8]) containing more than nine million DNA barcodes (Figure 1). The reference library is continuously expanding, with ~60% of the entries published during the last decade. This effort has been primarily driven by research projects promoted by the International Barcode of Life consortium (iBOL), such as Barcode 500K (completed in 2015) and BIOSCAN, launched June 2019 [9,10]. Other initiatives, such as the Earth Biogenome Project, aim to genome sequence all eukaryotic biodiversity in the upcoming decade, which will further expand DNA barcode coverage. The vast majority of organisms still lack DNA barcodes, and much of the current work has been carried out in Europe and North America, resulting in a bias in barcoded biota (Figure 1).

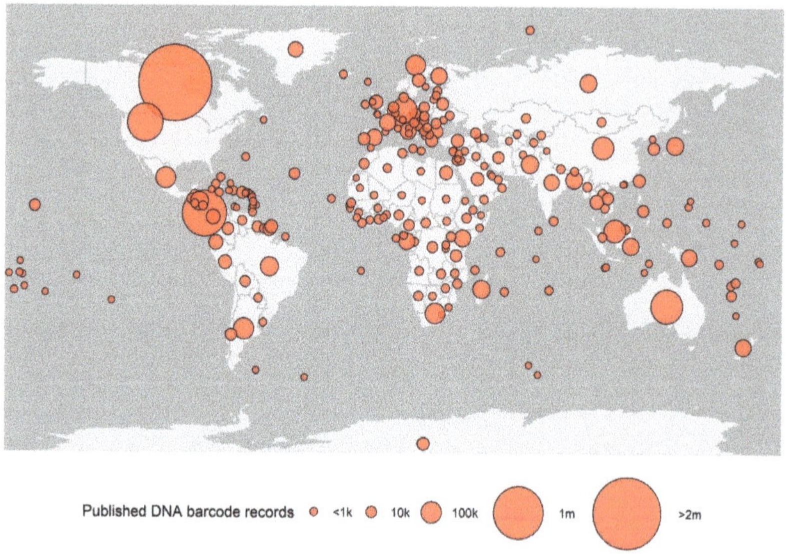

Figure 1. Number of public DNA barcodes available in Barcode of Life Data Systems (BOLD) by country (obtained 1 January, 2021).

A review by DeSalle and Goldstein [4] highlighted that DNA barcoding is a widely applied contemporary tool that has diversified paradigms and practices. The adoption of high-throughput sequencing (HTS) technologies has further decreased costs and increased the range of applications for DNA barcoding [5,11]. Despite differences in the choice of target DNA marker and challenges with generating barcodes for some taxa [12], DNA barcoding is now routinely used across the tree of life and functions as an integrated and standard methodology in biodiversity studies. The essential value of DNA barcoding as an identification tool is obvious: many species would remain unidentified, hidden, or cryptic by other means of identification. The added value of DNA barcodes for identification is that they elucidate species boundaries and provide information on relationships and interactions. A list of scientific advances was accomplished through the use of DNA barcodes and DNA barcoding (e.g., [6,13]).

Two major advances in DNA barcoding have been the development of approaches for sequencing mixtures of samples, and high-throughput sequencing of PCR amplicons with generic primers (DNA metabarcoding). The metabarcoding approach enables the analysis of entire communities in complex samples [14,15], and has expanded the utility of DNA barcoding and associated libraries to microbiomes (e.g., [16]), diets (e.g., [17]), bulk sample biomonitoring (e.g., [2]), sequencing environmental samples (eDNA, e.g., [18]), and paleogenomics (e.g., [19]). DNA metabarcoding contributes to the molecular toolbox for studying both temporal and spatial species dynamics [20].

Beyond the above-mentioned large-scale initiatives and widespread global engagement, a measure of the impact of DNA barcoding can be deduced from the steep curve of the annual number of scientific publications on this topic. Our search (21 January 2021) in Scopus® for publications with 'DNA barcod*' or 'DNA metabarcod*' in the title, abstract, or keywords for the period 2003–2020 returned 14,229 publications from a variety of journals, representing extensive scientific diversity and applications. The publication numbers on these topics have been steadily growing since their introduction (Figure 2).

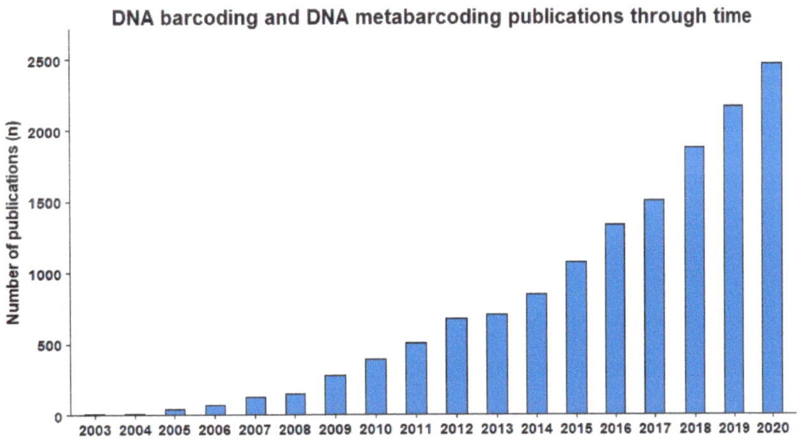

Figure 2. Publications per year registered in Scopus®, containing 'DNA barcod*' or 'DNA metabarcod*' in the title, abstract, or keywords (obtained 21 January 2021).

Nearly two decades after DNA barcodes were first proposed, we reflect upon their future utility and value. In a world of rapid scientific progress, technology has introduced not only novel research avenues, but also rapidly evolving scientific practices. We pose the following two key, overarching questions: (1) Will DNA barcoding stand the test of time as technological progress enables relatively easy access to large-scale genomic data? (2) Will DNA barcoding alter how we describe, assess, and investigate biodiversity?

To answer these questions and contemplate the future of DNA barcoding, we organized a discussion on this topic among early career researchers during the ForBio course, *DNA barcoding—from sequences to species*, held online 21–25 September 2020. The course covered multiple theoretical and practical aspects of the use of molecular tools to delimit and identify species. To prime the discussion, arguments were organized through a SWOT analysis, and were facilitated by tutors in the course. This allowed us to develop and collate opinions on the key aspects, current state (strengths and weaknesses), and future prospects (opportunities and threats) of DNA barcoding. This analysis served as a starting point for a comprehensive discussion with flexible category boundaries. For instance, an opportunity may be seen as a threat and vice-versa. The main objective of this opinion paper is to communicate views and perspectives on the future of DNA barcoding from early career researchers, following comprehensive discussions and literature reviews.

2. SWOT Analysis and Early Career Opinions

Our SWOT analysis, led by early career researchers, identified several current characteristics and prospects for DNA barcoding (Figure 3). In the following, we discuss the most important aspects related to the future of DNA barcoding in biodiversity research and management.

STRENGTHS

DNA barcoding:
- enables quick species identification
- increases the discovery rate of new species
- adds data to solve taxonomic uncertainty
- allows species identification with limited amounts of DNA
- increases cost and time efficiency compared to traditional identification methods
- objectively identifies species with scalable protocols
- produces references for matching DNA of unknown origin such as metagenomics and eDNA

WEAKNESSES

DNA barcoding:
- requires better quality control of reference libraries
- occasionally co-amplifies non-target organisms with non-specific primers
- does not provide species resolution in all groups
- sometimes requires use of multiple markers
- needs mutual agreement on marker choice
- standardization is difficult across organism groups
- demands resources (equipment, labs, funding)
- requires curated open access databases

THREATS

DNA barcoding:
- risks being surpassed by novel technologies
- needs a standardized international legal framework to prevent deepening the inequality gap
- may lead to misinterpretation of data without sufficient knowledge of taxonomic or genetic variation
- needs stable infrastructure and economic resources
- is vulnerable to instability of digital preservation and cybersecurity
- may create resource competition between barcoding and morphological approaches

OPPORTUNITIES

DNA barcoding:
- can evolve with new technology
- assists in revealing global biodiversity
- may automate biomonitoring
- provides molecular context to historic specimens
- can stimulate public engagement with genetic tools
- increases international access to DNA resources
- contributes to a reliable library of life
- can estimate species diversity from complex and environmental samples

Figure 3. Major strengths, weaknesses, opportunities, and threats for DNA barcoding, resulting from the SWOT analysis.

2.1. DNA Barcoding Offers Efficient, Affordable, and High-Throughput Solutions

For most biological diversity, DNA barcoding can be more efficient and require less expertise compared to traditional morphological methods for species identification. It can also be more affordable, particularly for large sample sizes, since the price of generating a single barcode sequence can be as low as USD 0.10 if workflows are scaled efficiently [21]. However, the present costs and efficiency of DNA barcoding vary depending on the research question, taxonomic target group, and project scale, and may not be optimal for all studies. The uneven distribution of infrastructure required for DNA barcoding must also be considered in cost calculations. For regions with limited access to DNA technology, the more realistic identification option for individual specimens may still be morphology. Such scenarios have spurred collaborations, in which local knowledge on species identification has been coupled with sequencing capacity, generating barcoding "hubs" (e.g., [22]). As the international DNA barcoding program continues to expand, key priorities include ensuring equitable global access to technologies, and that samples, knowledge, and benefits are treated in line with the Nagoya Protocol. Inclusive collaboration should be a priority for established and early career researchers alike.

With the advent of various HTS platforms, whole-genome sequencing and metagenomics have become more affordable in recent years. Such big data sequencing approaches have been considered a threat to DNA barcoding [23]. Although genomic data may provide deeper insights for some biodiversity-related questions, DNA barcoding remains more scalable when species identifications are needed. The analysis of genomic data is time-consuming, requires considerable bioinformatic competence compared to standardized DNA barcodes, demands more energy for data computation and storage, and is challenging for quality control when shared [24]. There is an inherent complementarity in DNA barcoding and genomics, with "sequencing a small amount of DNA from vast sample sizes", appropriate for species identification and biomonitoring, and "sequencing a vast amount of nucleotides from smaller sample sizes", appropriate for understanding genomic complexity, diversity, and function. Within this continuum between whole genomes and DNA barcoding, there are organism groups that benefit from deeper sequencing strategies to better address species-level diversity, such as plants [25]. In plant studies, whole-plastid sequencing has some potential to increase taxonomic resolution in species identifications [26,27], and the development of extended barcoding using the nuclear genome is underway [28]. Thus, the question is less about which method is better for discriminating between species and more about which is appropriate for a specific application. There are also clear mutual benefits between barcoding and genome sequencing, with DNA barcoding providing a framework for well-identified samples in genome sequencing projects, and genomic studies contributing insights that may identify new barcode regions in groups where the standard regions are suboptimal.

2.2. DNA Barcoding will Survive and Thrive with Accessible and Curated Reference Libraries

Public biological databases that contain sequence information (e.g., BOLD and GenBank) are pivotal for biodiversity science and equal opportunities in academia. The usefulness of open databases can, however, be compromised by erroneous or ambiguous sequence data [29]. For instance, certain primer sets can lead to accidental co-amplification of non-target organisms [30]. From the start, quality control measures have been implemented in BOLD, for instance, highlighting records that are not barcode compliant, display stop codons, or result from contamination or misidentifications [8,31]. Despite quality control measures, mistakes can still arise from specimen misidentification or errors during one of the many workflow steps [32,33]. Mislabeling of sequences, cross-contamination, low-quality sequences, and sequencing errors may be unnoticed and become potential liabilities for downstream applications [29,34].

The necessity of comprehensive and accurate reference libraries for DNA (meta) barcoding is well-understood, as is the importance of record curation [35]. Despite this understanding, there is admittedly little incentive for researchers producing the data to also curate their shared data. It is our view that there should be incentives (funding and/or recognition) to encourage the development and curation of reference libraries. BOLD is especially useful as it incorporates several pieces of information (trace files, metadata, photos, etc.) and cross-shares data with other repositories [36]. BOLD holds approximately 9 million barcode compliant sequences, although only ~2.2 million are publicly available (BOLD, accessed 11 February 2021). Private records can be made public at any time and shared among researchers in private projects. Researchers that publicly share DNA barcodes bolster the extent and quality of public databases, enabling use and quality control. Machine learning is already in use for the detection of technical and biological errors in sequence data [37] and has the potential to further enhance quality assessments of public data repositories. Another opportunity to strengthen DNA barcode reference libraries is to invest in the production of barcodes for vouchered specimens in curated natural history collections. Obtaining DNA barcode data from these well-curated samples offers the potential to increase the confidence and quality of reference libraries for many taxonomic groups. Another potential step would be to move to routine inclusion of reference barcodes as part of new species descriptions, although it would be premature to make this mandatory as it may prevent many new species from being described (due to lack of access to technology, failed sequencing, degraded DNA, or requiring destructive sampling methods).

2.3. DNA Barcoding Enhances Biodiversity Discovery and Monitoring

The importance of species discovery, species identification, and biodiversity monitoring cannot be overstated, as these are the only means to quantitatively and qualitatively measure the impacts of climate change, habitat degradation, ecosystem management, and other anthropogenic impacts on the biosphere. DNA barcode data can provide a comprehensive basis for organizing and recognizing species-rich groups in the tree of life, providing good starting points for taxonomy as well as biodiversity assessments and biomonitoring (e.g., [38]).

The application of DNA barcoding to species discovery and identification is well-established, including the ability of the methodology to cope with different life stages and provide insights into cryptic species diversity [39]. Since these initial applications, rapid species identification with DNA barcodes has been deployed in several fields, including forensic science [40], control of the food supply chain [6], and understanding disease [41]. Its use in biodiversity characterisation and descriptive taxonomy remains important [38], and acceleration of species discovery is increasingly crucial, given the current threats to biodiversity and elevated rates of extinction [42].

Biomonitoring is a major application of DNA barcoding, and although the term is most often used to refer to ecological assessments, it also encompasses biological identifications to support border control, food authenticity, pharmaceutical monitoring, etc., with sample characterization and identification as the common base task (e.g., [43–45]). Increased knowledge of community composition and species interactions can lead to more precise biomonitoring and allow for the tracking and tracing of particularly important taxa, including endangered and invasive species (e.g., [46]). For instance, DNA barcoding of a single specimen's symbiome, through targeted sequencing of all coexisting organismal DNA, may shed new light on species interactions (e.g., food webs, microbiomes, and parasites) and provide information for environmental management decisions. Detailed mapping of organisms' symbiomes may even be an effective tool to intercept future pandemics [47].

Biomonitoring is often performed at the species level, but DNA barcoding also enables population-level research, assessing, for example, intraspecific genetic structure, population segregation, and phylogeographic patterns (e.g., [48]). As reference databases are compiled, multiple sequences per species will accumulate. This represents a natural foundation for inquisition into population-level dynamics. Sequencing of barcode markers is often the starting point in a phylogeographic study design due to low initial commitment costs before focusing on additional nuclear DNA regions, which is the preferred target in systematics due to their biparental inheritance [48,49]. In recent years, metabarcoding approaches on environmental and fecal samples have yielded insights into population structure in multiple species [50–52]. Likewise, metabarcoding of stream water can help elucidate the ecological impacts of environmental stressors by analyzing the haplotype richness and perseverance of selected macroinvertebrate species [53]. The application of eDNA (meta)barcoding for biomonitoring at the population-level has just begun, and there is considerable expansion potential [54]. As distinct populations are typically handled as separate entities, for example, in estimating quotas and making stock assessments for commercial fish species (e.g., [55]), expanding the reference databases to include wider population coverage per species will also expand applications into population-level inferences.

There are also interdisciplinary avenues that use DNA barcoding and metabarcoding. For example, paleo-reconstructions utilize ancient DNA metabarcoding to better understand past biodiversity, climate boundary conditions and response, past ocean conditions, and even past species distribution (e.g., [19]). The use of paleo-records is well-established, but the inclusion of DNA (meta)barcoding provides more resolution for these past environments compared to traditional methods [56]. The potential for recovering soft-bodied biota typically lost in the geological record creates a compelling argument for the implementation of metabarcoding and eDNA methods.

2.4. DNA Barcoding Methodology Is the Foundation for Automation and Accelerated Biodiversity Assessments

Every methodology has its limitations and challenges. Those that utilize DNA barcoding for species identification must acknowledge the challenges in order to mitigate them. Some species may not be well-discriminated by standard barcodes due to the absence of a clear barcode gap (i.e., maximum intraspecific distance lower than minimum interspecific distance), and this can be particularly problematic in groups that have recently diverged, show extensive hybridization, and/or have slow mutation rates relative to rates of speciation [28]. To overcome the limited discriminatory efficiency for standard barcodes, multiple alternative markers or even approaches are suggested, exemplified by the conundrum of plant DNA barcoding where no single DNA barcode marker separates all or most plant species [26,57]. Moreover, established universal primers may bind to a variety of templates but fail to amplify a specific target group, hence establishing a need for either more degenerative or target-specific primers [58]. Yet another challenge includes barcode pseudogenes (i.e., non-functional copies of barcode regions), which can result in the overestimation of species diversity and misidentifications [59].

DNA metabarcoding has a particular set of challenges, as the outcome of studies is influenced by several variables and decisions made in the experimental setup; this includes the choice of primers, marker specificity, and taxonomic resolution [5]. The requirements of metabarcoding protocols have resulted in the use of additional or alternative DNA barcoding regions more suitable for specific taxa or applications (e.g., 12S for fish eDNA [60] and the *trnL* intron for plants [61]). This utilization of alternative barcoding regions can increase recoverability and resolution (and thus provides clear benefits) while maintaining similarity to a standardized system, using a common set of loci for the molecular identification of species.

Fully accepting the challenges and limitations outlined above, ongoing technological developments are considerably improving the efficiency of DNA barcoding and metabarcoding. One example of this is the use of the PacBio Sequel platform for extensively multiplexing samples and reducing costs [11]. Another innovation is where single-species biomonitoring techniques have been developed based on barcoding primers designed to detect target species in complex samples with a dip-stick. Doyle and Uthicke [62] designed the tool by combining a lateral flow assay with species-specific primers to successfully detect the presence of crown-of-thorn starfish on the Great Barrier Reef. This dip-stick method may potentially detect a wide variety of species from environmental samples, requiring little scientific training or laboratory access, making it well-suited for citizen science and remote conservation projects. Another future prospect is the potential for closed-tube PCR and automation, such as FASTFISH-ID [63], aiming to complete DNA barcoding in the field. When automated, these set-ups can become remote, real-time sensors. Deployment of such devices can efficiently provide unprecedented detail of real-time species movement, migration, and distribution. These tools, as well as other technological advancements for automatic sampling and processing, may be used for policy development, conservation, and biosecurity applications.

2.5. DNA Barcoding for Everyone, Everywhere

The DNA barcoding community contributes to networks, collaborative projects, data sharing, citizen science initiatives, and informed policy design. For instance, iBOL estimates 29,000 users of the Barcode of Life Data Systems database from 200 nations, which includes ~9 million barcodes, and the ambitious goal to expand by another 2 million barcoded taxa by 2026 (iBOL, http://ibol.org; access on 1 January 2021). Access to these and other reference barcodes is pivotal for well-rounded science and academic inclusivity. Researchers and organizations planning international collaborations should acknowledge funding bias and implement benefit-sharing with regions identified to have less barcoding capacity. In addition to academic projects, DNA barcoding is accessible to the public and suitable for citizen science. Citizen science projects such as the School Malaise Trap Program can result in data collection, education opportunities, and two-way collaboration between scientists and the general public [64,65].

The effectiveness of collaboration efforts relies on improved and continued open access to sequence information. However, freely accessible DNA barcode data can be targeted by commercial and exploitative research [66,67]. Thus, the delicate discussion of DNA barcodes as a form of digital sequence information (DSI, [68]), is needed. Digital sequence information is not yet regulated by the Nagoya Protocol on Access and Benefit-sharing that came into effect in 2014, despite ongoing discussions regarding DSI inclusion. Some support open-access DSI as a form of benefit-sharing, while others propose tighter restrictions [66]. The outcome of these discussions will have ramifications for DNA barcoding and metabarcoding and should be considered by anyone working directly or indirectly with DNA barcodes.

From our reflections, as long as a focus on data quality is prioritized and the methodological and technological advancements remain aligned, DNA (meta)barcoding will continue to impact, shape, and respond to changes in biological sciences, and DNA barcoding will continue to grow and increase our knowledge of global biodiversity. The scalability, accessibility, and automation potential of DNA (meta)barcoding methods strengthen biodiversity investigations. Beyond biodiversity monitoring, the knowledge provided by DNA barcoding can help mitigate threats to global biodiversity through improved environmental management and informed conservation measures.

Author Contributions: All authors contributed to the conceptualization, discussion and writing of this article. All authors have read and agreed to the published version of the manuscript.

Funding: The Research School in Biosystematics (ForBio) is supported by a grant from the Norwegian Biodiversity Information Centre.

Institutional Review Board Statement: Not applicable.

Data Availability Statement: Data sharing not applicable to this article as no research data were generated or analysed during the current study.

Acknowledgments: We are grateful to the Research School for Biosystematics (https://www.forbio.uio.no/, accessed on 13 June 2021), for organizing the course 'DNA Barcoding—from sequences to species'.

Conflicts of Interest: The authors declare no conflict of interest.

References

1. Hebert, P.D.N.; Cywinska, A.; Ball, S.L.; deWaard, J.R. Biological identifications through DNA barcodes. *Proc. R. Soc. Lond. Ser. B* **2003**, *270*, 313–321. [CrossRef]
2. Bush, A.; Monk, A.W.; Compson, G.Z.; Peters, D.L.; Porter, M.T.; Shokralla, S.; Wright, G.T.M.; Hajibabaei, M.; Baird, J.D. DNA metabarcoding reveals metacommunity dynamics in a threatened boreal wetland wilderness. *Proc. Natl. Acad. Sci. USA* **2020**, *117*, 8539–8545. [CrossRef]
3. Pennisi, E. DNA barcodes jump-start search for new species. *Science* **2019**, *364*, 920–921. [CrossRef]
4. DeSalle, R.; Goldstein, P. Review and interpretation of trends in DNA barcoding. *Front. Ecol. Evol.* **2019**, *7*, 302. [CrossRef]
5. Taberlet, P.; Bonin, A.; Zinger, L.; Coissac, E. *Environmental DNA: For Biodiversity Research and Monitoring*; Oxford University Press: Oxford, UK, 2018.
6. Galimberti, A.; Casiraghi, M.; Bruni, I.; Guzzetti, L.; Cortis, P.; Berterame, N.M.; Labra, M. From DNA barcoding to personalized nutrition: The evolution of food traceability. *Curr. Opin. Food Sci.* **2019**, *28*, 41–48. [CrossRef]
7. De Boer, H.J.; Ichim, M.C.; Newmaster, S. DNA barcoding and pharmacovigilance of herbal medicines. *Drug Saf.* **2015**, *38*, 611–620. [CrossRef] [PubMed]
8. Ratnasingham, S.; Hebert, P.D.N. BOLD: The Barcode of Life Data System (http://www.barcodinglife.org). *Mol. Ecol. Notes* **2007**, *7*, 355–364. [CrossRef]
9. Hobern, D. BIOSCAN: DNA barcoding to accelerate taxonomy and biogeography for conservation and sustainability. *Genome* **2021**, *64*, 161–164. [CrossRef] [PubMed]
10. Hobern, D.; Hebert, P.D.N. BIOSCAN-revealing eukaryote diversity, dynamics, and interactions. *Biodivers. Inf. Sci. Stand.* **2019**, *3*, e37333. [CrossRef]
11. Hebert, P.D.N.; Braukmann, T.W.; Prosser, S.W.; Ratnasingham, S.; DeWaard, J.R.; Ivanova, N.V.; Janzen, H.D.; Hallwachs, W.; Sones, E.J.; Zakharov, E. A Sequel to Sanger: Amplicon sequencing that scales. *BMC Genom.* **2018**, *19*, 219. [CrossRef]

12. Kress, W.J. Plant DNA barcodes: Applications today and in the future. *J. Syst. Evol.* **2017**, *55*, 291–307. [CrossRef]
13. Govender, A.; Willows-Munro, S. The utility of DNA barcoding as a tool to assess the success of ecological restoration using Hemiptera as a biological indicator. *Restor. Ecol.* **2019**, *27*, 1409–1419. [CrossRef]
14. Taberlet, P.; Coissac, E.; Pompanon, F.; Brochmann, C.; Willerslev, E. Towards next-generation biodiversity assessment using DNA metabarcoding. *Mol. Ecol.* **2012**, *21*, 2045–2050. [CrossRef]
15. Beng, K.C.; Corlett, R.T. Applications of environmental DNA (eDNA) in ecology and conservation: Opportunities, challenges and prospects. *Biodivers. Conserv.* **2020**, *29*, 2089–2121. [CrossRef]
16. Yatsunenko, T.; Rey, F.E.; Manary, M.J.; Trehan, I.; Dominguez-Bello, M.G.; Contreras, M.; Magris, M.; Hidalgo, G.; Baldassano, N.R.; Anokhin, P.A.; et al. Human gut microbiome viewed across age and geography. *Nature* **2012**, *486*, 222–227. [CrossRef]
17. Berry, O.; Bulman, C.; Bunce, M.; Coghlan, M.; Murray, D.C.; Ward, R.D. Comparison of morphological and DNA metabarcoding analyses of diets in exploited marine fishes. *Mar. Ecol. Prog. Ser.* **2015**, *540*, 167–181. [CrossRef]
18. Deiner, K.; Bik, H.M.; Machler, E.; Seymour, M.; Lacoursiere-Roussel, A.; Altermatt, F.; Creer, S.; Bista, I.; Lodge, M.D.; de Vere, N.; et al. Environmental DNA metabarcoding: Transforming how we survey animal and plant communities. *Mol. Ecol.* **2017**, *26*, 5872–5895. [CrossRef] [PubMed]
19. Alsos, I.G.; Lammers, Y.; Yoccoz, N.G.; Jørgensen, T.; Sjögren, P.; Gielly, L.; Edwards, M.E. Plant DNA metabarcoding of lake sediments: How does it represent the contemporary vegetation. *PLoS ONE* **2018**, *13*, e0195403. [CrossRef]
20. Creer, S.; Deiner, K.; Frey, S.; Porazinska, D.; Taberlet, P.; Thomas, W.K.; Potter, C.; Bik, H.M. The ecologist's field guide to sequenceh-based identification of biodiversity. *Methods Ecol. Evol.* **2016**, *7*, 1008–1018. [CrossRef]
21. Srivathsan, A.; Lee, L.; Katoh, K.; Hartop, E.; Kutty, S.N.; Wong, J.; Yeo, D.; Meier, R. MinION barcodes: Biodiversity discovery and identification by everyone, for everyone. *bioRxiv* **2021**, 434692. [CrossRef]
22. Janzen, D.H.; Hallwachs, W.; Pereira, G.; Blanco, R.; Masis, A.; Chavarria, M.M.; Chavarria, F.; Guadamuz, A.; Araya, M.; Smith, M.A.; et al. Using DNA-barcoded Malaise trap samples to measure impact of a geothermal energy project on the biodiversity of a Costa Rican old-growth rainforest. *Genome* **2020**, *63*, 407–436. [CrossRef]
23. Joly, S.; Davies, T.J.; Archambault, A.; Bruneau, A.; Derry, A.; Kembel, S.W.; Peres-Neto, P.; Vamosi, J.; Wheeler, T.A. Ecology in the age of DNA barcoding: The resource, the promise and the challenges ahead. *Mol. Ecol. Resour.* **2014**, *14*, 221–232. [CrossRef] [PubMed]
24. Obringer, R.; Rachunok, B.; Maia-Silva, D.; Arbabzadeh, M.; Nateghi, R.; Madani, K. The overlooked environmental footprint of increasing Internet use. *Resour. Conserv. Recycl.* **2021**, *167*, 105389. [CrossRef]
25. Hollingsworth, P.M.; Graham, S.W.; Little, D.P. Choosing and using a plant DNA barcode. *PLoS ONE* **2011**, *6*, e19254. [CrossRef] [PubMed]
26. Li, X.; Yang, Y.; Henry, R.J.; Rossetto, M.; Wang, Y.; Chen, S. Plant DNA barcoding: From gene to genome. *Biol. Rev.* **2015**, *90*, 157–166. [CrossRef] [PubMed]
27. Liu, Z.-F.; Ma, H.; Ci, X.-Q.; Li, L.; Song, Y.; Liu, B.; Li, H.W.; Wang, S.L.; Qu, X.J.; Hu, J.L.; et al. Can plastid genome sequencing be used for species identification in Lauraceae? *Bot. J. Linn. Soc.* **2021**, boab018. [CrossRef]
28. Hollingsworth, P.M.; Li, D.-Z. van der Bank, M. Twyford, A.D. Telling plant species apart with DNA: From barcodes to genomes. *Philos. Trans. R. Soc. B* **2016**, *371*, 20150338.
29. Fontes, J.T.; Vieira, P.E.; Ekrem, T.; Soares, P.; Costa, F.O. BAGS: An automated Barcode, Audit & Grade System for DNA barcode reference libraries. *Mol. Ecol. Resour.* **2021**, *21*, 573–583. [PubMed]
30. Siddall, M.E.; Fontanella, F.M.; Watson, S.C.; Kvist, S.; Erséus, C. Barcoding bamboozled by bacteria: Convergence to metazoan mitochondrial primer targets by marine microbes. *Syst. Biol.* **2009**, *58*, 445–451. [CrossRef]
31. Ratnasingham, S.; Hebert, P.D.N. A DNA-based registry for all animal species: The Barcode Index Number (BIN) system. *PLoS ONE* **2013**, *8*, e66213. [CrossRef]
32. Weigand, H.; Beermann, A.J.; Čiampor, F.; Costa, F.O.; Csabai, Z.; Duarte, S.; Geiger M., F.; Grabowski, M.; Rimet, F.; Rulik, B.; et al. DNA barcode reference libraries for the monitoring of aquatic biota in Europe: Gap-analysis and recommendations for future work. *Sci. Total Environ.* **2019**, *678*, 499–524. [CrossRef] [PubMed]
33. Weigand, A.M.; Jochum, A.; Pfenninger, M.; Steinke, D.; Klussmann-Kolb, A. A new approach to an old conundrum—DNA barcoding sheds new light on phenotypic plasticity and morphological stasis in microsnails (Gastropoda, Pulmonata, Carychiidae). *Mol. Ecol. Resour.* **2011**, *11*, 255–265. [CrossRef]
34. Pentinsaari, M.; Ratnasingham, S.; Miller, S.E.; Hebert, P.D.N. BOLD and GenBank revisited—Do identification errors arise in the lab or in the sequence libraries? *PLoS ONE* **2020**, *15*, e0231814. [CrossRef] [PubMed]
35. Rimet, F.; Aylagas, E.; Borja, A.; Bouchez, A.; Canino, A.; Chauvin, C.; Chonova, T.; Ciampor, F., Jr.; Costa, F.O.; Ferrari, B.J.D.; et al. Metadata standards and practical guidelines for specimen and DNA curation when building barcode reference libraries for aquatic life. *Metabarcoding Metagenom.* **2021**, *5*, e58056.
36. Porter, T.M.; Hajibabaei, M. Over 2.5 million COI sequences in GenBank and growing. *PLoS ONE* **2018**, *13*, e0200177. [CrossRef]
37. Krachunov, M.; Nisheva, M.; Vassilev, D. Machine learning models for error detection in metagenomics and polyploid sequencing data. *Information* **2019**, *10*, 110. [CrossRef]
38. Sharkey, M.J.; Janzen, D.H.; Hallwachs, W.; Chapman, E.G.; Smith, M.A.; Dapkey, T.; Brown, A.; Ratnasingham, S.; Naik, S.; Manjunath, R.; et al. Minimalist revision and description of 403 new species in 11 subfamilies of Costa Rican braconid parasitoid wasps, including host records for 219 species. *ZooKeys* **2021**, *1013*, 1–666. [CrossRef]

39. Montes-Ortiz, L.; Elías-Gutiérrez, M. Water mite diversity (Acariformes: Prostigmata: Parasitengonina: Hydrachnidiae) from karst ecosystems in southern of Mexico: A barcoding approach. *Diversity* **2020**, *12*, 329. [CrossRef]
40. Buenaventura, E.; Valverde-Castro, C.; Wolff, M.; Triana-Chavez, O.; Gómez-Palacio, A. DNA barcoding for identifying synanthropic flesh flies (Diptera, Sarcophagidae) of Colombia. *Acta Tropica* **2018**, *182*, 291–297. [CrossRef]
41. Ćelepirović, N.; Novak Agbaba, S.; Karija Vlahović, M. DNA Barcoding of Fungi in the Forest Ecosystem of the Psunj and Papuk Mountains in Croatia. *South-East Eur. For.* **2020**, *11*, 145–152. [CrossRef]
42. IPBES. *Summary for Policymakers of the Global Assessment Report on Biodiversity and Ecosystem Services of the Intergovernmental*; IPBES Secretariat: Bonn, Germany, 2019.
43. Madden, M.J.L.; Young, R.G.; Brown, J.W.; Miller, S.E.; Frewin, A.J.; Hanner, R.H. Using DNA barcoding to improve invasive pest identification at U.S. ports-of-entry. *PLoS ONE* **2019**, *14*, e0222291. [CrossRef] [PubMed]
44. Raclariu, A.C.; Heinrich, M.; Ichim, M.C.; de Boer, H. Benefits and limitations of DNA barcoding and metabarcoding in herbal product authentication. *Phytochem. Anal.* **2018**, *29*, 123–128. [CrossRef]
45. Solano, J.; Anabalón, L.; Figueroa, A.; Gangitano, D. ITS barcoding using high resolution melting analysis of *Cannabis sativa* drug seizures in Chile: A forensic application. *Forensic Sci. Int.* **2020**, *316*, 110550. [CrossRef] [PubMed]
46. Kuzmina, M.L.; Braukmann, T.W.A.; Zakharov, E.V. Finding the pond through the weeds: eDNA reveals underestimated diversity of pondweeds. *Appl. Plant. Sci.* **2018**, *6*, e01155. [CrossRef] [PubMed]
47. Kress, J.W.; Mazet, J.A.K.; Hebert, P.D.N. Intercepting pandemics through genomics. *Proc. Natl. Acad. Sci. USA* **2020**, *117*, 13852–13855. [CrossRef] [PubMed]
48. Vieira, P.E.; Desiderato, A.; Holdich, D.M.; Soares, P.; Creer, S.; Carvalho, G.R.; Costa, F.O.; Queiroga, H. Deep segregation in the open ocean: Macaronesia as an evolutionary hotspot for low dispersal marine invertebrates. *Mol. Ecol.* **2019**, *28*, 1784–1800. [CrossRef] [PubMed]
49. Hupało, K.; Teixeira, M.A.L.; Rewicz, T.; Sezgin, M.; Iannilli, V.; Karaman, G.S.; Grabowski, M.; Costa, F.O. Persistence of phylogeographic footprints helps to understand cryptic diversity detected in two marine amphipods widespread in the Mediterranean basin. *Mol. Phylogen. Evol.* **2019**, *132*, 53–66. [CrossRef]
50. Sigsgaard, E.E.; Nielsen, I.B.; Carl, H.; Krag, M.A.; Knudsen, S.W.; Xing, Y.; Holm-Hansen, H.T.; Møller, P.R.; Thomsen, P.F. Seawater environmental DNA reflects seasonality of a coastal fish community. *Mar. Biol.* **2017**, *164*, 128. [CrossRef]
51. Parsons, K.M.; Everett, M.; Dahlheim, M.; Park, L. Water, water everywhere: Environmental DNA can unlock population structure in elusive marine species. *R. Soc. Open Sci.* **2018**, *5*, 180537. [CrossRef]
52. Bohmann, K.; Gopalakrishnan, S.; Nielsen, M.; Nielsen, L.D.S.B.; Jones, G.; Streicker, D.G.; Gilbert, M.T.P. Using DNA metabarcoding for simultaneous inference of common vampire bat diet and population structure. *Mol. Ecol. Resour.* **2018**, *18*, 1050–1063. [CrossRef]
53. Zizka, V.M.; Geiger, M.F.; Leese, F. DNA metabarcoding of stream invertebrates reveals spatio-temporal variation but consistent status class assessments in a natural and urban river. *Ecol. Indic.* **2020**, *115*, 106383. [CrossRef]
54. Sigsgaard, E.E.; Jensen, M.R.; Winkelmann, I.E.; Møller, P.R.; Hansen, M.M.; Thomsen, P.F. Population-level inferences from environmental DNA—current status and future perspectives. *Evol. Appl.* **2020**, *13*, 245–262. [CrossRef] [PubMed]
55. Stransky, C.; Baumann, H.; Fevolden, S.E.; Harbitz, A.; Høie, H.; Nedreaas, K.H.; Salberg, A.-B.; Skarstein, T.H. Separation of Norwegian coastal cod and Northeast Arctic cod by outer otolith shape analysis. *Fish. Res.* **2008**, *90*, 26–35. [CrossRef]
56. Zobel, M.; Davison, J.; Edwards, M.E.; Brochmann, C.; Coissac, E.; Taberlet, P.; Willerslev, E.; Moora, M. Ancient environmental DNA reveals shifts in dominant mutualisms during the late Quaternary. *Nat. Commun.* **2018**, *9*, 139. [CrossRef]
57. Alsos, I.G.; Lavergne, S.; Merkel, M.K.F.; Boleda, M.; Lammers, Y.; Alberti, A.; Pouchon, C.; Denoeud, F.; Pitelkova, I.; Pușcaș, M. The treasure vault can be opened: Large-scale genome skimming works well using herbarium and silica gel dried material. *Plants* **2020**, *9*, 432. [CrossRef]
58. Lobo, J.; Costa, P.M.; Teixeira, M.A.; Ferreira, M.S.; Costa, M.H.; Costa, F.O. Enhanced primers for amplification of DNA barcodes from a broad range of marine metazoans. *BMC Ecol.* **2013**, *13*, 34. [CrossRef] [PubMed]
59. Leite, L.A.R. Mitochondrial pseudogenes in insect DNA barcoding: Differing points of view on the same issue. *Biota Neotrop.* **2012**, *12*, 301–308. [CrossRef]
60. Miya, M.; Sato, Y.; Fukunaga, T.; Sado, T.; Poulsen, J.Y.; Sato, K.; Minamoto, T.; Yamamoto, S.; Yamanaka, H.; Araki, H.; et al. MiFish, a set of universal PCR primers for metabarcoding environmental DNA from fishes: Detection of more than 230 subtropical marine species. *R. Soc. Open Sci.* **2015**, *2*, 150088. [CrossRef]
61. Taberlet, P.; Coissac, E.; Pompanon, F.; Gielly, L.; Miquel, C.; Valentini, A.; Vermat, T.; Corthier, G.; Brochmann, C.; Willerslev, E. Power and limitations of the chloroplast trnL (UAA) intron for plant DNA barcoding. *Nucleic Acids Res.* **2007**, *35*, e14. [CrossRef]
62. Doyle, J.; Uthicke, S. Sensitive environmental DNA detection via lateral flow assay (dipstick)-A case study on corallivorous crown-of-thorns sea star (*Acanthaster* cf. *solaris*) detection. *Environ. DNA* **2021**, *3*, 323–342. [CrossRef]
63. Naaum, A.M.; Cusa, M.; Singh, M.; Bleicher, Z.; Elliott, C.; Goodhead, I.B.; Hanner, R.H.; Helyar, S.J.; Mariani, S.; Rice, J.E.; et al. Validation of FASTFISH-ID: A new commercial platform for rapid fish species authentication via universal closed-tube barcoding. *Food Res. Int.* **2021**, *141*, 110035. [CrossRef] [PubMed]
64. Kobori, H.; Dickinson, J.L.; Washitani, I.; Sakurai, R.; Amano, T.; Komatsu, N.; Kitamura, W.; Takagawa, S.; Koyama, K.; Ogawara, T.; et al. Citizen science: A new approach to advance ecology, education, and conservation. *Ecol. Res.* **2016**, *31*, 1–19. [CrossRef]

65. Steinke, D.; Breton, V.; Berzitis, E.; Hebert, P.D.N. The School Malaise Trap Program: Coupling educational outreach with scientific discovery. *PLoS Biol.* **2017**, *15*, e2001829. [CrossRef]
66. Watanabe, M.E. The Nagoya Protocol: The conundrum of defining digital sequence information. *Bioscience* **2019**, *69*, 480. [CrossRef]
67. Bond, M.R.; Scott, D. Digital biopiracy and the (dis)assembling of the Nagoya Protocol. *Geoforum* **2020**, *117*, 24–32. [CrossRef]
68. Houssen, W.; Sara, R.; Jaspars, M. *Digital Sequence Information on Genetic Resources: Concept, Scope and Current Use*; Convention on Biological Conservation CBD/DSI/AHTEG: Montreal, QC, Canada, 2020.

MDPI
St. Alban-Anlage 66
4052 Basel
Switzerland
www.mdpi.com

Diversity Editorial Office
E-mail: diversity@mdpi.com
www.mdpi.com/journal/diversity

www.ingramcontent.com/pod-product-compliance
Lightning Source LLC
Chambersburg PA
CBHW061459210926
51850CB00008B/1767
9783725804870